The Welfare of Cattle

Animal Welfare

VOLUME 5

Series Editor

Clive Phillips, *Professor of Animal Welfare, Centre for Animal Welfare and Ethics, School of Veterinary Science, University of Queensland, Australia*

Titles published in this series:

Volume 1: The Welfare of Horses
Natalie Waran
ISBN 1-4020-0766-3

Volume 2: The Welfare of Laboratory Animals
Eila Kaliste
ISBN 1-4020-2270-0

Volume 3: The Welfare of Cats
Irene Rochlitz
ISBN 978-1-4020-3226-4

Volume 4: The Welfare of Dogs
Kevin Stafford
ISBN 978-1-4020-4361-1

Jeffrey Rushen • Anne Marie de Passillé
Marina A. G. von Keyserlingk • Daniel M. Weary

The Welfare of Cattle

Springer

A C.I.P. Catalogue record for this book is available from the Library of Congress

ISBN 978-1-4020-6557-6 (HB)
ISBN 978-1-4020-6558-3 (e-book)

Published by Springer,
P.O. Box 17, 3300 AA Dordrecht, The Netherlands

www.springer.com

Printed on acid-free paper.

Contents

Animal Welfare Series Preface

Animal welfare is attracting increasing interest worldwide, but particularly from those in developed countries, who now have the knowledge and resources to be able to offer the best management systems for their farm animals, as well as potentially being able to offer plentiful resources for companion, zoo and laboratory animals. The increased attention given to farm animal welfare in the West derives largely from the fact that the relentless pursuit of financial reward and efficiency has led to the development of intensive animal production systems that challenge the conscience of many consumers in those countries. In developing countries, human survival is still a daily uncertainty, so that provision for animal welfare has to be balanced against human welfare. Welfare is usually provided for only if it supports the output of the animal, be it food, work, clothing, sport or companionship. In reality, there are resources for all if they are properly husbanded in both developing and developed countries. The inequitable division of the world's riches creates physical and psychological poverty for humans and animals alike in many parts of the world. Livestock are the world's biggest land user (FAO, 2002) and the population is increasing rapidly to meet the need of an expanding human population. Populations of farm animals managed by humans are therefore increasing worldwide, and there is the tendency to allocate fewer resources to each animal.

Increased attention to welfare issues is just as evident for companion, laboratory, wild and zoo animals. Although the economics of welfare provision may be less critical than for farm animals, the key issues of provision of adequate food, water, a suitable environment, companionship and health remain as important as they are for farm animals. Of increasing importance is the ethical management of breeding programmes, now that genetic manipulation is more feasible, but there is less tolerance of deliberate breeding of animals with genetic abnormalities. However, the quest for producing novel genotypes has fascinated breeders for centuries, and where dog and cat breeders produced a variety of extreme forms with adverse effects on their welfare in earlier times, nowadays the quest is pursued in the laboratory, where the mouse is genetically manipulated with even more dramatic effects.

The intimate connection between animal and owner or manager that was so essential in the past is rare nowadays, having been superseded by technologically efficient production systems, where animals on farms and in laboratories are tended by increasingly few humans in the drive to enhance labour efficiency. With today's busy lifestyle, pets too may suffer from reduced contact with humans, although their value in providing companionship, particularly for certain groups such as the elderly, is increasingly recognised. Consumers also rarely have any contact with the animals that produce their food. In this estranged, efficient world man struggles to find the moral imperatives to determine the level of welfare that he should afford to animals within his charge. Some, and in particular many of the companion animal owners, aim for what they believe to be the highest levels of welfare provision, while others, deliberately or through ignorance, keep animals in impoverished conditions or even dangerously close to death. Religious beliefs and directives encouraging us to care for animals have often been cast aside in an act of supreme human self-confidence, stemming largely from the accelerating pace of scientific development. Instead, today's moral codes are derived as much from media reports of animal abuse and the assurances that we receive from supermarkets, that animals used for their products have not suffered in any way. The young were always exhorted to be kind to animals through exposure to fables, whose moral message was the benevolent treatment of animals. Such messages are today enlivened by the powerful images of modern technology, but essentially still alert children to the wrongs associated with animal abuse.

This series has been designed to provide academic texts discussing the provision for the welfare of the major animal species that are managed and cared for by humans. They are not detailed blue-prints for the management of each species, rather they describe and consider the major welfare concerns of the species, often in relation to the wild progenitors of the managed animals. Welfare is considered in relation to the animal's needs, concentrating on nutrition, behaviour, reproduction and the physical and social environment. Economic effects of animal welfare provision are also considered where relevant, and key areas requiring further research.

In this volume four of the world's leading scientists in the field, Drs Jeffrey Rushen, Anne Marie de Passillé, Marina von Keyserlingk and Dan Weary, present a challenging account of the welfare issues facing dairy and beef cattle. Drawing on their detailed knowledge of the behavioural and physiological correlates of welfare in cattle, they provide an account of the major issues facing one of the most important of agricultural species.

With the growing pace of knowledge in this new area of research, it is hoped that this series will provide a timely and much-needed set of texts for researchers, lecturers, practitioners, and students. My thanks are particularly due to the publishers for their support, and to the authors and editors for their hard work in producing the texts on time and in good order.

Clive Phillips, Series Editor,
Professor of Animal Welfare and Director, Centre for Animal Welfare and Ethics, School of Veterinary Science, University of Queensland, Australia

Reference

Food and Agriculture Organisation (2002). http://www.fao.org/ag/aga/index_en.htm.

Acknowledgements

We are very grateful to all of you who have contributed to our thinking on the issues discussed in this book, particularly the many collaborators, students, and technicians that we have had the very good fortune of working with. We especially thank the following whose thoughts, help or work we relied on extensively in writing this book: Margit Bak Jensen, Lorna Baird, Erin Bell, Isabelle Blanchet, Alain Boissy, Fernando Borderas, Naomi Botheras, Monika Budzyaska, Luisa Brusius, Art Ceballos, Beverley Chua, Lisa Croteau, Nuria Chapinal Gomez, Trevor De Vries, Michelle Drissler, Marcia Endres, Frances Flower, David Fraser, Jose Fregonesi, Chantal Gaboury, Maria Hötzel, Laura Paton, Amanda Grout, Vanessa Guesdon, Derek Haley, Laura Hänninen, Paul Hemsworth, Julianna Huzzey, Kiyomi Ito, Martine Janzen, Laura Kielbauch, Yuki Koba, Jan Ladewig, Lena Lidfors, Carlos Pinhiero Machado, Emmi Manninen, Pierre-Guy Marnet, Georgia Mason, Julie Ménard, Omar Mendo, Laura Mowbray, Lene Munksgaard, Audrey Nadalin, João Negrão, Sophie Neveux, Keelin O'Driscoll, Ed Pajor, Carol Petherick, Émilie Pombourcg, Katy Proudfoot, Guylaine Richer, Mairi Robertson, Pierre Rybarczyk, David Sanderson, Mitja Sedlbauer, Marie-Josée Sirois, Igor Spierts, Marek Spinka, Marjolaine St. Louis, Joe Stookey, Lizanne Stuenenberg, Karen Schwartzkopf-Genswein, Brigid Sweeney, Hajime Tanida, Tamiko Thomas, Sylvia ter Maat, Erin Tom, Cassandra Tucker, Geoff Urton, Reza Valizadeh, David Val-Laillet, Marcela Vankova, Elsa Vasseur, Isabelle Veissier, Karen Vickers, Doug Veira, Andreia Vieira, Tyler Vittie, Tristan Welp, Christoph Winckler, Fernando Wolf, Gosa Zobel, and Gosia Zdanowicz. If we have inadvertently left anyone off this list, we humbly beg forgiveness. We thank Frances Flower, Joe Stookey, Elissar Maidaa, and Denise Turner for permission to use their photographs. We are very grateful to Lori Vickers, Jeffrey Spooner and Nicole Fenwick for excellent help with the references. We also thank Canada's Natural Sciences and Engineering Research Council, the Dairy Farmers of Canada, Novalait, Agriculture and Agri-Food Canada and many others for research support.

Chapter 1
Introduction: What is Animal Welfare?

1 Introduction

Concern about the welfare of farm animals is nothing new – farmers and veterinarians have always been concerned about the condition of animals in their care and have tried to ensure that they are healthy and well nourished. In this older tradition of animal care, good welfare is seen largely as the absence of pain, illness or injury, and the focus is upon protecting the welfare of individual animals, especially ensuring that sick animals receive timely and effective care. The more recent interest in farm animal welfare, however, stems largely from the public concern about some modern farming techniques, especially the use of intensive husbandry (Figure 1.1).

In many modern farms, especially in the industrialized world, animals are housed indoors in apparently "unnatural" conditions, with limited space and often a limited ability to engage in social interactions and other natural behaviours. The more recent concerns of the public are with widespread and accepted industry practices rather than with individual acts of cruelty or neglect, and they focus upon whole "systems" of housing and management, as much as the health of individual animals. A convenient date to time the beginning of this more recent tradition in animal welfare, at least within the English-speaking parts of the world, is with the publication of *Animal Factories* by Ruth Harrison (1964). Similar developments occurred in other countries, for example, the writings of Astrid Lindgren (www.astridlindgren.com), a popular author of children's stories, were instrumental in encouraging the Swedish government to enact animal welfare legislation. One concern raised by Lindgren was indoor housing of cattle all year round rather than just in winter. Her position on this later influenced Swedish legislation providing cattle-grazing "rights" in the summer months.

In response to public expressions of concern about the welfare of animals in modern farming conditions, the British Government established the Brambell committee (Brambell, 1965) to "enquire into the welfare of animals kept under intensive husbandry systems". The Brambell report to the UK Government was one of the most influential writings on animal welfare. The views expressed on animal welfare, and the particular issues and concerns that were examined by the committee, had a great influence on the topics and the nature of the subsequent research into

Figure 1.1 Our ideas of what is best for animal welfare are affected by what we know (or think we know) of how they live under "natural" conditions. Keeping animals in conditions which appear unnatural, in which they cannot perform many of their normal behaviours, is one of the factors that raises concern about animal welfare. For cattle, a "natural life" generally involves grazing in fields or on pasture, with young calves suckling from their mothers. However, cattle are increasingly being housed indoors without access to pasture and dairy calves are typically separated from their mothers at birth. Many members of the public are disquieted by such types of housing systems. However, we should not assume that animals housed extensively or in apparently natural conditions do not have welfare problems. They do, even though these may be different from welfare problems facing animals housed indoors

animal welfare that was done. For example, the Brambell report drew particular attention to the problems of behavioural restriction that resulted from intensive, indoor housing systems. In the appendix to this report, Thorpe (an ethologist) wrote: "we must draw the line at conditions which completely suppress all or nearly all the natural, instinctive urges and behaviour patterns characteristic of actions…as found in the ancestral wild species and which have been little, if at all, bred out in the process of domestication" (Brambell, 1965, p. 79). This interest in behavioural deprivation, which was not a topic in traditional veterinary science or animal science education, led many researchers in animal welfare to focus attention on behavioural problems. This focus on behaviour and behavioural deprivation has been highly contentious, and we discuss some of the issues in using behavioural indicators of animal welfare in Chapter 4.

The Brambell committee identified a number of welfare concerns for cattle production. These focused mainly on calves and included the early separation of cow from calf, disease incidence in early weaned calves, and cross-sucking among milk-fed calves. The strongest concerns were with the methods used in veal production, especially the restriction of movement and of social contact imposed by the veal crate and the low iron diet, with the risk of anaemia. The committee concluded that "the methods of rearing calves in the 'white' veal industry do not conform to the principles of welfare" that the committee had adopted (p. 40). Recommendations were made to provide roughage daily, to prohibit close tethering of calves, except for short periods, to increase the size of pens used for individual calves and to provide bedding. The report also expressed some concerns with intensive beef production, notably the use of individual tethering and totally slatted floors, high stocking densities in pens, and the high incidence of liver disorders that results from the use of high-grain diets. Interestingly, the report did not include dairy cattle on the grounds that there had been few expressions of public anxiety over the welfare of dairy cows and that the evidence was that "no other kind of farm livestock is so well cared for" (p. 35). In the second section of this book, we focus upon some of the major challenges to the welfare of cattle, and find that many of the topics raised by the Brambell committee are still with us. We also discuss important welfare problems not recognized by the Brambell committee, including those facing dairy cattle today.

2 The Legislative Approach

Governments of many European countries responded to the public concerns about animal welfare by adopting legislation that prohibited certain practices. The Swedish and Swiss regulations were among the earliest and perhaps the most notable in explicitly dealing with the problems of behavioural deprivation and in laying down in detail which practices would no longer be tolerated. For example, The Swiss Animal Protection Ordinance of 1978/1998 states that animals "shall be kept in such a manner as not to interfere with…their behaviour" and, when dealing with cattle, states that calves must be kept in groups from 2 weeks to 4 months of age and that cows in loose-housing systems must not exceed the number of lying stalls available, among other provisions.

Similar animal welfare legislation was adopted in other European countries and formed the basis for subsequent European Union (EU) legislation. The European Convention for the Protection of Animals Kept for Farming Purposes of 1978 focused upon the importance of avoiding suffering and ensuring that housing, nutrition, and management systems should be appropriate to animal's "physiological and ethological needs in accordance with…scientific knowledge". These last requirements, especially the reference to "ethological needs", precipitated considerable scientific research aimed at better understanding such needs and how these differ among species. The involvement of scientists in the debate led to attempts to improve the definition of welfare to make it more amenable to scientific research. Unfortunately this resulted in some definitions that suited scientists by, for example, redefining animal welfare to make it easier to measure (discussed in more detail by Rushen, 2003). However, definitions often failed to properly address the full range of societal concerns.

In EU legislation, cattle have generally received less attention than other species, such as pigs and poultry, and legislation dealing specifically with cattle has focused most upon calves, especially veal calves. The Council of the European Communities Directive "laying down minimum standards for the protection of calves" of 1991 and its amendment in 1997, specified space allowances, types of flooring, iron content of diets etc. and, perhaps most controversially, prohibited the individual housing of calves over the age of 8 weeks, except in the case of veterinary treatment. Cattle are also covered by EU directives covering the transport and slaughter of animals.

It is beyond the scope of this book to describe and evaluate all laws regarding the welfare of cattle. Fraser (2006) provides a good discussion of the advantages and disadvantages of the legislative approach to dealing with animal welfare (Table 1.1). We mention them because the legislative approach to dealing with animal welfare set the agenda for much of the research that was done, either in preparation for the legislation or subsequently (Rushen, 2003). For example, much research effort was directed at developing a scientific conception and methodology

Table 1.1 Animal welfare issues have been dealt with at a societal level in a number of ways. Most European countries have adopted animal welfare legislation, but this has been resisted in North America. Many food retailers have implemented quality assurance (QA) schemes to assure their customers that the products they buy have come from animals that have been humanely handled. There are a variety of labelling schemes that attempt to differentiate products from farms that use special rearing methods thought to be better for animal welfare. Finally, there are a variety of voluntary codes that farmers can follow. However, these different approaches all have their own advantages and disadvantages (Based on the discussion in Fraser, 2006.)

Degree to which programmes are	Legislation	Corporate QA schemes	Labelling	Voluntary codes
Supported by agricultural industry	No	Yes or No	Yes or No	Yes
Easy to implement	No	Yes	No	Yes
Enforceable	Yes	Yes	Yes	No
Comprehensive in application	Yes	No	No	Yes

for dealing with "ethological needs", as these were controversial aspects of the welfare legislation. This research has proved to be of broad value in improving our understanding of the effects of husbandry practices and housing systems on cattle welfare, topics which are addressed in detail in this book.

The effect of legislative action on animal welfare research was not always positive. Much of the research was based upon comparisons of different types of housing systems, for example, group housing of veal calves versus the individual stalls, since this was a focus of legislation. Unfortunately, such studies often involved a comparison using only one farm of each type. The underlying assumption seemed to be that the type of housing "system" was the predominant factor influencing animal welfare. Consequently, it was assumed that one farm could serve as an example for all farms of that type of housing system, and that it was meaningful to talk of the average level of animal welfare within each type of housing system. However, it has become apparent that the level of animal welfare within each type of housing system is strongly dependent upon the type of management and stockmanship, and upon the details of the housing system. Consequently, it is often meaningless to talk of "average" levels of welfare with different housing systems, especially when such comparisons are based on just one or few farms. It may be more useful to talk of the capacity of different housing systems to provide an acceptable level of animal welfare. We discuss the difficulties in comparing different housing systems for their effect on animal welfare in Chapter 6.

3 Animal Welfare as a Consumer Concern

A complementary approach to improving animal welfare comes from the recognition that animal-welfare concerns can affect the buying habits of consumers. Surveys undertaken in the EU show that consumers often state that animal welfare issues are important to them in making purchasing decisions, although sometimes these are of secondary importance compared to food safety, taste, and nutrition (e.g. Weatherell et al., 2003; Grunert et al., 2004). Many consumers feel that information about the production system, including animal welfare, should be part of product labelling (Bernues et al., 2003). Stated preferences do not necessarily translate into changes in consumer behaviour, however, at least when there is a price differential among products (Webster, 2001). For example, there is little evidence that interest or concern with animal welfare issues affect general consumption of meat products such as beef (Mannion et al., 2000). Webster (2001) makes the telling point that radical improvements in the welfare of farm animals could be achieved if consumers were willing to accept an almost imperceptible increase in the price of food. However, while animal welfare issues may not have a marked effect on the day-to-day buying habits of the majority of consumers, animal welfare can be a "sleeping issue" that has potential to affect buying habits at moments of crises, somewhat like food-safety issues (Grunert et al., 2004). Furthermore, it is clear that subgroups of consumers are emerging who are concerned with a range of "civic" issues and that these concerns can influence purchasing decisions for this (sometimes sizeable) segment of the population (Weatherell et al., 2003). This has led to the development of niche-markets, of which that for organic, ecological, or biological products is the most successful. Interestingly, concern about animal welfare issues appears to be one of the main reasons that consumers do buy organic animal products, especially in the UK (Grunert et al., 2004) and is one factor leading consumers to prefer food considered to be locally produced (Weatherell et al., 2003).

The recognition of these civic concerns has led to a proliferation of "quality assurance" schemes that try to assure consumers that the products they buy are produced according to practices that do respect the environment, animal welfare, etc. Quality-assurance (QA) schemes that specifically deal with animal welfare now exist in a number of countries, many having been developed by food retailers (Fraser, 2006). In addition, the majority of standards for organic animal production contain provisions regarding animal welfare (Vaarst et al., 2003). The use of animal welfare standards developed by food retailers is now the main way of dealing with animal welfare issues in North America, and has resulted in definite improvements in some limited aspects of the ways that animals are housed and slaughtered (Mench, 2003; Fraser, 2006). The most successful independent scheme is probably the Royal Society for the Prevention of Cruelty to Animal's (RSPCA) Freedom Foods that provides standards for beef, and dairy cattle. Such schemes have led to the development of animal welfare audits (e.g. Grandin, 2002) and indices (e.g. Bartussek, 1999; von Borell et al., 2001) that attempt to assess animal welfare on-farm or during transport and slaughter.

The development of these QA schemes and animal welfare audits and indices has affected the way that research in animal welfare is done. More specifically, there is interest in finding ways to assess the welfare of animals in "real life", i.e. on commercial operations, such as on individual farms or abattoirs. The aim is to ensure that the animal welfare standards are adequate and are being respected by individual farmers, truckers, or slaughter plants. In some cases, these QA schemes deliberately avoid dealing with contentious issues, such as whether or not individual housing of calves should be allowed, and instead seek to define the conditions under which maximum welfare can be achieved within each type of housing system.

Unfortunately, assessing animal welfare on commercial operations is more difficult than assessing welfare in more controlled, experimental conditions. Often the focus is upon "how" something is done (e.g. the way in which calves are dehorned and the care taken in doing the procedure) as much as upon "what" is done (e.g. the fact that calves are dehorned at all).

4 What is Animal Welfare?

Animal welfare advocates are often vocal about their concerns, making these relatively easy to describe. Table 1.2 illustrates some of the aspects of dairy, beef, and veal production that are often referred to.

Table 1.2 Some of the main welfare concerns that have been expressed by animal-welfare groups about the welfare of cattle. The list is not exhaustive, but is meant to illustrate the wide-ranging nature of the issues

Veal calves
– Inability to turn around or lie down comfortably due to small size of veal crates
– Diets lacking adequate roughage or iron
– Lack of opportunity for social contact
– Early separation from mother
Beef cattle
– Metabolic problems from high-grain diets (acidosis, laminitis, liver abscesses)
– Rough handling
– Pain from dehorning, castration, branding
– Slaughter and stunning techniques (possibility of consciousness during slaughter)
– Transport (long distance, live transport)
– Feed lots (stocking densities, heat stress)
Dairy cattle
– Lack of opportunity to graze
– Metabolic problems and infectious disease following parturition
– Lameness
– Pain from dehorning
– Use of BST to increase milk yield

We do not wish to prejudge the extent that these concerns are justified: these will be evaluated in the second part of this book. The table simply illustrates the wide range of concerns that have been expressed. Consumers, legislators, and the general public have taken an interest in animal welfare, and any scientific approaches to the improving welfare must address the full range of concerns. In this way animal welfare science differs from many other branches of science: it is not driven primarily by curiosity (like astronomy), or primarily by economic opportunities (like applied electronics), but rather by ethical and societal concerns. Fraser and Weary (2004) present a more complete discussion of the unusual aspects of this branch of scientific enquiry.

Three types of concern about animal welfare are typically heard: those that involve the biological functioning of the animal, those that involve how the animal is "feeling", that is with the animal's affective or emotional state, and those that involve the ability of the animal to live a "natural" life (Fraser, 2003, 2006). People concerned with the biological functioning of the animal (most often veterinarians and farmers) generally focus on disease, injury, poor growth rates, and reproductive problems. There is little disagreement about whether such problems are of welfare concern, and much research in animal welfare has focused on these issues as described in Chapters 2 and 3. People concerned more with the affective state, or the emotions of the animal, focus upon whether the animals are suffering from unpleasant feelings, such as pain, fear, or hunger. These obviously aversive experiences have received much attention from researchers (discussed in Chapters 4 and 5), but some effort has been made to come to grips with more subtle emotions, such as boredom or frustration, as well as pleasure and other positive states. It is now widely accepted that the topic of animal sentience is a key one in understanding animal welfare, and that our concept of animal sentience will change continuously as a result of scientific and philosophical research in this area (Duncan, 2006).

For at least some people interested in animal welfare, a key concern is whether the animal is able to live a relatively natural life, and the issue of natural behaviour has been central to discussions of animal welfare (discussed further in Chapter 4). Natural living includes both allowing animals to live in a manner to which they are adapted and to develop in a manner that is normal for the species (Fraser and Weary, 2004).

Clearly these different types of concern about animal welfare can and do overlap (Fraser et al., 1997). A lactating dairy cow unable to seek shade on a hot day (natural behaviour), will likely feel uncomfortably hot (affective state), and may show signs of hyperthermia (biological functioning). In such cases, research directed at any or all the levels can help address the welfare problem. In other cases, overlap may be less obvious or the different concerns may even be in conflict. For example, housing dairy calves in groups allows them to engage in natural social interactions, but when poorly managed can lead to increased incidence of certain diseases or aggressive interactions (see Chapter 7). Different scientists can thus reach opposite conclusions about the relative advantages of different housing systems by favouring different welfare indicators (for a case study, see Fraser, 2003). Clearly the best research solutions will be those that address all concerns, for example, by creating group-housing systems for calves that avoid competition and allow calves to stay healthy. In this way these three types of concerns can be considered as a checklist to help researchers identify and solve the various welfare issues.

A widely cited "inclusive" approach is provided by the "Five Freedoms" (Webster, 2001; Mellor and Stafford, 2001). These are freedom from hunger and thirst, discomfort, pain, injury or disease, fear and distress, and the freedom to express normal behaviour. The five freedoms do not describe the criteria that must be fulfilled if any housing or management system is to achieve an acceptable level of welfare. It is unrealistic to think that animals (in any environment) could attain complete "freedom" in this respect. Rather, the five freedoms indicate a way of identifying welfare problems, and a direction in which we should move if we wish to improve welfare. An advantage of a clear statement of the main threats to animal welfare is that it gives some direction as to what needs to be measured. That is, we need indicators that address all of the five freedoms, not just one.

In contrast to the inclusive approach exemplified by the Five Freedoms, some authors have argued that the "essence" of animal welfare is best captured by a single type of concern. For example, Duncan (2002) argues that an animal's welfare depends on the animal's affective state, so that a threat to animal welfare can be measured in terms of how much suffering it causes the animal (through pain, fear, frustration etc.), or how much it limits the animal's ability to experience positive affect (i.e. pleasurable states). From this perspective, problems with biological functioning (e.g. ill health or injury) or with natural behaviour (e.g. restricted ability to move or interact socially) are important to animal welfare only to the extent that they lead to negative affect or restrict positive affect. However, such an approach ignores those whose concerns about animal welfare focus more on the animal's ability to lead a natural life or on the biological functioning criteria. For example, organic producers argue that the ability of farm animals to lead reasonably natural lives is of inherent concern, not simply a way of avoiding certain types of suffering (Lund and Weary, 2003). Similarly, many veterinarians would argue that disease and injury are inherent threats to animal welfare, even in cases where these lead to little or no apparent pain or discomfort. In this book we take all three types of concern seriously, although much of the research (and hence much of our coverage) is focused on biological functioning (including measures of health and production covered in Chapters 2 and 3) and emotional or affective state (Chapter 4).

The published discussions about the different definitions of animal welfare often give the impression that there is little agreement on what constitutes good welfare. However, despite occasional disagreements between scientists, considerable consensus does exist. Some recent examples show how consensus can be achieved using methods such as the Delphi technique. Two studies (Main et al., 2003a; Whay et al., 2003a) found considerable agreement between experts (veterinarians and behavioural scientists) as to which of two dairy farms had the higher level of welfare, when they were presented with a variety of information about the state of the cows on the two farms. The results show that with appropriate techniques, people with experience in animal welfare are capable of integrating a variety of sources of information about animal welfare and can achieve a fair consensus on the level of welfare on the farm. Such techniques have heuristic values in making decisions about animal welfare even in the absence of a single definition of welfare.

Where genuine disagreement does exist, it often occurs when poorly understood measures of welfare are used. Most often this involves disagreements about the relative importance of behavioural, physiological, and immunological measures of welfare. These we discuss further in Chapters 3 and 4. In contrast, there is little disagreement about the negative effects of poor health on animal welfare (Broom, 2006). Finally, disagreements tend to occur most often over the most complicated issues in animal welfare, for example, the relative advantages of different types of housing systems. In Chapter 6, we discuss why these housing system comparisons are particularly difficult to interpret. Finding ways of improving routine procedures, such as branding or tail docking, or of making small improvements in animal housing, such as better flooring, tends to provoke far less disagreement.

5 Criteria for Judging Animal Welfare

One of the main issues addressed in this book is how to judge animal welfare. The criteria for assessing animal welfare are conventionally divided into those associated with the environment of the animal (input-based, engineering, or design criteria) versus those associated with the state of the animal (outcome-based or animal-based criteria) (e.g. Rushen and de Passillé, 1992; Mench, 2003). Input-based criteria generally describe the environment of the animals including the way that animals are kept, fed and managed, and include such aspects as the use of battery cages, space allowances, group size, and use of tethering. Outcome-based criteria attempt to directly assess the state of the animals, and include behavioural, physiological and immune measures, incidence of health problems, and production levels. Input-based criteria are usually favoured in animal welfare standards since they are easier to audit (Mench, 2003). For example, of the several hundred items in the RSPCA welfare standards for beef cattle, the majority describe the environment of the animals or the management practices. Only a handful of items describe the actual state of the animals, such as haemoglobin levels in calves, body-condition score and the incidence of various diseases. Similarly, the welfare standards being developed by US retailers are also heavily based on input-based criteria (Mench, 2003).

A potential advantage of well-chosen input-based criteria is that these should prevent welfare problems from occurring. Unfortunately, little is known about how specific input criteria actually relate to animal welfare although some research has attempted to address this link. For example, Main et al. (2003b) described how animal-based measures of welfare differed on dairy farms that did or did not conform to the RSPCA Freedom Food standard that was based primarily on input criteria.

Input-based criteria make it difficult to establish the equivalence of animal welfare standards in different countries because different countries may use quite different housing systems or production techniques as well as different breeds of animals. It would be difficult, for example, to establish animal welfare standards for dairy cattle, based solely on input-based criteria, which would allow a comparison of the level of animal welfare in the pasture-based system of New Zealand, the large-scale indoor housing systems of North America, and the smaller-scale mixed indoor/pasture systems of some European countries. In such cases, outcome-based criteria are preferable, since it should be possible to measure the actual state of the animals, irrespective of how they are housed or managed. Outcome-based criteria also come closer to demonstrating the actual level of animal welfare, and allow for corrective actions to be taken if welfare problems arise. Unfortunately, outcome-based criteria for animal welfare are sometimes impractical to measure in an audit, and there is controversy over which measures are most appropriate. It is increasingly clear that many endemic health problems represent some of the most serious welfare problems, especially for high-producing animals, and the incidence of these illnesses may be one of the most effective outcome-based indicators of animal welfare (Chapter 2). These include the incidence of lameness in dairy cattle, and rates of mortality and morbidity in dairy calves, veal calves, and feed-lot cattle. Part 1 of this book addresses the various outcome-based criteria for judging animal welfare.

6 Scope of the Book

In this book we review and discuss the research that has been done on the welfare of cattle. According to Webster (2001), a full understanding of animal welfare issues requires that we have (1) a scientific understanding of the factors that affect the welfare of farm animals, (2) an understanding of the ethical or moral reasons why we should respect the welfare of farm animals, and (3) an understanding of the economic forces that lead us to treat animals the way we do. This book focuses exclusively on the first of these. We only address the other two issues when necessary to understand how or why research was done. Unlike some basic research, the economic and ethical context of the debate about animal welfare has had a marked effect on how research on animal welfare is done and where the research is directed.

We cover dairy, veal, and beef production, although the fact that all four authors are more familiar with dairy production will be apparent in our coverage. Our coverage is international, but more focused on issues related to production in the "industrialized" world, since this is where most research has been done. In addition, we focus more on intensive housing, not because extensive systems are free of welfare problems (Petherick, 2005), but because research has tended to concentrate on more intensive systems.

We have divided the book into two parts. Part 1 discusses various methods of assessing animal welfare, including behavioural, physiological, health, and production measures. Part 2 discusses the main threats to the welfare of cattle, including those arising from the way the cattle are housed, the types of procedures we inflict on them, the interactions between the animals and the people that handle them, and the way the cattle are fed.

Part I
Indicators of Animal Welfare

Chapter 2
Health, Disease, and Productivity

1 Introduction

It seems evident that poor health can be a major cause of poor welfare in farm animals and that the occurrence of illness can be used to assess animal welfare. Cattle of all ages, both in traditional and modern housing systems, and under intensive and extensive management, suffer from a variety of endemic diseases, both infectious and non-infectious, as well as outbreaks of epidemic diseases, such as foot and mouth and BSE. Cattle also suffer from various forms of physical injury as a result of poor housing or management.

Health measures have long been recognized as potentially useful indicators of animal welfare and have figured prominently in assessments of cattle welfare (Broom, 2006). However, researchers in animal welfare have often paid little attention to animal health problems (Rushen, 2003), perhaps because of difficulties obtaining data about the incidence of health problems in the different types of cattle production (Figure 2.1); for example, information on animal health is more available and more easily obtained for intensively housed cattle than for extensively managed ones.

In this chapter, we focus on the issues that arise when we try to use measures of the incidence or prevalence of illness as indicators of animal welfare. While it may seem obvious that illness or injury reduces welfare, we argue that there are a number of unresolved problems in using health measures to assess animal welfare. The main issues that we deal with are the difficulty in judging the relative impact on animal welfare of different forms of illness or injury, and the difficulty in obtaining reliable and valid information on the occurrence of illness and injury, especially where welfare is being assessed in commercial settings.

As well as animal health, this chapter briefly touches on two, more problematic and controversial welfare indicators: productivity and reproductive success. Farmers and others involved in the industry often claim that, since dairy cows are producing enormous quantities of milk or that beef cattle are growing at prodigious rates, then their welfare must be satisfactory. In contrast, animal-welfare critics point out that increases in productivity often result from specific practices, such as the use of growth enhancers or BST, rather than reflecting the general welfare of the animal. Furthermore, there is increasing concern that the high level of productivity in modern

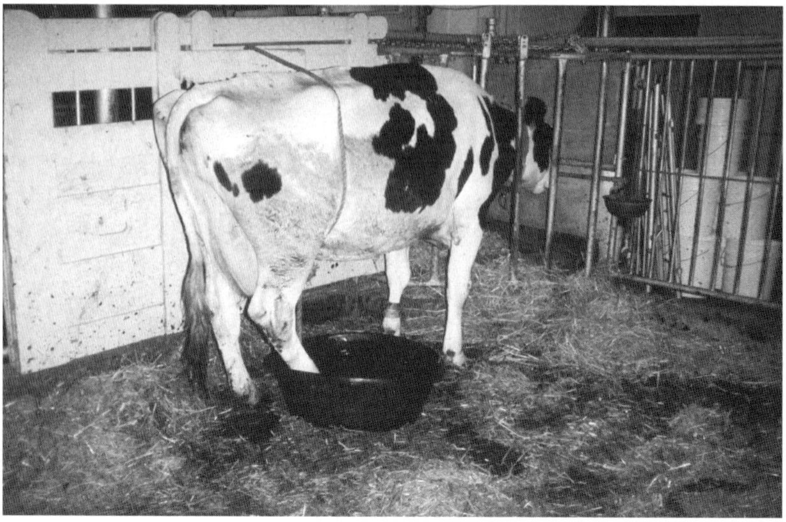

Figure 2.1 Cattle suffer from a number of painful diseases, such as lameness, which potentially can be used in welfare assessment. However, many such diseases are not treated by a veterinarian and questions about the accuracy of the diagnosis and difficulties in obtaining accurate and reliable records of their incidence limit their usefulness in assessing animal welfare under commercial conditions

cattle production is itself a risk factor for welfare problems. This issue we discuss in Section 6 of this chapter.

Similarly, reduced reproductive success would seem promising in providing information about poor welfare. There is much evidence that animals in poor condition (e.g. those that are ill or suffering from chronic stress) are less likely to reproduce successfully. Critics of animal agriculture often argue that the low reproductive rate of dairy cattle is an indicator of poor welfare. However, as we argue later, while poor welfare may indeed lead to lower reproductive success, it does not follow that high reproductive success indicates a lack of welfare problems.

2 Relationship of Animal Health to Animal Welfare

In animal agriculture, the importance of a disease is often judged by its direct economic impact, but a broader view requires a better understanding of how different diseases affect animal welfare (Wells et al., 1998). Animal health problems are best related to animal welfare to the extent that they are associated with the animal's suffering (Wells et al., 1998), either in the past, the present, or the future. While it may seem obvious that good welfare is dependent on good health, in order to be able to use measures of the incidence of different diseases to assess overall welfare, we need to be able to estimate the relative impact of these diseases.

Ideally, we would be able to directly measure the amount of suffering caused to the animal, but at present this is not feasible. An indirect approach is to compare different diseases by the severity of the symptoms. Severity can be assessed by the duration of the disease and the likelihood of the disease causing death. Another approach is to examine any common symptoms that different diseases may have. For example, reductions in feed intake and general activity and increased time spent resting are behavioural changes that accompany a wide number of illness in many species (see Section 5 of this chapter and Chapter 4), such that the relative severity of illnesses might be judged by comparing the relative magnitude of these changes.

Some data is available to let us judge the relative impact of different diseases of dairy cattle. For lactating cows, many illnesses reduce milk production and feed intake. By examining detailed records of milk production and feed intake from a research farm, Bareille et al. (2003) were able to estimate both the likely durations and the effects on milk production and feed intake of various diseases, as illustrated in Table 2.1.

The diseases are ranked according to a rough measure of their severity, based on the combined loss of milk and the reduction in feed intake. For example, foot lesions associated with lameness were found to affect milk production for a total of 117 days, leading to a cumulative loss of milk of almost 77 kg and a cumulative reduction in feed intake of almost 28 kg (dry matter). Lameness associated with hock lesions affected milk production for over 130 days, reducing milk yield by 109 kg and feed intake by 48 kg. In contrast, localized mastitis (not shown in Table 2.1) affected the cow for 54 days, reducing milk yield by 13 kg and feed intake by only 2 kg. Thus, according to this approach, localized mastitis may have fewer detrimental effects than lameness on cow welfare.

Table 2.1 Effects of different diseases of dairy cows on milk production and feed intake. The diseases are ordered by averaging their ranking for the effect on milk production and the ranking for the effect on food intake

	Duration[1]	Lost Milk[2]	Reduced feed[3]
Hock injury	131	109	48
Systemic mastitis	143	160	30
Acute metritis	145	57	47
Ketosis	131	20	72
Very difficult calving	70	52	43
Hoof lesions	117	77	28
Milk fever	52	45	38
Teat injury	63	155	5
Chronic metritis	104	39	18
Difficult calving	98	6	37
Retained placenta	56	33	10
Local mastitis	54	13	2

[1]Duration of effect based on data in Bareille et al., (2003). This estimate is the largest of the effects on either milk production or feed intake.

[2]Estimate of the cumulative loss of milk production (kg) due to the disease. Based on data presented in Table 3 of Bareille et al. (2003).

[3]Estimate of the cumulative reduction in feed intake (kg dry matter) due to the disease. Based on data presented in Table 4 of Bareille et al. (2003).

This type of data has some potential in evaluating the relative effects of different diseases on cattle welfare, but there are difficulties that must be considered. First, certain diseases, such as afflictions of the udder, may have pronounced local effects but little overall effect on the cow. For example, in the study reviewed above, teat injuries had a very large effect on milk production but relatively little effect on feed intake (Table 2.1). In contrast, metabolic diseases such as ketosis have relatively little influence on milk production but large effects on feed intake (Table 2.1). Second, the reliability of estimates of the effects of the diseases on milk production is uncertain. Other estimates (e.g. Ostergaard and Sorensen, 1998; Gröhn et al., 2003) of the impact of diseases on milk production differ somewhat from those of Bareille et al. (2003). Third, we need a better way of weighting the different symptoms. Table 2.1 is derived simply by ranking each disease and giving equal weight to the effects on milk production and feed intake with no justification for this weighting.

An alternative means of assessing the impact of illness or injury upon animal welfare is to ask the experts. While this may seem an unhappily subjective approach to some (who are the "experts"?), this provides a means forward when it is difficult or impossible to obtain more objective information, and a number of procedures have been developed to determine what consensus there is among experts.

Whay and colleagues conducted a Delphi exercise, involving repeated consultation with a panel of animal-welfare researchers and veterinarians who were asked to rank various animal-welfare parameters, including health problems, in terms of their importance for animal welfare. The results of this exercise were then used to assess a number of dairy farms in the UK (Main et al., 2003a; Whay et al., 2003a). The results were then given to a further panel of 50 experts who in turn were asked to determine the severity of each welfare parameter and whether some intervention was required to rectify the situation. This gives some idea of the relative importance the experts felt each welfare indicator had for animal welfare. A large majority (at least 75%) of the experts felt intervention was needed in the case of mastitis when the incidence was between 27% and 29% (which involved the worst 60% of farms). In the case of lameness, the majority of experts felt that some intervention was called for when the incidence was around 18% (i.e. on the worst 80% of farms). This indicates that, generally, the experts felt that lameness represented a more severe threat to welfare than mastitis.

These results are, of course, limited by what the experts know; the importance of some diseases is likely to be underestimated. Also judgements will be affected by what the experts are used to. For example, mastitis was judged by the experts to be less important than lameness in terms of its impact on animal welfare perhaps because many experts now consider a high incidence of this disease to be the norm.

Clearly, to understand the relative impact of different animal disease on animal welfare, we need better data on the duration and severity of the full range of symptoms of each disease. Until then, we may need to rely on techniques for obtaining "expert opinion" on the importance of various diseases for animal welfare.

3 Reliability and Validity of Measures of Animal Health

Even where we have good grounds for believing that a disease has a marked effect on animal welfare, to use the occurrence of this disease to assess animal welfare, we must be able to obtain valid and reliable measures of its occurrence. Work on limited numbers of research herds allows easy and continuous access to consistently managed animals, which greatly increases the chance of obtaining reliable data. In epidemiological research, large numbers of herds and animals are available, but continuous access to these animals is difficult and many differences among farms cannot be controlled or even assessed by the researcher.

3.1 Incidence and Prevalence

The frequencies of different diseases or health problems are usually assessed by measuring either the prevalence or the incidence. Prevalence is assessed by determining the number of cases (i.e. the number of animals that suffer from a disease) at any one point in time. Incidence is assessed by the number of new cases that develop over a given period of time. The period of time over which incidence is assessed will vary according to the nature of the ailment; for example, this can be over a year, lactation, or other stage of production. For chronic illness, where the duration of the illness is long, the prevalence is often higher than the incidence, since the latter is based only on new cases, whereas measures of prevalence will include all cases (new and ongoing). However, many illnesses of cattle are relatively short lasting and in such cases measures of incidence will be higher than those of prevalence. In terms of assessing welfare, measures of prevalence and incidence each have their own advantages and drawbacks.

The main advantage with a measure of prevalence is that it can be taken during one or a few visits. This is important when researchers do not have continuous access to animals, for example, when animals are housed extensively and are rounded up only a few times a year. The main problem with measures of prevalence concerns the adequacy of sampling, since the occurrence of illness fluctuates over time. The prevalence of an illness at any one point in time may be affected by a number of factors that can vary from visit to visit. The adequacy of a single-time sample will depend upon the duration of the illness; short-lasting illnesses are more likely to be missed, and prevalence measures of mortality are not possible. Furthermore, most illness is not randomly distributed over time: there is often marked seasonal or monthly variation in the incidence of diseases (e.g. Whitaker et al., 2000).

Measures of incidence involve the collection of data on the number of animals that become ill over a period of time and hence represent a more complete sample than do measures of prevalence. However, there is a practical difficulty in collecting such data: researchers must often rely on data collected by others, such as herdsmen, and any problems in record keeping or diagnosis will reduce the value of the data.

3.2 Validity and Reliability of Diagnosis

The reliability of recognition of disease and injury will depend greatly upon the skill and training of the observer, and performance can be improved by using well-defined diagnostic procedures. Some researchers rely upon records kept of veterinary treatment, but the quality of these records depends on the accuracy of the diagnostic procedure and the consistency with which it was applied, and these records likely underestimate the incidence of many diseases. For example, in Denmark, the incidence of veterinary treatment for hoof problems (hoof lesions or digital dermatitis) is around 1–2% (Alban et al., 1996) and for lameness 7% (Alban, 1995) even though direct recording of sole disorders at hoof trimming found an average incidence of 38% (Vaarst et al., 1998). One study of respiratory diseases in unweaned dairy calves showed that only 28% of cases were treated by a veterinarian (Svensson et al., 2003).

The decision whether to treat animals suffering from the more common diseases is most often made by the farmer and so depends greatly on the farmer's ability to correctly diagnose the disease and develop a treatment plan. This will vary greatly from farm to farm and from illness to illness. For example, because mastitis has immediate effects on milk revenue, cows with mastitis are more likely to be treated than cows suffering from lameness. Thus measures of treatment likely underestimate the incidence of some diseases and likely reflect the perceived economic return on the cost of treatment.

Even when treatment occurs accurate records may not always be kept. Scandinavian countries typically have a centralized data bank recording all veterinary treatments of farm animals (Ekesbo et al., 1994), and these have been used to assess the incidence of various diseases of dairy cattle and to relate these to various management practices in Denmark (Alban et al., 1996; Bruun et al., 2002), Finland (Schnier et al., 2002), Sweden (Manske et al., 2002), and Norway (Waage et al., 1998). However, such systems are not in use in other countries. Some countries require only that any treatments involving antibiotics be recorded. Some alternative health-recording systems are available, such as the National Animal Health Monitoring System of the US Department of Agriculture (Loneragan et al., 2001) and the various dairy-herd systems, but these usually rely on voluntary membership by the farmer. In most cases, data on the incidence of illness relies on the farmer making the diagnosis and keeping records.

Given the relative rarity of veterinary treatments for many diseases, most on-farm assessments of welfare rely upon the farmer's diagnosis. However, the accuracy of these will depend upon the farmer's skill, diligence, the amount of time available and the effectiveness of the recording method used. Farmers themselves underestimate the occurrence of some illness. For example, Whay et al. (2003b) reported that UK dairy farmers estimated the prevalence of lameness on their farms at 5.7%, whereas estimates of prevalence based on systematic gait scoring was 22%. Farmers may also change how they perceive a disease and even how they manage their cattle once they are asked to report on health problems (Ducrot et al., 1998).

As well as the difficulties in diagnosis of illness, farmers will vary in the quality of the records they keep. Studies typically report that up to 20% of records have to be excluded because of doubts about the validity of the data or because of poor or incomplete record keeping (Wells et al., 1996; Fourichon et al., 2001; Loneragan et al., 2001). The effectiveness of farmers' records will likely vary according to how they are kept, e.g. daily record keeping is likely to produce more valid data than when farmers are asked retrospectively to describe causes of death or illness. It is particularly troublesome if the quality of the diagnosis is related to the incidence of the illness on the farm. Loneragan et al. (2001) found that feedlots that reliably and regularly supplied data on mortality rates tended to have a lower rate of mortality than feedlots that only irregularly supplied data, and Schnier et al. (2002) found a trend for higher rates of mastitis among the farms that did not agree to participate in a study of dairy cow health problems. Thus, farms excluded because of poor or unreliable records may have poorer levels of animal welfare than those included.

Moreover, farmers may be reluctant to provide records of health treatments on farms. This "self-selection" by farmers is a common complicating factor in research done on farms. Studies that rely on farmers volunteering to comply do not represent a random sample of farms. Studies typically report that between 25% and 50% of farmers contacted will not participate in studies of animal health (Wells et al., 1996; Frei et al., 1997; Whay et al., 2003b). The effect of this bias on the results will depend on the extent to which the decision to participate or not was related to the incidence of illness. All of these factors greatly complicate the collection of reliable data on the incidence of illness using the farmer's diagnoses of illness.

Some information on animal health can be obtained from culling records. This is often easier in dairy production where milk-recording companies ask farmers for records and reasons for culling. However, again the usefulness of the records will depend upon the farmer's ability to correctly recognize the cause or causes of death. For example, reproductive failure is often cited as the leading reason why dairy cows are culled (e.g. Whitaker et al., 2000), although many cases of reproductive failure are rooted in lameness (Sprecher et al., 1997). Furthermore, farmers are not always consistent in whether or not to cull a cow. Certain diseases are more likely than others to result in cows being culled: although the incidence of mastitis is similar or only slightly higher than for lameness, nearly four times as many cows are culled because of mastitis (Seegers et al., 1998). Accurate records of illness and mortality are more easily obtained from indoor intensive production, where there is regular, daily contact between animals and the caretakers, than from outdoor, extensive production, where cattle may not be seen for days, weeks, or months at a time.

Such factors complicate the use of measures of illness, injury, and mortality, and make comparisons between different studies difficult.

4 Types of Diseases

Below we discuss some of the more common health ailments useful for assessing animal welfare. We discuss evidence that relates to the impact of these diseases on animal welfare, and the problems in detecting and measuring their incidence or prevalence in dairy, beef, and veal production.

4.1 Mortality

High rates of mortality on a farm can be a direct indicator of welfare problems. Death itself provides an end of suffering, but the illnesses that eventually lead to death are often the source of considerable suffering to animals. However, mortality figures do not take into account those diseased animals that do not die and, perhaps for this reason, mortality rates are not well correlated with the incidence of various diseases. For example, Busato et al. (1997) found only a low correlation between mortality and morbidity on cow–calf farms in Switzerland. Records of mortality also must be treated cautiously. Estimates can be complicated by epidemic diseases that kill relatively large numbers of animals over a short period. Mortality figures are also affected by culling, which may be done for humane, health, or production reasons. The importance of culling differs between different production systems: many more beef-suckler calves die than are culled, whereas the opposite is true of dairy cows. A high rate of cull can provide welfare advantages, as animals are slaughtered before disease becomes life-threatening, but cattle are frequently culled for reasons other than illness, such as low milk production, so cull rates are likely to be variable and difficult to interpret.

In assessing animal welfare, measures of mortality are more useful if we know the cause, but accurately identifying the cause of death can be difficult, especially under extensive husbandry conditions or other situations where contact with herdsmen is infrequent. In many cases a post-mortem is necessary to accurately reveal cause. For example, feedlot cattle initially thought to have died from respiratory disease were found upon post-mortem to have suffered from a variety of ailments (Loneragan et al., 2001).

Table 2.2 provides estimates of mortality rates that have been obtained for dairy, beef, and veal cattle in various production systems.

It is clear that mortality is highest for unweaned calves. Mortality is especially high soon after birth: Wells et al. (1996) report that of the calves that died during the 8 weeks before weaning, 30% died during the first week of life. The substantially higher death rates of dairy-beef calves suggests problems associated with the care of these animals, such as inadequate colostrum intake and protection from cold weather during and after transport (Moore et al., 2002). Mortality rates would therefore seem particularly useful in examining the welfare of younger animals.

Table 2.2 Some estimates of mortality rates in dairy and beef production

Type of animal	Type of production	Mortality	Mortality rate (%)
Calf	Beef cow-calf[1]	Pre-weaning	6.3
Calf	Dairy[2]	Pre-weaning	9.4
Calf	Dairy[3]	24-h post-calving	7
Calf	Dairy beef[4]	4-week post-calving	13.5
Calf	Dairy[5]	Post-weaning to 1st calving	2.2
Calves and adults	Beef feedlot[6]	Total	1.3
Cows	Dairy[7]	Annual	1.2

[1]Busato et al., 1997. [2]Losinger and Henrichs, 1997. [3]Fourichon et al., 2001. [4]Moore et al., 2002. [5]Wells et al., 1996. [6]Loneragen et al., 2001; Seegers et al., 1998.

The relatively low rates of mortality among older animals means that a sizeable number of animals are needed to make valid comparisons, suggesting the need for epidemiological rather than experimental research.

Mortality rates vary markedly from farm to farm. For example, Fourichon et al. (2001) found that calf mortality during the first 24 h after birth varied from 0% to 31% across dairy farms in France. A certain amount of this variability may be due to differences in record keeping or accuracy of estimates but it is difficult to believe that this accounts for most of the differences reported. That mortality rates vary so much between farms suggest that many of the deaths on the high mortality farms are avoidable and that mortality rates provide useful information about the relative level of animal welfare on different farms.

4.2 *Lameness*

Lameness in cattle can result from a number of causes including infectious disease (such as digital dermatitis and foot rot) and hoof lesions (e.g. ulcers, haemorrhages, white line separation) that are associated with both metabolic challenges and physical injury to the hoof. Lameness is widely regarded as a major welfare problem for dairy cows (Farm Animal Welfare Council 1997) and the incidence of lameness has been included in a number of on-farm animal-welfare assessment schemes. In their "Delphi exercise", Whay et al. (2003a) found that experts consistently ranked lameness as one of the most serious welfare problems for cattle. The majority of experts felt that the incidence on all dairy farms was sufficiently high to justify some intervention to reduce the incidence of lameness.

Obtaining more objective information to assess the magnitude of the impact of lameness on animal welfare has proved difficult. Lameness reduces animal welfare to the extent that animals are experiencing pain when walking or even standing still, and research using local anaesthetics illustrates such effects (e.g. Rushen et al., 2007b). However, this does not allow us to compare the effect of lameness with the effects of other diseases, such as metritis or mastitis. The effects of lameness upon milk production and feed intake reviewed earlier (Bareille et al., 2003) indicate that lameness has a large effect on welfare (Table 2.1).

4.2.1 Difficulties in Detection and Assessment

Given the effect of lameness on the welfare of the cattle, it would seem logical to use measures of the occurrence of lameness to assess the degree of animal welfare. However, this raises the issue of the accuracy of such measures. Unfortunately, studies show that dairy farmers find it difficult to identify lame animals, especially those at the early stages of lameness (Whay et al., 2003b; Espejo et al., 2006), casting doubt upon the validity of farm records. Given the challenges in identifying lame cows, it is not surprising that lameness is often not treated by a veterinarian and that few cows are culled because of lameness: within the UK only 1.7% of cows are culled because of lameness even though the annual incidence is over 25% (Whitaker et al., 2000). Thus records of veterinary treatment, farmers' diagnoses, and culling underestimate the incidence of lameness.

Even when lameness or hoof problems are assessed by trained veterinarians or researchers, there seems to be little consensus as to the best way of scoring the problem. The most direct way of detecting lameness is to score the gait of the cows while walking. Lameness researchers have developed and used locomotion-scoring systems to assess the severity, duration, and prevalence of lameness (e.g. Sprecher et al., 1997; Manson and Leaver, 1988). However, these gait-scoring systems suffer from a lack of standardization, often a low degree of reliability and have not been adequately tested for validity. Flower and Weary (2006) describe a gait-scoring system based on observable changes in the way cows walk when they become lame. Figure 2.2 and Table 2.3 illustrate this gait-scoring system.

The main advantage of this approach is that it describes a wide range of changes in gait that can occur when cows become lame, rather than relying too heavily on one or a few features. Furthermore, the system allows observers to rate components of the gait separately, perhaps improving the ability to detect gait problems.

The validity of this gait-scoring system has been tested in a number of ways. Since the effect of lameness on animal welfare occurs because lameness makes

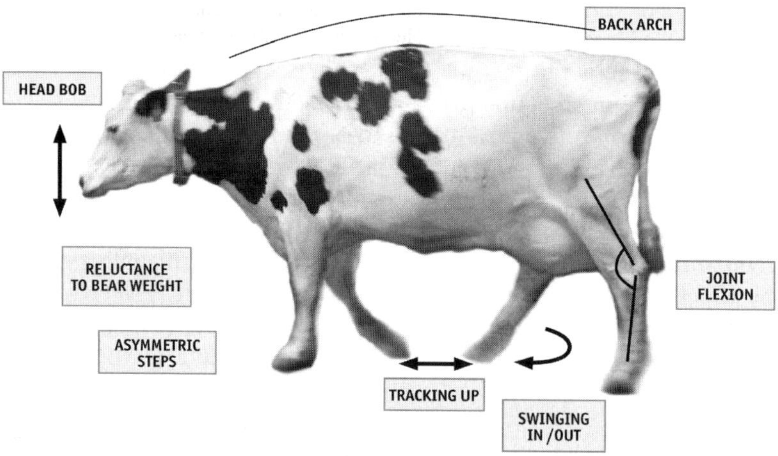

Figure 2.2 Cows that are lame or becoming lame show a variety of changes in the way they walk. Most obviously they can show a reluctance to bear weight on one or more legs or show some asymmetry in the length or timing of their stride. There can be reduced flexion of joints, making the legs appear stiff. In some cases, lame cows show an obvious arch to their back, their heads may bob up and down markedly and their back legs will either swing out or swing in. Normally when cows walk, they place their back hooves at the same place that they place their front hooves (tracking up). Lame cows, however, may place their back hooves further back behind where the front hooves were. Scoring the occurrence of each of these changes in gait is a promising way of detecting lameness (Flower and Weary, 2006) (Figure copyright Frances Flower, University of British Columbia.)

Table 2.3 An example of a gait-scoring system based on the observed changes in cattle's gait when they are becoming lame (illustrated in Figure 2.2) (Flower and Weary, 2006)

Locomotion score	Clinical description	Assessment criteria
1.0	Sound	Smooth and fluid movement. Flat back posture when standing and walking. Symmetrical gait without abduction/adduction. All legs bear weight equally and joint flex freely. Hind-claws land on or in front of fore-claw prints (tracking up). Head carriage remains steady as the animal moves.
2.0	Imperfect Locomotion	Ability to move freely not diminished. When standing and walking, back posture is flat or mildly convex – in the absence of other gait abnormalities this stance is likely attributed to normal posture. Gait slightly asymmetrical due to minimal adduction/abduction. All legs bear weight equally but joints show slight stiffness. Hind-claws do not track up perfectly but shortened strides are uniform. Head carriage remains steady.
3.0	Lame	Capable of locomotion but ability to move freely is compromised. Flat or mildly convex back posture when standing, but obviously arched when walking. Gait is asymmetrical due to adduction/abduction. All legs bear weight equally but a slight limp can be discerned in one limb. Joints show signs of stiffness but do not impede freedom of movement. Hind-claws do not track up and strides may be shortened. Head carriage remains steady.
4.0	Moderately Lame	Ability to move freely is obviously diminished. Obvious arched back posture when standing and walking. Gait is asymmetrical due to adduction/abduction and one or more strides obviously shortened. Reluctant to bear weight on at least one limb but still uses that limb in locomotion. Strides are hesitant and deliberate and joints are stiff. Head bobs slightly as animal moves in accordance with the sore hoof making contact with the ground.
5.0	Severely Lame	Ability to move is severely restricted. Animal must be vigorously encouraged to stand and/or move. Extreme arched back posture when standing and walking. Gait is asymmetrical due to adduction/abduction, one or more strides obviously shortened and/or inability to bear weight on one or more limbs. Obvious joint stiffness characterized by lack of joint flexion with very hesitant and deliberate strides. Head obviously bobs as sore hoof makes contact with the ground.

walking painful, we would expect that a valid gait-scoring system would be sensitive to the effects of pain-reduction methods. Recently, Rushen et al. (2007b) showed that the gait scores given to lame cows using this system were reduced when the cows were injected with a local anaesthetic in the leg that was responsible for the lameness. Since lameness in dairy cows is most often a consequence of lesions in the hoof, we would expect that a valid gait-scoring system should be able to identify cows that are suffering from these. Flower and Weary (2006) showed that this gait-scoring system was also successful identifying animals with severe lesions.

Gait-scoring systems are useful but require dairy farmers to spend time watching their animals walk. With the marked increase in the size of dairy herds, there has been a lot of interest in ways of automatically detecting lameness. Several such methods rely on how cows distribute their weight among their legs both when walking and standing (Rajkondawar et al., 2002; Neveux et al., 2006; Pastell et al., 2006). These studies also provide objective measures that provide a better basic understanding of how cows respond to the pain and injuries associated with lameness. Another objective approach has been to take quantitative measures of cow's gait by digitizing video records of cows walking with and without known injuries. In one study Flower et al. (2005) used motion-analysis software to calculate six stride variables for each hoof. Compared with cows with sole ulcers, healthy cows walked faster, had shorter stride durations, and longer strides. The percentage of time when cattle were supported by three legs (called triple-support) more than doubled for cows with sole ulcers compared with healthy cows. These results show that such "kinematic" gait analysis is a promising approach to understanding how the gait of dairy cows change with the onset of lameness.

Observing the incidence of hoof problems might seem to be a more reliable approach than gait analysis (Figure 2.3). However, there is little standardization in how hoof lesions are scored. Alban (1995) included all cases of "contusion, foul in the foot, sole ulcer, foot rot, digital and interdigital dermatitis, laminitis, swollen hock, arthritis", basing these on undefined diagnoses and without distinguishing between the problems. Vaarst et al. (1998) counted acute haemorrhage, sole ulcers, enlarged white line, heel horn erosion, and interdigital dermatitis, and analysed data on acute haemorrhages separately from that for sole ulcers. Manske et al. (2002) assessed severity on a 4- point scale of dermatitis (distinguishing three types), sole haemorrhages, separations (distinguishing between sole and white line), and sole ulcer, analysing the data by aggregating the first three categories separately from sole ulcers. A complete description of all of the different ways that researchers have measured hoof lesions or attempted to analyse their frequency is beyond the scope of this book, but these examples illustrate the lack of standardization in the characterization of hoof injuries. Many hoof injuries can only be assessed during hoof trimming, limiting the frequency of assessment. Thus observations of the hoof injuries are best measured as prevalence, rather than of incidence. Prevalence is known to vary with cow parity and stage of lactation (e.g. Offer et al., 2000), so these factors should be taken into account in any analysis.

In general there is a lack of knowledge about how painful different types of foot lesions are, or how foot lesions can be scored in terms of their relative effect on animal welfare. Different types of foot injuries are likely to cause different degrees of pain (Whay et al., 1998). Clearly, we need more standardized and better

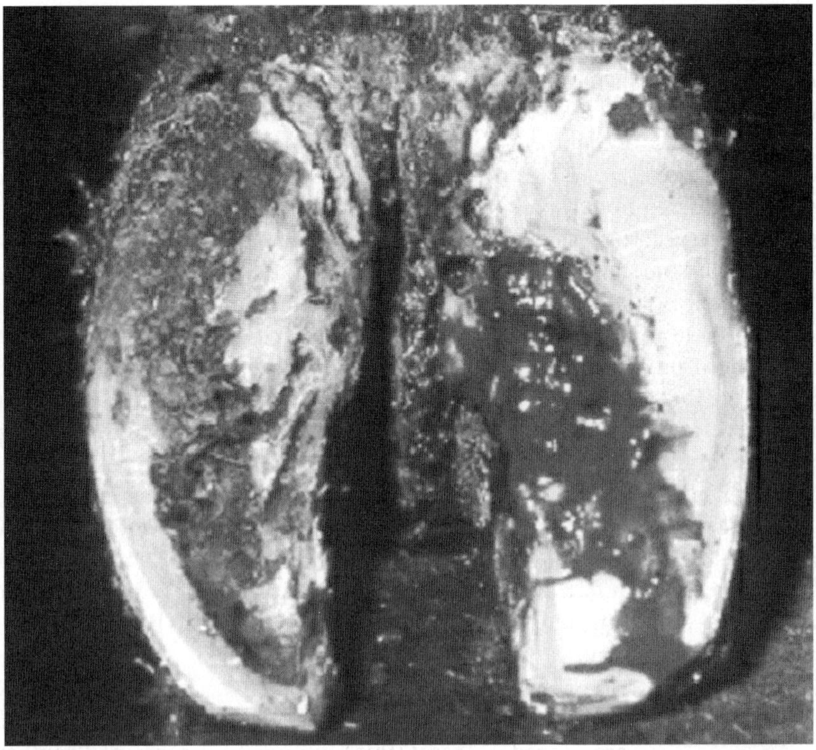

Figure 2.3 Lameness is a serious welfare problem in dairy cows and most cases of lameness arise from problems in the hooves. When hooves are trimmed they often show a variety of lesions, which can range from small areas of erosion to large, painful, bleeding ulcers. Studies in a number of countries suggest that a majority of dairy cows will have some sort of hoof lesion. Scoring the types and severity of hoof lesions can be useful in welfare assessment

validated approaches to assessing gait and hoof and leg pathology and a better understanding of the pain associated with different types and severities of injury.

4.2.2 Incidence and Prevalence of Lameness and Related Injuries

Despite the problems of detecting lameness, a number of studies have reported a high incidence of lameness among dairy cows (Frei et al., 1997; Whitaker et al., 2000; Fourichon et al., 2001; Espejo et al., 2006). Lameness is not restricted to the large intensive dairy farms of the industrialized world: small-scale dairy farms in Kenya report a lameness prevalence of nearly 12% (Gitau et al., 1996). Variation between farms is consistently reported to be high, with estimates varying from 0% up to 100% (Fourichon et al., 2001). Within the UK, where the most frequent and comprehensive surveys have been done, the average percentage of cows becoming lame during a lactation varied from 5.8% for the best quartile of farms to 50% for the worst quartile of farms (Whitaker et al., 2000). Much of this variability likely relates to a range of management and housing factors that have been identified as risks (e.g. Whitaker et al., 2000; Cook et al., 2004) and will be discussed in later chapters.

The occurrence of injuries to the hoof is even higher than the incidence of lameness: Manske et al. (2002) reported that on average 70% of cows on Swedish dairy farms suffered from some form of hoof lesion but this varied from 18% to 98% between different farms; Vaarst et al. (1998) reported a prevalence of 38% of cows suffering from some hoof disorder at the time of hoof trimming in Danish dairy cows. Erin Bell (2004) reported that in British Columbia, 86% of dairy cows suffered from some form of hoof lesion. Generally, lameness in dairy cows is more related to hoof problems than upper-leg problems (e.g. Gitau et al., 1996; Offer et al., 2000). However, lameness associated with upper-leg problems seems to have a larger effect on the cows' feed intake and milk production than does lameness associated with hoof problems (Bareille et al., 2003).

Although incidence is highest in dairy cows, lameness can also affect other cattle, e.g. 1.1% of unweaned dairy calves are lame or judged as having joint problems (Wells et al., 1996). Nearly 2% of feedlot cattle suffer from lameness (USDA, 2000) and over 73% of beef cows and over 64% of beef bulls arriving at an abattoir were observed to be suffering from lameness (Roeber et al., 2001).

4.3 Mastitis

Mastitis is an infection of the mammary gland caused by a number of different bacteria. *Eschericia coli* and *Staphylococcus aureus* generally account for the majority of cases with *Streptococcus* and *Klebsiella* responsible for a smaller incidence (e.g. Barkema et al., 1999a, b). Because of the direct effects of mastitis on milk quality, records of the incidence of mastitis are fairly good for dairy cows and mastitis is often treated by veterinarians. Even in countries where antibiotics can be administered by farmers, records of mastitis treatment with antibiotics are usually good, especially for dairy farmers enrolled in milk recording or dairy-herd improvement schemes.

Mastitis is usually detected by clinical examination of the udder or by visual inspection of milk samples. Milk samples from mastitic cows can be tested to identify the actual bacteria involved. When mastitis occurs, the cells from the immune system enter the mammary gland and can be detected in milk. Counts of such somatic cells (SCC) are regularly taken from the milk in bulk tanks to assess milk quality of a farm. However, the relationship between bulk-milk SCC of a farm and the actual incidence of mastitis tends to be weak (Whitaker et al., 2000; Barkema et al., 1999a, b), and varies with the cause of the infection, so bulk-milk SCC is of limited value in monitoring the incidence of mastitis for animal welfare purposes.

As with other afflictions, it is difficult to know how mastitis affects the welfare of the animal. As discussed earlier, experts appear to rate mastitis as having less of an effect on animal welfare than lameness (Whay et al., 2003a). In their attempt to "prioritize" diseases of dairy cows, Wells et al. (1998) did not consider mastitis to have important consequences for animal welfare. However, the effect of mastitis on the animal depends on the form of the disease. For instance, systemic mastitis has a longer duration of effect than localized mastitis (Bareille et al., 2003; Table 2.1) and may have greater welfare consequences.

Mastitis is one of the most common diseases affecting lactating dairy cows, with measures of incidence typically ranging from 25 to 40 cases per 100 cows per year in most western countries (Frei et al., 1997; Rajala and Gröhn, 1998; Barkema et al., 1999a, b; Whitaker et al., 2000; Fourichon et al., 2001). Mastitis is also an important cause of culling (Whitaker et al., 2000). The incidence of mastitis varies greatly between farms: Fourichon et al., (2001) reported that the number of cases of clinical mastitis varied between 3.4 and 137.5 per 100 cows per year. Whitaker et al. (2000) reported that the poorest quartile of UK dairy farms had 70 cases per 100 cows and the best quartile had only 13.

Although most often studied in dairy cows, mastitis can also occur among lactating beef cows, although here much less is known about the incidence or the causes. Small-scale studies report an average prevalence of 30–50% of lactating beef cows suffering from mastitis (Simpson et al., 1995; Duenas et al., 2001).

Numerous epidemiological studies have shown that the incidence of clinical mastitis is strongly related to housing and management factors (e.g. Waage et al., 1998; Barkema et al., 1999a, b). Furthermore, the incidence of mastitis has been used in on-farm welfare assessment schemes for dairy cattle (Whay et al., 2003a, b). However, the importance of management and housing varies according to the type of bacteria responsible for the infections. For example, cases due to *E. coli* are likely related to housing conditions, but those due to *S. dysgalactiae* are more likely related to milking procedures and equipment (Barkema et al., 1999a, b). Cows housed in tie stalls have higher rates of mastitis than cows housed in free stalls (Valde et al., 1997). Moreover, cows in confinement housing show the highest incidence of environmental mastitis in the warm and humid months of the year, since moisture and elevated temperatures support microbial growth. Consequently, measures of the overall incidence are of less use than those that identify the pathogens responsible.

4.4 Calving Difficulties

Calving difficulty (sometimes known as dystocia) is another common ailment with clear implications for animal welfare. Calving difficulties can be a leading cause of calf death; Nix et al., (1998) reported that 20–30% of beef calves were likely to die within the 24 h following calvings that required even mild assistance, compared to a baseline mortality rate of only 3%. Caesarean sections resulted in 50% of calves dying. Among dairy cattle, dystocia is a major cause of stillbirths (Meyer et al., 2001). The welfare of surviving calves is also affected: calves that needed assistance during delivery developed enteritis at an earlier age than calves that did not need such assistance (Sivula et al., 1996). Dairy herds with a high incidence of dystocia also tend to have a higher incidence of health problems in calves (Sanderson and Dargatz, 2000). Calving difficulties appear to have less of an impact on the cow than on the calf; although dystocia can cause reproductive problems in the cows, this appears to have only moderate effects on milk production or feed intake (Bareille et al., 2003). However, dystocia is associated with increased incidence of metritis and retained placenta (Huzzey et al., 2007; Gröhn et al., 2003), as it increases trauma to the uterine wall and increases the susceptibility to disease by increasing the risk of harmful bacteria entering the reproductive tract (Bruun et al., 2002) and it increases the likelihood that the cow will be culled (Rogers et al., 2004).

One barrier to research in this area is of measuring calving difficulty. Dystocia is often defined simply as cases requiring assistance or surgery (Fourichon et al., 2001). However, farms likely vary greatly in when they feel that some assistance is needed. Consequently, measures of calving assistance may reflect the farmer's attitudes rather than anything else. Measures of the degree of difficulty (e.g. Nix et al., 1998) may be more useful.

Many studies show that calving difficulties occur at a relatively low incidence; for example 2.2% of births on Swiss dairy farms (Frei et al., 1997), 2.1% of births in Finnish dairy farms (Rajala and Grohn, 1998), 5% of births on French dairy farms (Fourichon et al., 2001), and 4% of births on beef farms (Busato et al., 1997). However, Frei et al. (1997) found that as many as 50% of calvings were assisted on some Swiss dairy farms. Whitaker et al. (2000) reported that 8.7% of calvings on UK dairy farms were assisted, although this varied from 0% to 57% across farms. Dystocia appears to occur more often on US dairy farms: around 20% of calvings of primiparous cows and 6% of calvings of multiparous cows required some assistance (Meyer et al., 2001). One research herd in the US reported that 23.7% of calvings involved dystocia (Johanson and Berger, 2003).

Because of the welfare consequences to both the cow and calf associated with dystocia, as well as the great variability among farms in rates of calvings judged to require assistance, we feel that this is an urgent area for more research. Specifically, research is required to properly document the potential welfare risks associated with different degrees of assistance, and to determine more objectively when assistance is truly beneficial. New work is also required to understand the role of pain and how this may affect the cow and calf's recovery following difficult calvings (see Chapter 5).

4.5 *Illness at Calving*

The high incidence of morbidity in dairy cattle around the time of calving is of great current interest to both the dairy industry and to dairy researchers worldwide. The "transition phase", generally accepted as the period beginning 3 weeks prior to calving and ending 3 weeks following calving, is recognized as a critical phase in the cow's lactation (Drackley, 1999; Ingvartsen, 2006).

During the transition period cows face a number of stressors including several diet changes and social regroupings, as well as physical, hormonal, and physiological changes associated with calving and the onset of lactation. One of the main challenges for transition dairy cows is a sudden increase in nutrient requirements to support the onset of lactation at a time when feed intake lags behind (Drackley, 1999). The constraints imposed by the decline in feed intake, coupled with other stressors associated with the transition period, likely contribute to the high incidences of metabolic and infectious diseases.

4.5.1 Periparturient Diseases

There is a suite of diseases that afflict cows during transition, including metritis, ketosis, fatty liver, displaced abomasum (DA), and milk fever. Ketosis is a metabolic condition that occurs when a cow is in negative-energy balance immediately after calving. Peak prevalence of subclinical ketosis occurs in the first 2 weeks of lactation (Duffield et al., 1998). To support the cows' energy demands, the body must mobilize fat reserves resulting in the production of ketone bodies by the liver. This condition increases the cows' risk of DA, fatty liver, retained placenta, reduces milk production, and decreases reproductive performance (Duffield, 2000). The reported prevalence of subclinical ketosis ranges from 8.9% to 34% in various studies (Dohoo and Martin, 1984; Duffield et al., 1997). Duffield et al. (1998) reported that the cumulative incidence of subclinical ketosis over the first 9 weeks of lactation in 507 untreated cows from 25 Holstein dairy farms was 59% and 43% depending on the assessment method. It is difficult to compare these numbers across studies since numerous factors beyond cow- and herd-level risk influence the rates. Dohoo and Martin (1984) found that cows with subclinical ketosis had an increased risk of metritis or clinical ketosis 4 days later. However, these authors argued that since metritis is a condition that normally develops at calving, subclinical ketosis is more likely a result rather than a cause of metritis. Milk fever (hypocalcemia) can result due to the increased demands for calcium with the onset of milk production (Østergaard and Larsen, 2000). Hypocalcemia affects muscle function, heart rate, and can lead to rumen bloat, decreased ruminal contractions, and suppressed appetite (Østergaard and Larsen, 2000).

Overall estimates combining a suite of transition diseases indicate that up to 30% of dairy cows in North America are afflicted. A Minnesota study reported that 25% of cows leaving the herd do so in the first 60 days after calving, and suggested that many of these culls are due to these diseases (Godden et al., 2003). These disorders have not figured prominently in on-farm welfare-assessment schemes (e.g. Main et al., 2003a, b; Whay et al., 2003a, b) and unfortunately there is little data to judge the relative impact of these diseases on animal welfare. Ketosis appears to have only moderate effects on milk production (although it has marked effects on feed intake) (Bareille et al., 2003; Table 2.1). Experts recommend intervention when the incidence of milk fever (hypocalcemia), in particular, is quite low (Main et al., 2003a) but this may reflect the relatively low incidence of the diseases compared to mastitis or lameness. As veterinary examination of post-partum cows is relatively infrequent on most dairy farms, typically only during routine herd health checks that commonly occur once every two weeks, many cases of periparturient disease may go unnoticed. Producers can use urine or milk tests to monitor the health of their animals, but frequent administration of tests on a herd-wide scale can be costly and time consuming. Moreover, no such tests are available for diagnosing inflammatory uterine disease (metritis or endometritis), one of the most common disorders after calving.

4.5.2 Metritis

The most common reproductive disease in dairy cattle is metritis, an inflammation of the uterine wall caused by bacterial infection and usually diagnosed by elevated body temperature, vaginal discharge, and a large, flaccid uterus. The latter is usually determined by rectal palpation, although it appears to be a poorer diagnostic tool than examining vaginal discharge directly (LeBlanc et al., 2002). Metritis is normally associated with calving, and is most likely to be diagnosed in the first few weeks after parturition.

Metritis can have long-term effects on dairy cows, substantially reducing both milk production and feed intake (Bareille et al., 2003; Table 2.1) and indicating a relatively severe impact on animal welfare. It is therefore unfortunate that metritis tends not to be included (as a separate disease) in on-farm welfare assessment protocols (e.g. Whay et al., 2003a, b). Estimates of the incidence of this disease tend to be variable. Some studies report only a low incidence, often of less than 5% per lactation (Rajala and Grohn, 1998; Fourichon et al., 2001; Bruun et al., 2002). However, other studies have found a much higher incidence: Frei et al. (1997) reported a mean incidence of 37.2% of metritis in Swiss dairy cows, Fleischer et al. (2001) reported a mean lactational incidence of 23.6% in German dairy farms while estimates in US dairy cows are 65% (Hirvonen et al., 1999) and 38% (LeBlanc et al., 2002). These differences likely reflect differences in how well the disease is diagnosed, especially when diagnosed by farm staff.

Even where reported incidence is low, there is usually considerable variation between farms, with some farms reporting a high incidence. For example, the median incidence of treatment on Danish farms was 0.7%, but the highest incidence of treatment recorded on any one farm was 21% (Bruun et al., 2002). On French dairy farms, the median incidence was found to be only 3%, but the maximum incidence was nearly 17% for acute metritis and 50% for chronic metritis (Fourichon et al., 2001).

The risk factors for metritis are not well known. Many studies (e.g. Curtis et al., 1985; Correa et al., 1993) have found that metritis is associated with retained placenta. However, the nature of the relationship between housing and management factors and metritis is not clear. Kaneene and Miller (1995) found that "larger herds, problem calvings, overconditioning, and underconditioning" were associated with increased incidence of metritis although their discussion of these relationships was speculative. Much remains to be learned about if and how this disease can be prevented by changes in housing, nutrition, and management. It also seems likely that metritis has been underestimated as a cause of poor welfare in dairy cattle and measures of its incidence underused in welfare assessment.

Interestingly, infectious diseases such as metritis generally receive less attention than the metabolic diseases that are prevalent after calving. This is presumably because ketosis and milk fever have dramatic impacts on milk production and on an animal's appearance, while metritis has few overt symptoms (Lewis, 1997). Interest in prevention and treatment of metritis has increased due to concerns that metritic cows suffer from reduced fertility.

4.6 Bovine Respiratory Disease

Probably the most serious welfare problem affecting beef cattle in feedlots is bovine respiratory disease (BRD). Although it is an infectious disease, BRD is thought to result from an interaction between the presence of the pathogens, and suppressed immunity resulting from the stress of transport to and arrival at the feedlot. Respiratory disease accounts for 57.1% of deaths of feedlot cattle (or 0.72% of all cattle in feedlots) in the US (Loneragan et al., 2001). As discussed earlier in this chapter, post-mortem results show the number of animals that die from the disease (Loneragan et al., 2001), but these numbers underestimate the numbers of animals that suffer from the disease. Pulmonary lesions can be found in up to 70% of steers in feedlots, although these are not detected through standard clinical observations (Wittum et al., 1996). In feedlots, sick cattle are usually identified by behavioural changes, such as a reduced feed intake. Unfortunately, relatively little is known about how behaviour changes in response to respiratory disease. Research to date has indicated that feedlot cattle suffering from BRD increase their number of drinking bouts, reduce time spent feeding and reduce the number of visits to the feed bunk (Sowell et al., 1999; Buhman et al., 2000; Quimby et al., 2001), but much remains to be learned about how changes in behaviour can be used to improve early identification of animals with this and other diseases (Section 5 of this chapter).

4.7 Respiratory and Gastrointestinal Diseases in Calves

For young calves, the main diseases of importance are respiratory and gastrointestinal (GI) disorders. They are also major causes of death, responsible for 52% and 16% respectively of pre-weaning deaths in beef calves (Busato et al., 1997) and 46% and 14% of deaths of dairy calves (Agerholm et al., 1993). The relative importance of the two as a cause of death varies between regions; in contrast to the figures from Western countries provided above, GI-tract disorders accounts for more deaths (31%) on dairy farms in Kenya than do respiratory problems (17%) (Mulei et al., 1995). These two diseases are among the major causes of antibiotic use in the dairy industry (Ortman and Svensson, 2004).

The incidence of both afflictions is high, especially in the first few weeks after birth, with several large-scale national surveys showing diarrhoea affecting between 10% and 35% of dairy calves and respiratory disorders affecting between 8% and 15% (Wells et al., 1996; Svensson et al., 2003; Frei et al., 1997). Clearly there is much potential in using measures of respiratory and GI-tract disorders in assessing the welfare of unweaned calves.

4.8 Body Injuries

Due to the pain that results from most physical injury, occurrence of injury would seem to provide a relatively clear indication about the existence of a welfare problem. Problems arise, however, in finding a reliable and accurate means of scoring injuries, especially more minor ones.

Difficulties in defining an injury can lead to marked differences between studies. For example, in a survey of UK dairy herds, Whitaker et al. (2000) reported that only less than 2% of cows suffered from injuries and that on most farms no injuries were seen. In this case, injuries were defined as "damage to teats and injuries from equipment, buildings and obtruding objects". Enevoldsen et al. (1994) similarly report that between 1% and 5% of dairy cows have injuries. However, other studies report a much higher incidence, suggesting that there were marked differences in how serious an injury had to be before it was counted. Detailed observation of injuries by Whay et al. (2003b) found that on average well over half the cows on dairy farms had some form of injury. Weary and Taszkun (2000) found that over 75% of dairy cows had some form of hock lesion (Figure 2.4). Much of this variation results from the fact that an "injury" can range from a major trauma to a minor skin abrasion.

Finding a reliable and standardized measure of the incidence and severity of smaller injuries remains one of the obstacles to using these in welfare assessment. Injuries to the leg have been evaluated using qualitative methods of assessment (e.g. Weary and Taszkun, 2000) or quantitative measurements such as surface area of hair loss (Mowbray et al., 2003). The quantitative measurements have the advantage of being more repeatable, and more amenable to parametric statistical analyses, but taking such measures is often much more time consuming. The choice of the method of assessment should ultimately depend on how well it reflects the way that the injury actually affects the animal, either in terms of the pain experienced, or in predisposing the animal to other injuries, infections, or physical impairments such as abnormal gait. Unfortunately, for many injuries, little or no research is yet available to establish these links.

For dairy cattle, injuries to the legs and the udder are most common. Experts judge that the presence of injuries, particularly swollen or ulcerated hocks, is one of the most serious threats to the welfare of dairy cattle (Whay et al., 2003a), with the majority of experts recommending some corrective action when the incidence of swollen hocks was less than 10%. Hock injuries that are severe enough to result in lameness reduce milk production and feed intake (Bareille et al., 2003; Table 2.1). Less severe hock injuries are common in dairy cattle (Weary and Taszkun, 2000) and, along with swollen knees, are an indicator of problems in the

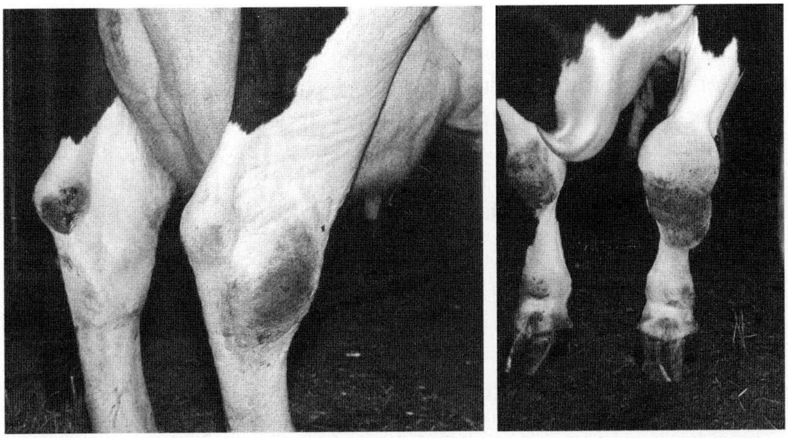

Figure 2.4 Dairy cows housed indoors show a variety of minor injuries to their legs, such as hock abrasions (left) and swollen knees (right). These can be painful and may be a route of infection. To use such injuries to assess animal welfare requires that we have reliable and accurate means of measuring their occurrence and severity

design of housing for dairy cows (Chapter 6). For dairy cattle, udder and teat lesions appear to be a particular problem. One study of teat injuries on Finnish dairy cows that were treated by a veterinarian reported an incidence of 3.4% (Rajala and Grohn, 1998), and a second study found that teat injuries were responsible for 1.2% of the cullings on French dairy farms (Seegers et al., 1998). Teat injuries are also of economic importance because they have a marked effect on milk production.

Measures of bruising have been used to assess the effects of transport and pre-slaughter management on the welfare of beef cattle. Examination of animals arriving at market can provide useful information about the extent of bruising or injury. The National Beef Quality Audit in the US (McKenna et al., 2002) examined the condition of over 12,000 animals or carcasses at 30 abattoirs. Over 45% of the cattle had some bruising, 16% had multiple bruises and about 5% of carcasses exhibited evidence of extreme or "critical" bruising. A similar audit in Canada found bruising on 54% of the carcasses with 17% of carcasses having major or severe bruising (Van Donkersgoed et al., 2001).

In summary, physical injury is an obvious concern for animal welfare, but difficulties in measuring the occurrence and severity of some common injuries has limited their use in assessing animal welfare.

4.9 Gastrointestinal Ulcers

Cattle of all ages can suffer from ulcers in the GI tract, but abomasal ulcers in young milk-fed calves have attracted most attention in terms of assessing animal welfare. A very high incidence of such ulcers has been reported among calves in veal production, with often one-third to three-quarters of calves being affected (e.g. Gottardo et al., 2002). The presence of abomasal ulcers can generally only be detected after slaughter, and various methods have been used to assess their severity. Sometimes this involves a simple 3-point scale (e.g. Bokkers and Koene, 2001) while in other cases ulcers are distinguished from stomach "erosion" and inflammation (e.g. Mattiello et al., 2002). It seems likely that such ulcers are a source of pain for the calves, but as yet there is little direct evidence to judge their impact upon animal welfare. Abomasal ulcers also occur among nursing beef calves and can result in death (Jelinski et al., 1996).

5 Early Indicators of Illness

A key message of this chapter is that if we use illness to assess animal welfare, we must have accurate methods for detecting the diseases. To date, most on-farm assessments of cattle health have relied on farmers' or veterinarians' records, or on measures of prevalence based on single or a few visits to farms. As discussed earlier, this approach is hindered by the low rate of veterinary treatment of some diseases and doubts about the reliability of diagnosis and of record keeping. In addition, there is often a delay between when the disease affects the animal and when it is detected. Green et al. (2002) found that lameness in dairy cows was affecting milk production several months before it was diagnosed and treated. Because of the increasing incidence of various diseases, and the large economic cost for both dairy and beef production, there is a great deal of interest in early detection of illness.

A promising approach is the automatic monitoring of animal health (Ingvartsen et al., 2003). Automated detection of animal-health problems provides a complement, and perhaps eventually an alternative to the current use of farmers' records or records of veterinary treatment to assess animal welfare.

5.1 Physiological and Biochemical Signs of Illness

For research purposes, a number of physiological parameters have been found to correlate with health problems. For example, measures of plasma haemoglobin or iron concentrations can detect anaemia in veal calves (Welchman et al., 1988; Bokkers and Koene, 2001), plasma concentrations of biotin in milk and plasma may give some information about hoof quality (Higuchi and Nagahata, 2001; Higuchi et al., 2003), electrical conductivity of milk may be used to automatically detect mastitis (de Mol and Ouweltjes, 2001), plasma gastrin concentrations are a sign of bleeding ulcers (Ok et al., 2001), and negative energy balance in postpartum cows can be detected by a change in concentration of metabolites in the blood (Aeberhard et al., 2001). These are just a few examples. It is likely impractical (and arguably inhumane) to take blood samples regularly from a large number of animals, although this could be done at critical moments, such as entry into veal-calf operations, entry into a heifer-growing facility or feedlots or just after calving. At present, routine use of physiological indicators to monitor health is likely to be most feasible for dairy cattle, where milk samples could be used (Ingvartsen et al., 2003). Recently, Schaefer et al. (2004) has shown that illness in beef cattle can be detected by thermographic measures of body temperature, several days before the traditional clinical examinations. Infrared thermography can also detect joint inflammation that may lead to lameness (Cockcroft et al., 2000). However, the most practical way of monitoring health is to look for changes in animals' behaviour.

5.2 Behavioural Indicators of Illness

Animals respond to illness with a consistent and predictable pattern of behavioural changes, which typically include reduced feeding, increased rest and sleep, increased thermoregulatory behaviours and reduced social behaviour (Hart, 1988). Clinical diagnosis of illness often involves some objective assessments of these behaviours. These behavioural changes occur simultaneously with physiological and metabolic changes, most notably the fever response, and some of the behavioural changes serve to maintain the fever (Johnson, 2002). The physiological components of the acute-phase response to illness, such as fever, are now thought to be host defences, i.e. evolved, adaptive responses that help the animals recover from the illness (Stearns and Ebert, 2001; LeGrand and Brown, 2002). Consequently, it seems likely that many of these behavioural responses to illness are adaptive responses, in that they help animals recuperate (Hart, 1988; Johnson, 2002). The behavioural changes that occur when animals are ill are not simply a by-product of the debilitating effects of illness, but organized behavioural changes that help animals recuperate from the illness (Aubert, 1999). Interfering with the behavioural responses to illness, for example, by forced feeding or preventing sleep, will reduce the chance of the animal recovering (Johnson, 2002; Irwin, 2002) so that the behavioural changes can be considered as part of the immune response. Most of the behavioural responses to illness can be stimulated by injections of bacterial endotoxin, such as lipopolysaccharide (LPS) (Dantzer, 2001; Larson and Dunn, 2001), and cytokines from the immune system are now seen as the major controllers (Johnson, 1998; Dantzer, 2001; Larson and Dunn, 2001; Inui, 2001).

Some studies have documented how automatically detected changes in behaviour may help early detection of disease, especially in beef cattle in feedlots. Cattle in feedlots suffer from a variety of illness of which respiratory diseases and metabolic illness are the most prevalent. Changes in the amount of feeding, drinking, and activity are typical symptoms of all of these illnesses. Beef-cattle suffering from BRD show an increased number of drinking bouts, a reduced time spent feeding and a reduced number of visits to the feed bunk (Sowell et al., 1999;

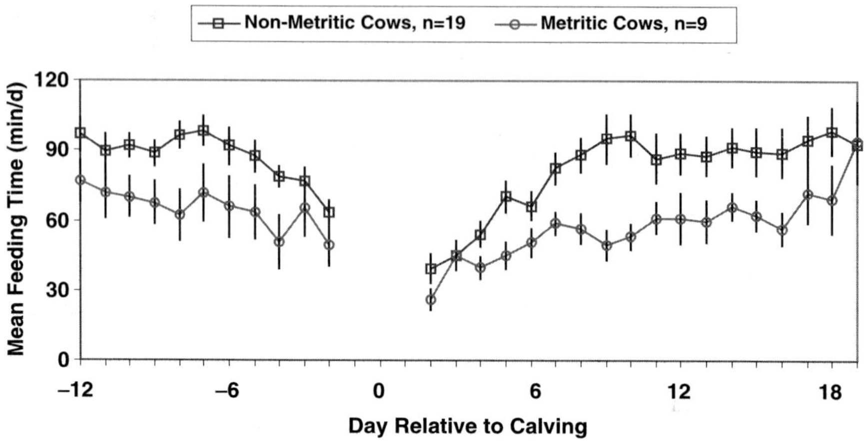

Figure 2.5 Daily feeding behaviour activity (min/day) of metritic and non-metritic Holstein cows during the period beginning 2 weeks prior to calving and ending 3 weeks post-calving (Data redrawn from Urton et al., 2005)

Buhman et al., 2000) and these behavioural changes could be detected before the steers were diagnosed clinically. Quimby et al. (2001) found that automated monitoring of feeding behaviour of calves in feedlots could detect calf morbidity several days earlier than the conventional methods employed in commercial feedlots. Recently, Urton et al. (2005) and Huzzey et al. (2006) were able to identify dairy cows that developed metritis several days before calving by observing reductions in feeding time (Figure 2.5). Unweaned dairy or veal calves are increasingly fed with automatic milk-dispensing systems that have the possibility of automatically collecting data on the sucking behaviour of the calves. This information may be useful as a means of detecting illness and hence serve in a welfare-monitoring system. Svensson and Jensen (2007) found a number of changes in calves' sucking behaviour 2 days before the calves were clinically diagnosed as being ill.

This type of information is critical for developing procedures that minimize the risk of disease and minimize animal suffering and the automatic monitoring of early indicators of illness, including behavioural changes, is an important and growing area for research in animal welfare.

6 Productivity

The use of measures of productivity (e.g. growth rates of dairy, veal or beef calves, or milk production by dairy cows) to assess animal welfare is controversial. Our position is that measures of productivity can be useful indicators of the welfare of the animals, but only under some circumstances. More specifically, we must understand the precise cause of the change or difference in productivity, and be able to relate this, with some confidence, to animal pain or suffering. There are good reasons why problems causing poor welfare can lead to lower productivity. Activation of the immune system during an illness uses a considerable amount of metabolic energy (Colditz, 2002) and illness often results in reduced feed intake, so that during illness, resources may be limited and diverted into immune function rather than milk production, growth, or reproduction. The important point is that changes in productivity can serve as an indicator only to the extent that there are clear and validated links with known problems. On their own, overall levels of productivity are of little value in welfare assessment.

For lactating cows, reduced milk production has been shown to reflect the condition of the animals, at least under some circumstances. As discussed earlier, a number of studies have estimated the magnitude of the reduction in milk yield that occurs when dairy cows suffer from various diseases (e.g. Bareille et al., 2003; Table 2.1). Short-term changes in milk yield have proven useful in assessing cows' responses to stressful events. For example, a variety of acute stressors such as novel surroundings can reduce oxytocin secretion, leading to blocked milk ejection and hence reduced milk yield (e.g. Bruckmaier and Blum, 1998). The increase in residual milk, which is obtained when the cows are injected with oxytocin after normal milking, can be an effective measure of the response of lactating cows to acute stressors (Bruckmaier et al., 1993, 1997; Rushen et al., 2001b; Figure 2.6). In these circumstances, the observed drop in milk yield from what is normal can be seen as an indicator of reduced welfare.

However, this does not mean that all or even most variation in milk yield is related to differences in animal welfare: few animal-welfare experts consider milk yield to be a useful measure in on-farm welfare assessments of dairy cattle (Main et al., 2003a; Whay et al., 2003a, b). Variation in milk yield can be affected by a host of factors (nutritional, genetic, and environmental) that are welfare-neutral. Similarly, measures of growth are not necessarily associated with illness. For example, Busato et al. (1997) found that the morbidity rates on cow-calf beef farms in Switzerland were not correlated with average weight gains of the calves.

A limiting factor in using measures of productivity to assess welfare is that assessing productivity of farm animals is not always straight forward. In part, this occurs because the measures depend as much on economics as biology. Animal productivity cannot be assessed except in terms of why the animal is kept: the productivity of a beef calf will be assessed largely in terms of its growth, whereas the productivity of a dairy cow is assessed mostly by the quantity of milk produced. Furthermore, the precise measures of productivity can depend on local economic factors: in some parts of the world, the income of dairy farmers is assessed by milk fat or protein, rather than by the quantity of milk itself. Beef producers can be paid according to the actual weight of the calf sold, or by its carcass weight at slaughter. At a more practical level, measures of productivity may not be that easy to obtain. Most dairy farmers in Western countries have data on daily milk production of individual cows, but in other countries farmers may only know the total milk

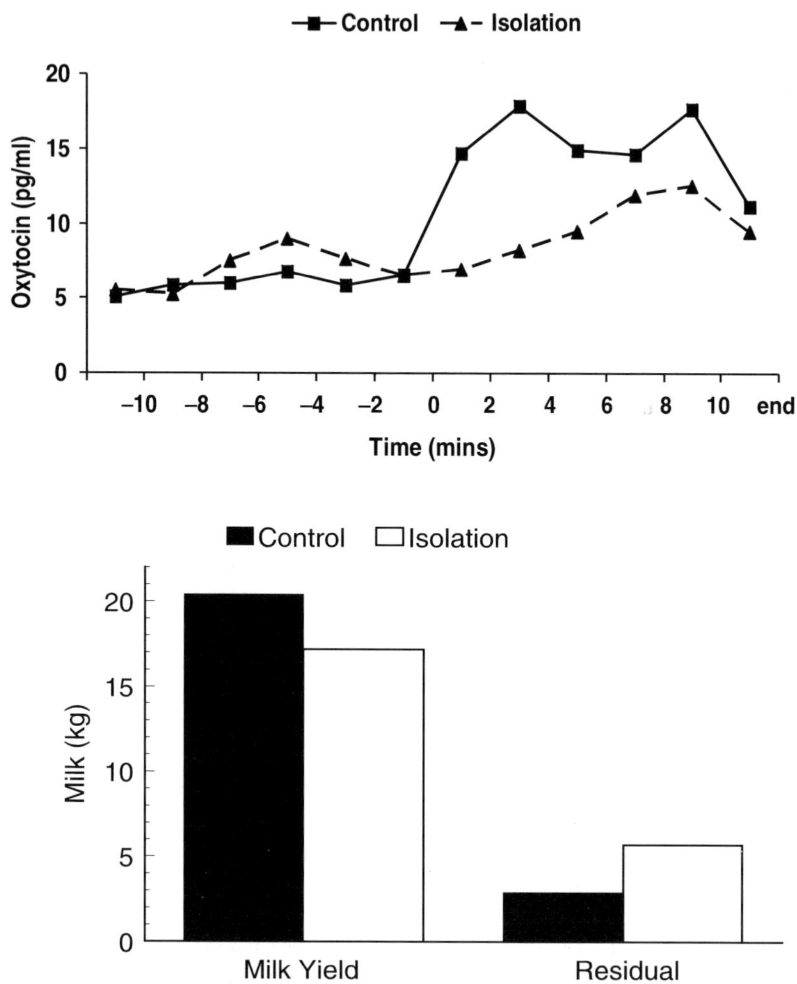

Figure 2.6 Under some circumstances sudden changes in milk yield can be used to assess acute stress in cows. When cows are milked under normal circumstances (CONTROL), oxytocin (Upper figure) increases dramatically leading to milk ejection. When they are milked while isolated in an unfamiliar place (ISOLATION), the stress on cows during milking reduces the release of oxytocin, which prevents milk ejection, and thus reduces milk yield and increases residual milk (Lower figure) (Based on data from Rushen et al. 2001b.)

produced by the herd. Many beef farmers do not have detailed data on the weight or growth rates of their animals.

Finally, there is increasing evidence that a high level of production may itself produce a threat to animal welfare. High growth rates among pigs and poultry are associated with an increased occurrence of health problems (reviewed in Rauw et al., 1998). Improved genetics and nutrition have resulted in a 2–3% increase in milk production per cow per year in Western countries. However, this increased production has put extra demands on the cow, leading to an increased incidence of disease and higher rates of involuntary culling. A number of studies have reported that high levels of milk production in dairy cattle is associated with an increased incidence of health problems (e.g. Kelm et al., 2000; Fleischer et al., 2001; Fourichon et al., 2001), although the nature of this relationship is not always clear (Ingvartsen et al., 2003). Furthermore, the cause of this relationship remains a matter of debate. The specifics of this issue are beyond the scope of our book, but Ingvartsen et al. (2003) and Rauw et al. (1998) both provide useful discussions. For high-producing dairy cattle, the problem appears to be not so much with the amount of milk produced, but rather the degree of negative-energy balance and concomitant changes in the endocrine and immune systems in the weeks following calving (Goff and Horst, 1997; Mallard et al., 1998; deVries et al., 1999; Ingvartsen et al., 2003). Since cows respond to the negative-energy balance by mobilizing body tissue, changes in body condition may be useful to detect the reduced energy status of the cows (Broster and Broster, 1998; Ingvartsen et al., 2003).

We conclude that there is no simple relationship between good productivity and good welfare. Sudden and inexplicable drops in milk production, growth rates, or body condition, may indicate illness, and hence can be used to assess welfare where the causes of these changes are understood. However, levels of production of individual animals and especially of farms cannot be used with any confidence to assess welfare.

7 Reproduction

Decreased reproductive success has often been proposed as a measure of welfare. Furthermore, reproductive success is decreasing in fast growing lines of beef cattle and high-yielding dairy cows (Lucy et al., 2001) and some cite this as evidence of welfare problems in modern dairy and beef production. These reproductive problems of dairy cows can themselves have an indirect effect on animal welfare. Increasingly, dairy producers are relying on a series of interventions to increase pregnancy rates through improved oestrus synchronization. Synchronization protocols consist of a series of timed injections (Pursley et al., 1998), each one likely a source of stress for the cows.

Many forms of illness, such as calving difficulties, lameness, and metritis can lead to reproductive failure (Section 4.6 of this chapter), and so measures of reproductive success may be indirect measures of these health problems. Furthermore, considerable research on a number of mammalian species has also shown the depressive effects of stress on reproductive capacity of both males and females and has described the physiological mechanisms underlying these effects. Such suppressive effects of stress have been found in cattle (Dobson and Smith, 2000) and a number of studies show that stressors reduce reproductive efficiency in cows. This is true not only of physical stressors, but also stressors that are more emotional or psychological in nature. Cows that increased their dominance status within a herd had fewer days between calving and conception than those that lost dominance status (Dobson and Smith, 2000). A 30-min truck journey has been shown to be sufficient to block the luteinizing hormone surge in response to oestradiol benzoate injections in early post-partum cows (Dobson and Smith, 2000). The effects of stress on the reproductive performance of bulls have been much less studied.

However, as is the case with productivity, the fact that poor welfare can sometimes affect reproduction does not mean that reproductive success can be used to assess welfare in practical circumstances. This is especially true when considering differences between farms. Differences among farms in reproductive failure could be due to many factors that are not related to the welfare of the cows, such as success at oestrus detection, effective artificial insemination strategies, as well as general reproductive management. Thus poor reproductive success of cattle can only be used to assess welfare in cases where there is evidence to link changes with stress or disease, and in these cases it would likely be more appropriate to measure the degree of stress or occurrence of illness directly. In short, although poor welfare may reduce reproductive success of cattle, measures of reproductive success alone, especially at the herd level, appear to have little potential in welfare assessment.

8 Conclusions

In this chapter, we have focused on some of the difficulties that limit our ability to use measures of disease incidence to assess animal welfare, especially under commercial conditions. Poor health obviously is associated with poor welfare, but there are still many uncertainties about the relative impact of different disease on animal welfare. In some cases, such as metritis, these uncertainties may result in an underestimation of the importance of the disease for animal welfare. Difficulties in obtaining accurate diagnoses and accurate and reliable records of disease incidence are other problems, especially for extensive beef production where there is less contact between caretakers and animals. One area of promise for future research is the use of automated monitoring of health problems.

Chapter 3
Stress and Physiological Indicators of Animal Welfare

1 Introduction

Placed in difficult and demanding situations, we have become accustomed to using the word "stress" to describe how we feel. Personal familiarity with the word leads us to assume that it has a clear definition, and when faced with the necessity of assessing the welfare of animals, it seems logical to try to determine if they too are "stressed". Scientists even claim to have physiological indicators of stress, apparently making the concept concrete. Unfortunately, the easy use of the word in everyday conversation obscures the semantic tangle that awaits those who venture in too deeply. In this chapter, we discuss the concept of stress as a justification for the various physiological measures that have been proposed as indicators of animal welfare. For a detailed historical exposition of the various concepts or models of stress, and a comprehensive review of the physiological systems involved, we refer the reader to Toates (1995), Sapolsky (1998), Moberg and Mench (2000), and Tsigos and Chrousos (2002). Lane (2006) and Mormède et al. (2007) present very good discussions of the relationship between animal welfare and the physiological responses to stress.

In this chapter, we discuss some of the potential and difficulties in using physiological measures to assess animal welfare. We focus primarily on the issues associated with validating such measures as indicators of animal welfare, and the complexities and difficulties associated with their measurement. Since our subject is animal welfare, we limit our discussion to aspects of the stress response that are the most clearly linked to animal welfare; space constraints prevent us from discussing other topics, such as the effects of stress on reproduction or meat quality. Although such effects can be of considerable economic importance, they are not directly relevant to animal welfare.

2 The Concept of Stress

One of the more influential models of what stress is and how it can affect animal welfare is that of Gary Moberg (1985, 1996, 2000). The model is outlined in Figure 3.1. Moberg distinguishes between three key components of the stress

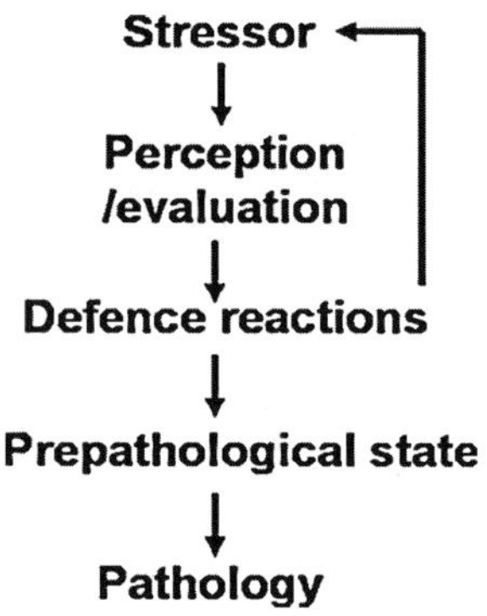

Figure 3.1 Outline of model of stress based on that proposed by Moberg (2000). To elicit a stress response, the stressor must first be perceived and evaluated as such by the animal. The behavioural and physiological responses are defence reactions aimed at dealing with the stressor. If these use too much of the animal's biological resources a pre-pathological state can occur, which if prolonged, eventually results in some long-term pathology

response: (1) the recognition or the perception of the stressor, (2) the biological defence against the stressor, and (3) the long-term consequences of the defence response for the animal. Most research on the physiological responses to stress has focused on the second component. However, we argue below that it is the first and the last components, that is, the causes and the consequences of the defence responses that are of most relevance for assessing animal welfare.

It is the animal's perception of threat, rather than the threat per se, that results in a stress response. For our purposes, the important fact is that to assess stress it is essential to understand how animals perceive events. The second component consists of the immediate response of the animal to the perceived stressor. The animal's behavioural, neurophysiological, and peripheral physiological responses to stressors are, in most cases, defence reactions – attempts by the animal either to deal with or avoid the stressor. Since different stressors will differ in many respects, the ways that the animal tries to deal with stressors will also vary from one stressor to another. Thus, we should not expect to find many general stress responses that occur irrespective of the type of stressor. Recent reviews of the various responses animals make to stress can be found in Mormède et al. (2007), Tsigos and Chrousos (2002), Moberg and Mench (2000), and Balm (1999). The two main physiological systems that respond to stress are the sympathetic nervous system (SNS) and the hypothalamic-pituitary-adrenal (HPA) axis. Increased activity of the HPA axis is one of the most commonly reported physiological responses an animal makes in response to stress. Indeed, early in stress research it was assumed that increased HPA activity was a common response to all stressors, which led to the idea that increased HPA activity was the defining indication that an event was perceived as a stressor. We now know that this assumption is not correct: not all stressors lead to increased HPA activity (Pacák and Palkovits, 2001) and there are other peripheral physiological responses to stress besides HPA axis activity. However, since many different stressors do elicit increased HPA activity, measures of HPA activity have figured prominently in the assessment of stress by researchers.

Finally, the third component of Moberg's model recognizes that the defence responses that animals make in dealing with stressors can have long-term consequences that may be deleterious for the animal's welfare (Figure 3.1). The distinction between the response itself and the biological consequences of the response is important for assessing animal welfare. The changes in biological functioning that follow the response to a stressor are often more important than the actual response itself. This is important to remember because often, when assessing animal welfare using physiological measures, it is the defence response that is measured rather than the long-term consequences.

Moberg suggests that the total biological cost to the animal of mounting a stress response is the most important determinant of welfare. When such costs are great the animal can enter a "pre-pathological" state, meaning that the animal comes under increased risk of developing some clinical pathology. An essential aspect of this concept is that animals have only limited biological resources (e.g. time available, energy, etc.), and these must be distributed among many activities essential for life (e.g. growth, reproduction, immune responses, etc.). Responding to stressors uses these resources, making them unavailable for the other activities. One of the primary resources is the animal's energy supply (Wingfield and Ramenofsky, 1999). Many of the stressors to which animals respond alter metabolism and increase energy consumption above basal levels (Wingfield and Ramenofsky, 1999; Steffens and de Boer, 1999; Elsasser et al., 2000). Furthermore, the behavioural responses to stress can take up the animal's time (Rushen, 2000) as well as alter the use of specific nutrients such as amino acids, minerals, etc. (Elsasser et al., 2000). Thus, the long-term effects of stress on animal welfare may be estimated by the biological resources that the stress response requires (Moberg, 2000).

We suggest that there are two ways that the physiological components of the stress response, outlined in Figure 3.1, can be used as a basis of welfare assessment. The first is to use these physiological responses as a sign that the animal is experiencing some kind of negative emotional response. As we argue later, that assessment is most appropriate when we are dealing with acute stressors. However, the measures of physiological responses may also be used to predict the long-term negative consequences for the animal, and this approach is likely most appropriate when dealing with long-term or chronic stressors.

3 Physiological Responses to Stress

It is important to repeat that not all of an animal's physiological responses to stress can be described simply by measures of HPA axis activity. However, many of the principles we wish to discuss can be illustrated with reference to the HPA axis, even though they apply to other physiological responses. Thus, in the following section we focus primarily on the HPA axis.

3.1 The HPA Axis as a Multilevel System

The physiological responses that we typically measure when an animal is exposed to a stressor are the result of changes in activity of a complex, multilevel physiological system, and this complexity can hinder our ability to correctly interpret any changes that do occur.

The anatomical components and the functioning of the HPA axis in mammals has been well described (Toates, 1995; Tsigos and Chrousos, 2002; Mormède et al., 2007), but most of the research has been done using laboratory rodents, and our knowledge of how the HPA axis is regulated in cattle is lacking. Increased activity of the HPA axis following stress is generally taken to begin with increased secretion of corticotrophin-releasing-hormone (CRH) (sometimes called corticotrophin-releasing-factor (CRF)) from the neurons of the paraventricular nuclei of the hypothalamus. Recent research has now shown that CRH and CRH receptors are implicated in the effect of stress upon many other physiological systems, including the SNS, and upon the behaviour of animals (e.g. Tsigos and Chrousos, 2002). The CRH enters the hypothalamic-hypophyseal portal blood vessels and induces secretion of adrenocorticotrophic hormone (ACTH) and beta-endorphin from the anterior pituitary gland into the general blood circulation. ACTH, in turn, increases the secretion of corticosteroids or glucocorticoids from the adrenal cortex. The corticosteroids are cortisol and corticosterone, which in cattle have been reported to occur at relative concentrations of 2.4:1 (Willet and Erb, 1972). Through a process of negative feedback, increased circulating cortisol can reduce secretion of CRH by acting on corticosteroid receptors in the hypothalamus.

The multi-step nature of the HPA axis response has implications for how we can interpret the response in terms of animal welfare. For example, if our purpose is to use changes in HPA activity as an indicator of the emotional response of animals to the stressor, then in theory we are best off directly measuring CRH activity, since this has the fewest intervening steps from the neurophysiological events that underlie the emotional response itself. In practice this is difficult to do since this activity occurs within the brain. Therefore we tend to rely on more downstream measures, such as cortisol, which can be detected in the blood stream. To further complicate matters, the different components of the response generally occur over different time periods. Release of CRH triggers the secretion of ACTH within a matter of seconds, and, in cattle, plasma concentrations of ACTH typically reach maximum values within 10 min of injection of CRH and return to baseline values within 60–180 min (Veissier et al., 1999). In contrast, the effects of ACTH on corticosteroid release occur within minutes, and, in cattle, plasma concentrations of cortisol usually reach peak values within 30–90 min of the injection of ACTH and return to baseline values within 150–300 min (Lay et al., 1996). Thus, assessment of plasma cortisol concentrations generally requires multiple blood samples to be taken every 15–20 min, while assessment of ACTH concentrations requires the multiple blood samples to be taken at even shorter intervals.

Many other neural and endocrine factors can influence the degree of HPA activity. There is clear evidence of an opioid receptor-based inhibition of HPA activity (Nanda et al., 1992; Rushen et al., 1999a; Tancin et al., 2000) and recently, there has been interest in the possibility that central oxytocin may reduce HPA responses (Uvnäs-Moberg et al., 2001). ACTH secretion is also affected by vasopressin, oxytocin, and epinephrine (Matteri et al., 2000), while corticosteroid release from the adrenal cortex is affected directly by CRH and by vasopressin, epinephrine, and various other hormones, neurohormones, and immune system factors (Matteri et al., 2000). In cattle, there is evidence that the increase in circulating cortisol that occurs during milking is not a result of ACTH secretion (Tancin et al., 2000). These other controlling factors can modulate the influence of stress on HPA activity; for example, in cattle it has been shown that vasopressin can potentiate CRH-induced ACTH secretion (Veissier et al., 1999). Furthermore, the glands described are not the only source of these hormones, for example, in cattle ACTH can also be released from lymphocytes (Dixit et al., 2001). The important point for our purposes is that changes in concentrations of cortisol in the plasma cannot be taken as a simple reflection of the secretion of CRH or ACTH; the route from application of a stressor to the release of cortisol into the plasma is a complicated one.

3.2 Basal Activity of the HPA Axis

An important aspect of the HPA axis is that stressor-induced HPA activity occurs against a background of changing basal activity. Plasma cortisol concentrations in adult cows follow a circadian rhythm but this does not appear to be solely related to the light–dark cycle: lowest values occurred between 1700 h and 0100 h, with the minima at 1800 h and a maximum at 0530 h (Lefcourt et al., 1993). However, these circadian changes are relatively small: Lefcourt et al. (1993) reported an average peak to trough difference of only 1 ng/ml, although for individual cows this difference can be larger. Lefcourt et al. (1993) suggested that the small size of the circadian rhythm may be related to the unusual sleep pattern of adult ruminants. Similar diurnal variation in plasma cortisol has been reported for bulls (Thun et al., 1981) but in young calves the situation is somewhat different. Hänninen et al. (2006) found highest cortisol concentrations between 1100 h and 1600 h, with lower concentrations during the night time, although a second peak in concentrations was seen between 0400 h and 0700 h. The peak to trough difference was also larger than for adult cows, around 5 ng/ml (Figure 3.2). In contrast to the circadian rhythms noted in plasma corticosteroid concentrations, milk cortisol concentrations were reported to be lower during afternoon milkings (1500 h) than morning milkings (0700 h) (Verkerk et al., 1998).

There is also evidence of larger, high-amplitude peaks in plasma cortisol concentrations although the intervals between peaks and the amplitude of the peaks varies according to the age and sex of the animals. In adult cows, these peaks occurred at regular intervals of 120 min with an amplitude of 16 ng/ml (Lefcourt et al., 1993). In growing bulls, peaks tend to occur at a mean interval of 135 min with a mean amplitude of 5 ng/ml (Ladewig and Smit, 1989). For young male calves, Hänninen et al. (2006) reported that peaks in cortisol concentrations occurred at intervals of 4–5 h with a mean amplitude of 7–8 ng/ml. The circadian

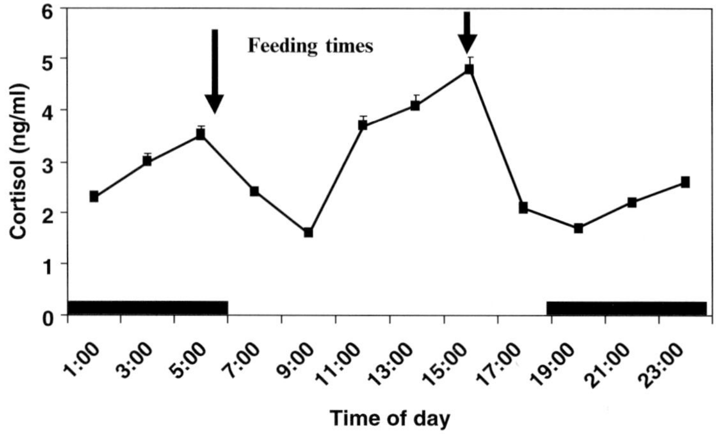

Figure 3.2 Plasma concentrations of cortisol in young calves show clear circadian rhythms. Concentrations are slightly lower at night time than during the day, but the main factor affecting cortisol concentrations is the timing of feed delivery. Cortisol concentrations were highest immediately after the calves were fed. The cortisol concentrations shown are mean values based on repeated samples taken at 20-min intervals over 2-h periods of time, beginning at the time of day noted on the *x*-axis (Data from Hanninen et al., 2006.)

and ultradian rhythms and the highly pulsatile nature of cortisol secretion have important implications for the measurement of plasma cortisol levels.

3.3 *The Sympathetic Nervous System*

In contrast to the largely endocrine HPA axis, the SNS is primarily neural, being a branch of the autonomic nervous system that governs much of the body's visceral responses. Arising in the hypothalamus, the SNS makes direct neural connections with many internal organs, of which the cardiovascular system is perhaps the most widely studied within the context of stress. The majority of the connections involve norepinephrine as the neurotransmitter. Complications arise from the fact that the SNS also has direct neural control of the adrenal medulla, which secretes the catecholamines epinephrine and norepinephrine (and some neuropeptide Y) into the general circulation. Thus, the majority of the internal organs can be influenced both by direct neural connections and through secondary stimulation via circulating catecholamines. Measures of circulating catecholamines therefore give only limited information as to the degree of activation of the SNS, and consequently SNS activity is often measured through changes in the activity of the organs themselves, for example, through increased heart rate, blood pressure, etc. In cattle, the adrenal medulla appears to be the only significant source of epinephrine, but norepinephrine may also be released directly from the central nervous system (Hard et al., 2001).

4 Measurement of Physiological Changes Following Acute Stress

The physiological changes that result when an animal is exposed to a stressor are most clearly seen following short-term exposure to a stressor that has a well-defined beginning and end. When the stress is prolonged or repeated, the nature of the physiological response changes (Mormède et al., 2007).

4.1 The Hypothalamic-Pituitary-Adrenal (HPA) Axis

Measures of HPA axis activity in cattle following acute stress have generally relied on endocrine measures, most commonly of cortisol or ACTH. These can be detected in plasma, but also in other bodily fluids, such as saliva, milk, or urine.

4.1.1 Plasma Hormonal Concentrations

Studies that have taken repeated blood samples with minimal disturbance show that basal cortisol concentrations in adult cattle and young calves are generally between 2 and 5 ng/ml (Ladewig and Smit, 1989; Lefcourt et al., 1993). Application of a short-term stress such as social isolation (Figure 3.3), a brief period of transport, branding, or dehorning results in a relatively quick increase in plasma concentrations which reach peak values of 12–40 ng/ml within 20–45 min, and which usually return to baseline values within a few hours, although some treatments can lead to cortisol concentrations being elevated for longer (Table 3.1).

Accurate assessment of cortisol concentrations requires that sufficient blood samples be taken over the whole period. Typically, blood samples are taken every 15–20 min. In general, measurements of plasma cortisol would seem to give fairly reliable information about the changes in HPA activity following an acute stress.

However, there are a number of methodological issues that have not been fully resolved. First, what actual measure should be used – plasma concentrations at some defined point in time after the application of the stressor, peak concentrations regardless of what time after the stress they occur, or some estimate of the integrated HPA activity over a period of time, such as the area under the curve? Often, studies will use a mix of these but without any clear rationale for choosing the best. Peak values in cortisol concentrations may not provide the best estimates of HPA activity because of a "plateau" effect where cortisol concentrations soon reach a maximum and cannot increase further. For example, Veissier et al. (1999) injected low doses of CRH (0.01–0.03 μg/kg) into cattle and noted that increases in the dose of CRH injected were apparent in increased maximum plasma cortisol concentrations as well as increased length of time that the concentrations remained above baseline. However, when larger doses were used (0.1–1 μg), which produced higher

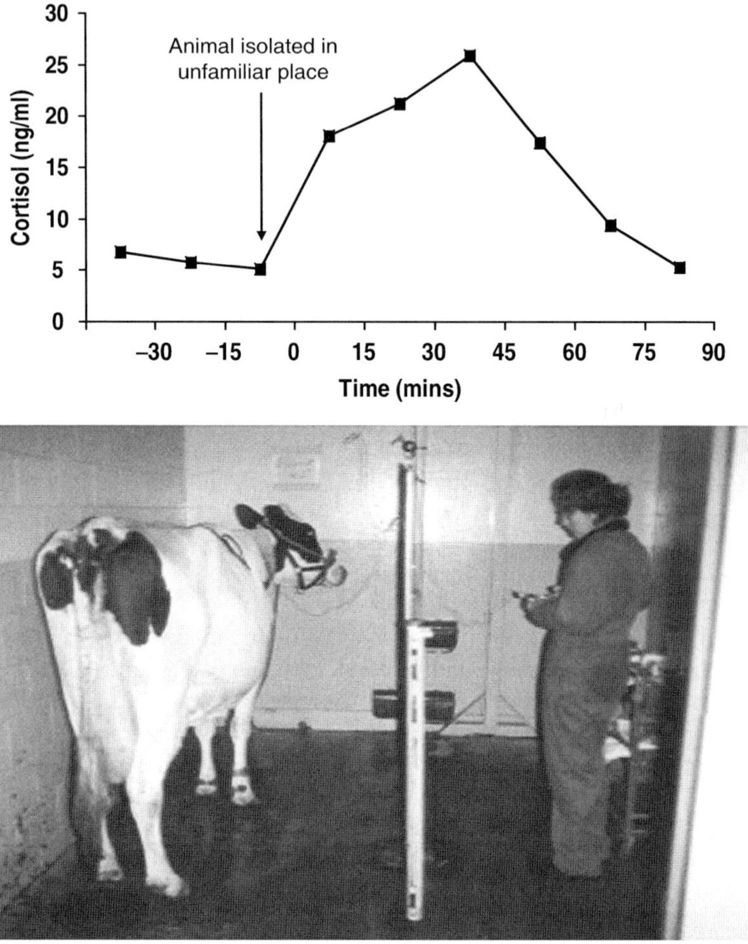

Figure 3.3 Plasma concentrations of cortisol often rise quickly when cattle are subject to acute stress, such as being isolated in unfamiliar surroundings (lower figure). However, blood sampling can be invasive, often requiring the animal to be catheterized and handled repeatedly to obtain blood samples (Data from the experiment described in Rushen et al., 2001b.)

plasma cortisol concentrations, maximum cortisol concentrations soon reached a plateau and increasing the doses of CRH did not result in further increases in maximum concentrations. Rather there was an increase in the duration of time that the concentrations remained above baseline. In contrast, peak concentrations of plasma ACTH did continue to reflect the dose of CRH used. Similar results are found when ACTH is injected (Lay et al., 1996). This plateau effect suggests that large increases in HPA activity may not be apparent in the maximum plasma cortisol concentrations achieved but in the length of time that plasma cortisol concentrations remain elevated above baseline. This is best determined from an

Table 3.1 Effects of various acute treatments on plasma cortisol concentrations of cattle. Concentrations are given either as nanograms per millilitre (ng/ml) or nanomoles per litre (nmol/L)

Treatment	Type of animal	Baseline	Peak values	Time of peak	Return to baseline
Branding[1]	Yearling beef heifers	10–15 ng/ml	35–37 ng/ml	20 min	60 min
Branding[2]	Beef calves	10–20 ng/ml	35 ng/ml	NA	NA
Dehorning[3]	3–4-month-old calves	20–40 nmol/L	130–140 nmol/L	30 min	7 h
Dehorning + analgesia[4]	3–4-month-old calves	20–40 nmol/L	65 nmol/L	6–7 h	~12 h
Castration[5]	5-month-old bulls	5–9 ng/ml	46 ng/ml	30 min	8–10 h
Castration + anesthesia[6]	5-month-old bulls	5–9 ng/ml	35 ng/ml	30 min	8–10 h
Castration[7]	13-month-old bulls	8 nmol/L	100 nmol/L	30–60 min	2–6 h
Castration + analgesia[8]	13-month-old bulls	8 nmol/L	67 nmol/L	30 min	2 h
Transport[9]	Pregnant heifers	<5 ng/ml	25–35 ng/ml*	NA	NA
Isolation/new surroundings[10]	Lactating cow	5–7 ng/ml	25 ng/ml	45 min	90 min
Milking[11]	Lactating cow	2–3 ng/ml	12 ng/ml	15 min	2 h
ACTH injection[12]	Pregnant heifers	10 ng/ml	50–60 ng/ml	15–90 min	>5 h

[1] Schwartzkopf et al., 1997a; [2] Lay et al., 1992b; [3,4] Sutherland et al., 2002; [5,6] Fisher et al., 1996; [7,8] Ting et al., 2003; [9,12] Lay et al., 1996; [10] Rushen et al., 2000; [11] Negrao et al., 2004.
*Value obtained at the end of transport.
NA = not available.

integrated measure of area under the curve based on a number of blood samples. In summary, the issue of what is the best measure of changes in plasma concentrations of cortisol remains unresolved; however, it is clearly unwise to rely on single sample measures of cortisol secretion.

4.1.2 Hormonal Concentrations in Other Bodily Fluids

Measurement of cortisol concentrations in the blood requires blood sampling, which entails either use of a catheter or some form of venipuncture. HPA activity is known to be sensitive to the handling associated with catheterization and blood sampling, and there is always the risk that stress at the time of handling might influence HPA activity to the extent that the effect of other stressors is masked (Figure 3.3). Thus, there is interest in sampling cortisol concentrations in other bodily fluids, such as saliva, urine, or milk, which might be obtained with fewer disturbances to the animals (reviewed in Mormède et al., 2007).

For lactating dairy cows, milk is an obvious alternative to blood, since milk collection is a routine procedure with which the cows are familiar. Steroid hormones that permeate cell membranes should pass directly from the blood into the alveolar secretory cells of the mammary gland and thus into the cisternal milk. The concentration of these hormones in milk should thus be a direct function of the concentrations in the plasma integrated over a period of time (Bremel and Gangwer, 1978). A number of studies have found correlations between the concentrations of cortisol in milk and in plasma following injections of ACTH and other stresses such as transport (Bremel and Gangwer, 1978; Verkerk et al., 1998). Thus, the concentrations of milk cortisol can be used to estimate the concentrations in plasma. However, there are a number of factors that limit the usefulness of milk cortisol concentrations.

First, cortisol concentrations in milk tend to be much lower than those in plasma. Bremel and Gangwer (1978) injected a high dose of ACTH into lactating cows that resulted in peak concentrations of plasma cortisol of 60 ng/ml in 8–10 h; cortisol concentrations remained elevated for up to 48 h. Cortisol concentrations in the milk collected 8 h later were significantly elevated to 6–12 ng/ml (compared to a baseline of around 2 ng/ml). The increase in milk concentrations was therefore much smaller and is likely to be less sensitive to differences in the magnitude of the stress. Furthermore, there was a very large variation between cows, with some cows consistently having milk cortisol concentrations above 10 ng/ml even when there was no apparent stress. Verkerk et al. (1998) obtained similar results using a smaller dose of ACTH, and found cortisol concentrations in milk decreased fairly quickly, so that when milking occurred 4 h after the injection of ACTH, milk cortisol levels were not elevated above those of the control group. The concentrations of cortisol in foremilk were also found to differ from those in the remainder of the milk, which the authors suggested was due to the distribution of milk between the cistern and the alveolar cells prior to milking. When plasma cortisol levels begin decreasing, the cortisol in the milk that is still within the alveolar cells is thought to diffuse back into the blood stream more quickly than cortisol in milk that has already entered the cistern. If this is true then cortisol concentrations in milk may be markedly affected by the ratio of cisternal and alveolar milk. This will introduce an extra source of variance making it difficult to draw firm conclusions about the magnitude of the stress response. Finally, although milking is a regular event, we should not assume that it is not stressful to the cows (Rushen et al., 1999b). In short, there are several disadvantages of using milk cortisol concentrations to assess acute stress in cattle.

Assessment of cortisol in saliva has been used successfully in other species such as pigs, where cannulation is difficult. Figure 3.4 shows that salivary cortisol levels in cattle closely follow those in the plasma following ACTH injection (Negrao et al., 2004). However, as with milk cortisol concentrations, the values are substantially lower, suggesting a lower sensitivity to the magnitude of the stressor. It is easier to control the timing of saliva samples than of milk samples, but in our experience it is more difficult (and probably more stressful) to take saliva samples rather than blood samples from cattle. Unless it is very difficult to

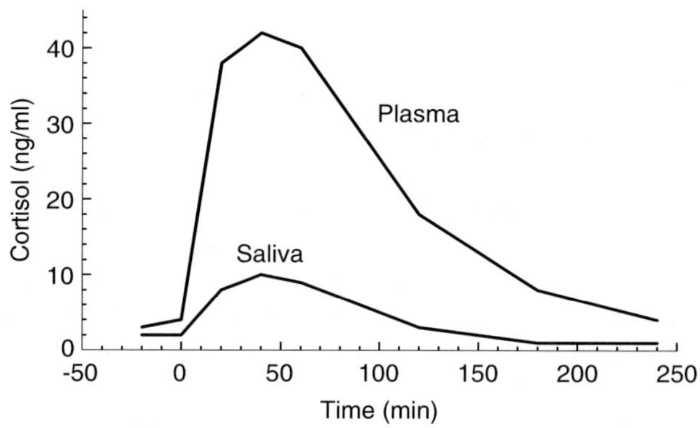

Figure 3.4 Cortisol concentrations in both plasma and saliva following injections of ACTH. Data taken from Negrao et al. 2004

take blood samples, use of salivary cortisol measures do not appear to hold any advantages.

Möstl and Palme (2002) have suggested that the concentrations of cortisol or its metabolites in urine or faeces may be useful under some circumstances. However, insufficient research has been done on cattle to judge this. Palme et al. (2000) did find an increased concentration of cortisol metabolites in the faeces of cows following transport. Again, given the difficulties of controlling the time at which urine or faecal samples can be taken, use of urine or faeces would seem limited to cases where it is impossible or very difficult to handle the animals sufficiently to take either blood or saliva samples. Free-ranging beef cattle might be one example. In such cases, however, the development of automated blood sampling techniques may be a more accurate alternative.

4.2 Measures of the Sympathetic Nervous System

The responses of sympathetic nervous system activity to stress have been measured far less often in cattle than have changes in HPA activity. Some studies have measured plasma concentrations of epinephrine and norepinephrine (e.g. Lefcourt et al., 1986; Locatelli et al., 1989; Petty et al., 1994; Hydbring et al., 1999; Hopster et al., 2002; Mellor et al., 2002). Plasma epinephrine and norepinephrine concentrations rise markedly after painful events such as dehorning (Mellor et al., 2002). For epinephrine, the increase is rapid, with peak values being reached within 5 min and a return to baseline at 10 min. If the stress is prolonged for 30 min, plasma concentrations of epinephrine stay elevated, but still drop below peak values fairly quickly (Locatelli et al., 1989). For norepinephrine, the increase following dehorning is more gradual, peak values being reached within 15–20 min, with a return to baseline values about 60 min after dehorning (Mellor et al., 2002). However, changes in norepinephrine were not noted after a period of simulated transport, even though plasma cortisol and epinephrine concentrations were elevated (Locatelli et al., 1989). The rapid rise and fall of epinephrine concentrations indicate the difficulty of using measures of these hormones in practical situations. Accurate assessment of the peak values would require frequent blood sampling at short intervals (5 min or less).

An alternative approach to measuring SNS responses has been to look at changes in the cardiovascular system. Heart rate has been the measure most often used. With modern techniques, heart rate can now easily be measured with telemetry. The most common approach has been to measure average heart rate over a period of time. Average heart rate in adult cattle tends to be around 70–90 beats per min (e.g. Boissy and LeNeindre, 1997; Rushen et al., 1999a; Waiblinger et al., 2004). Heart rates increase to about 90–120 beats per min during acute stress, for example, during a period of social isolation (Boissy and Le Neindre, 1997; Rushen et al., 2001b). The problem with this approach is that heart rate will closely reflect the posture and activity of the animal, and this may mask any changes due to the stressor. For example, de Passillé et al. (1995) found that heart rate increased when calves were placed in an open-field test. However, the calves also increased their activity, and multiple regression showed that the increased activity accounted for the entire rise in heart rate. More recent research has investigated whether measures of the variability of heart rate may be more effective in detecting responses to acute stress, with some limited success (Després et al., 2002; Mohr et al., 2002; Hagen et al., 2005). However, Després et al. (2002) did not detect any measures of heart rate variability that clearly reflected sympathetic tone. More research is needed to determine which measures of cardiovascular activity are the most appropriate to assess animals' responses to acute stress.

In general, measures of sympathetic nervous system activity appear to reflect fairly rapid responses of animals to a challenge, and mainly because of measurement difficulties, appear to have limited value in the assessment of animal welfare at this time.

4.3 Residual Milk and Reduced Oxytocin Secretion

Inhibition of oxytocin secretion during milking or lactation is a common response of lactating cows to acute stress. A variety of acute stressors, such social isolation in novel surroundings (Bruckmaier et al., 1993), or fear of the people present at milking (Rushen et al., 1999b) lead to inhibition of milk ejection in cattle. This inhibition can either be a centrally mediated inhibition of oxytocin secretion or a peripheral inhibition of milk ejection even in the presence of normal levels of oxytocin (Bruckmaier and Blum, 1998; Wellnitz and Bruckmaier, 2001). A central inhibition of oxytocin secretion can be detected by an absence of peak increases in circulating oxytocin concentrations or by an increased volume of residual milk, which is obtained by giving oxytocin injections after normal milking (Figure 2.5 in the previous chapter). The neurophysiological causes of the central inhibition of oxytocin are not clear, but they do not appear to be opioid based or involve catecholamines (Wellnitz and Bruckmaier, 2001; Macuhova et al., 2002). In contrast, the peripheral inhibition of milk ejection does appear to be a consequence of sympathetic nervous system activity, since it is mediated by adrenergic receptors (Wellnitz and Bruckmaier, 2001). Suppression of oxytocin secretion and the resulting increase in residual milk is an effective way of detecting acute stress at the time of milking. While a failure of milk ejection is undoubtedly important for production efficiency, it is difficult to judge whether this represents a welfare problem, except perhaps in the case of nursing cows where milk ejection failure may lead to a food shortage in the calf. Furthermore, we know little about the nature of the underlying emotional responses and the central inhibition of oxytocin secretion is not well correlated with behavioural indicators of fearfulness, such as increased vocalization or defecation (Rushen et al., 2001b).

5 Detection of Chronic or Prolonged Stress

Blood sampling for cortisol or catecholamine assays may be an appropriate means of detecting HPA or SNS responses to acute stress that have a well-defined beginning and end. However, this method is far more problematic when dealing with prolonged stress (such as occurs with the effects of housing) or where the end of the stress is not well defined (e.g. where there may be prolonged effects, such as chronic pain following dehorning or tail docking) (Mormède et al., 2007). Because of the difficulties in assessing prolonged or chronic stress in terms of plasma concentrations of hormones, there has been interest in developing tests in which the HPA axis is challenged in some way (see Mormède et al., 2007).

5.1 Plasma Hormonal Concentrations During Chronic Stress

There are three important issues that affect our ability to assess chronic changes in HPA activity from changes in blood cortisol concentrations.

The first is a methodological issue and involves the pulsatile release of cortisol and the number of samples required to detect long-term changes in pulsatile release. Due to the pulsatile secretion of cortisol from the adrenal gland (discussed in Section 3.2 of this chapter), plasma cortisol concentrations can change by as much as 5–20 ng/ml in a few minutes or a few hours (Ladewig and Smit, 1989). This makes it difficult to assess effects of different conditions on HPA activity by taking single blood samples. The concentrations of cortisol in a single sample can vary greatly depending on when the sample was taken relative to a pulse. Since these pulses do not appear to be closely related to external events such as feeding or milking, it is difficult if not impossible to know when they occur relative to the sample. Thus, assessment of chronic changes in plasma cortisol concentrations will require sufficient number of samples taken over the day to control for both circadian and ultradian rhythms.

The second issue is a theoretical one: even if sufficient samples are taken to determine average plasma concentrations over the day, there remains the question as to whether or not the average concentration is the most appropriate measure or whether an alternative measure of the nature of the pulsatile release or of the circadian rhythm might be more appropriate. In examining the effects of lying deprivation on HPA activity in cows, Munksgaard and Simonsen (1996) took blood samples every 30 min for 7.5 h. They found no treatment effects on daily average ACTH concentrations, but there was an increased ACTH concentration at certain times of the day. Ladewig and Smit (1989) noted that when young bulls were tethered, average plasma concentration of cortisol increased but so did the nature of the pulsatile release: they reported an increase in the duration and frequency of the pulses, but not in their amplitude. Munksgaard and Simonsen (1996) also reported a reduction in the number of peaks in ACTH concentrations in cows that had been prevented from lying down for 14 h per day for 3 weeks. Thus, measures of the nature of the pulsatile release of cortisol may provide a more appropriate way of assessing HPA responses to chronic stress than average plasma concentrations.

A third issue concerns the nature of the adaptation in HPA activity that occurs when stress is prolonged. Ladewig and Smit (1989) found that the increased pulsatile secretion of cortisol that occurred when bulls were first tethered was absent 1 month later. This may, at first, be taken as evidence that the bulls had fully adapted to the stress. However, when the researchers measured the amount of cortisol that was released in response to ACTH injections they noticed that the apparently adapted bulls showed a smaller response. Thus the reduced plasma secretion of cortisol might have been due purely to a reduced sensitivity of the adrenal cortex to ACTH, without necessarily any change in the amount of ACTH or CRH being released (which was not measured). Thus, prolonged or chronic stress may alter the sensitivity of various components of the HPA axis, which may not be apparent in the plasma levels of cortisol.

Because of these difficulties in using plasma concentrations of cortisol, there has been marked interest in using various measures of HPA responsiveness ("challenge tests") to detect the effects of chronic stress (see Mormède et al., 2007).

5.2 *Challenge Tests*

For cattle, three types of challenge test have been used or proposed. These have involved examining changes in adrenal cortex responsiveness to ACTH stimulation (ACTH test), changes in pituitary responsiveness to CRH stimulation (CRH test), and changes in glucocorticoid control of CRH secretion (the dexamethasone test).

5.2.1 The ACTH Test

The ACTH test involves injecting the animal with a standard dose of ACTH and measuring the changes in plasma cortisol concentrations that follow. These changes reflect the sensitivity of the adrenal cortex to ACTH stimulation. An alternative approach is to measure the ratio of plasma cortisol to plasma ACTH following application of a stressor or injection of CRH. The principle behind this variant is that, if the adrenal cortex is more sensitive to ACTH stimulation, then the ratio of cortisol: ACTH should be higher.

In one of the first uses of the ACTH test on cattle, Gwazdauskas et al. (1975) noted that cattle exposed to heat stress did not show changes in plasma cortisol concentrations but secreted lower amounts of cortisol when injected with ACTH. They suggested that the continued stimulation of the adrenal gland by ACTH, which occurs when the stress is prolonged, results in reduced responsiveness of the adrenal cortex. However, subsequent studies found that prolonged stress associated with social competition for lying stalls resulted in an increased, rather than a reduced, cortisol secretion to ACTH (Friend et al., 1977, 1979). In an attempt to unravel the time course of the effect, Gwazdauskas et al. (1980) gave repeated sequential injections of ACTH (seven injections over 3 days) and found that the increase in cortisol concentrations following ACTH was highest on the second day of treatment and that this began to decline on the third day. They suggested that prolonged stress would result in an initial increase in sensitivity of the adrenal gland followed by a decrease. Support for this was provided by Friend et al. (1977), who noted that increased density of cows in a free stall barn resulted in an enhanced cortisol response to ACTH on day 2, which was diminished on day 3.

Other research, however, has suggested that the picture may not be so simple. Ladewig and Smit (1989) did detect a reduced adrenal response 1 month after tethering bulls. In contrast, Munksgaard et al. (1999) prevented bulls from lying down for 14 h per day and noted an increased cortisol response to ACTH injection 53 days after the deprivation began. However, a similar experiment with cows found no such effect after 3 weeks (Munksgaard and Simonsen, 1996). Fisher et al. (2002) found that reducing the amount of time that cows could lie down resulted in an increase in plasma cortisol and also increased cortisol: ACTH ratio in plasma concentrations (presumably reflecting an increased adrenal sensitivity to ACTH) at least 9 days after the beginning of the deprivation period.

The time course of this initial increase in adrenal sensitivity followed by a decrease will likely depend on the magnitude of the stress and the interval between subsequent stressors. For example, studies that have used different doses of ACTH suggest that a greater stimulation of the adrenal gland by larger doses of ACTH will result in a quicker suppression of the cortisol response (Gwazdauskas et al., 1980). However, when ACTH is injected repeatedly at intervals of 20 days, no change in adrenal sensitivity is detected (Lay et al., 1996). Thus it is unclear whether a prolonged or chronic stress will reduce or increase adrenal sensitivity. The inconsistent findings make the ACTH test difficult to interpret. Furthermore, cortisol responses to ACTH injection do not appear to correlate well with cortisol responses to an actual stressor (e.g. Van Reenen et al., 2005). Our ability to use the ACTH test as a measure of responsiveness will be enhanced once we know more about the mechanisms by which ACTH does stimulate cortisol synthesis and secretion and how these mechanisms change in response to repeated activation of the HPA axis.

5.2.2 The CRH Test

Research with people and other animals suggests that the anterior pituitary may be less sensitive to CRH stimulation following exposure to long-term stress. This can be assessed by measuring the size of the increase in ACTH concentrations in the plasma following an injection of CRH. Veissier et al. (1999) established a dose-response curve for young calves and found a threshold dose of CRH to be around 0.03 μg/kg body weight, while all animals were responding with increased ACTH following 0.1 μg/kg body weight. Fisher et al. (2002), working with cows, and Munksgaard et al. (1999) working with bulls found that preventing animals from resting for many hours a day resulted in a reduced ACTH response to CRH that was apparent during the first 6 days after the treatment commenced; Munksgaard et al. (1999) found that this affect was absent 49 days after treatment commenced. This provides evidence for desensitization (at least in the short term) of the pituitary to CRH in response to a stressor.

Unfortunately the ability of the CRH test to detect responses to stress in cattle has not yet been adequately evaluated for its use to be recommended.

5.2.3 The Dexamethasone Test

Circulating corticosteroids exert a negative feedback on the secretion of CRH and ACTH primarily acting on corticosteroid receptors in the hypothalamus. Injections of corticosteroids (usually the synthetic glucocorticoid, dexamethasone) therefore reduce circulating concentrations of ACTH. Numerous studies on stressed rodents or depressed humans have shown that in these cases, injections of dexamethasone result in a smaller-than-usual reduction in ACTH concentrations, suggesting a reduced negative feedback of corticosteroids. The only study to use this test in cattle (Fisher et al., 2002) found no effects of feed deprivation or reduced lying time, perhaps because the stressors were not sufficiently severe. Thus, evaluation of this test as a means of detecting chronic stress in cattle awaits further studies.

5.3 Conclusions

The assessment of chronic stress in cattle through physiological measures remains highly problematic partly because of the difficulty in detecting the changes that occur in physiological systems during a prolonged stress, but also because we still lack a good picture of how the functioning of the HPA axis in cattle changes with chronic stress. This places a major limitation on our ability to use such data to assess the effects of long-term challenges, such as housing, on animal welfare.

6 Validating Physiological Measures of Welfare: Signs of Suffering

So far we have focused on the technical difficulties in measuring physiological responses. However, another obstacle to the use of physiological data to assess animal welfare involves the extent that changes in physiological variables have been validated as indicators of animal welfare.

The measures of physiological activity, such as changes in HPA axis activity, have been promoted and used to assess animal welfare many times, but others have questioned the extent to which such changes really do reflect the level of animal welfare (e.g. Rushen and de Passillé, 1992). Throughout this book, we have argued that indicators of animal welfare need to be validated, that is a link between the measures of welfare chosen and an acceptable definition of animal welfare must be shown. Unfortunately, this has rarely been done for many physiological measures of stress. Based on the model of Moberg (Figure 3.1), we suggest that physiological changes, such as altered activity of the HPA axis, can be taken to indicate changes in animal welfare if these physiological changes indicate that the animal is suffering from some aversive emotional experience, for example, experiencing pain, fear, anxiety, etc.

Relating a change in a physiological variable to the animal's emotional experience is not easy. Evidence for a link between HPA axis activity and long-term or chronic suffering tends to be indirect, and is often circumstantial and requires that we accept an extrapolation from human experience.

Our confidence that the physiological changes do indicate suffering would be enhanced if we could show that such changes occurred when we submitted the animals to a procedure that we have good reason to believe is aversive. In the case of pain, this argument is further supported if the use of analgesics or anaesthetics reduces or prevents the physiological changes from occurring. Numerous studies have shown increases in plasma cortisol concentrations following apparently painful procedures such as branding, castration, and dehorning in cattle (see Chapter 5; Table 3.1). Furthermore, these increases are effectively reduced by the use of analgesics or local anaesthetics. Thus, there seems little dispute that pain can result in marked increases in HPA activity in cattle.

However, a number of apparently painful treatments do not increase HPA activity. When given a brief electric shock, cattle will show marked behavioural signs of distress and increased heart rate but do not show increased cortisol concentrations (Lefcourt et al., 1986). In heifers, cortisol concentrations increase during parturition, but do not differ between heifers that calve without difficulty and those that have a prolonged parturition requiring human assistance (Hydbring et al., 1999). Furthermore, plasma concentrations of corticosteroids often appear to reach an upper boundary or "ceiling" (see Section 4.1.1 of this chapter) such that they are little value in discriminating between pains of different intensities (Mormède et al., 2007). For sheep, measures of plasma corticosteroids are less effective than behavioural changes at detecting pain responses (Molony et al., 2002). Most importantly, increased plasma concentrations of cortisol also occur following events that do not, on the surface, appear to be painful or cause suffering, for example, after sexual behaviour, feeding, and milking (Borg et al., 1991; Lindstrom et al., 2001; Bruckmaier et al., 1993; Negrao et al., 2004). Similarly, increases in heart rate have been reported to occur in situations associated with positive emotions (e.g. Hagen and Broom, 2004). In other cases, treatments known to be stressful, such as increased ambient temperatures, can result in reduced cortisol concentrations (Ronchi et al., 2001). In short, increases in plasma cortisol concentrations alone cannot be taken as sufficient evidence that the animal is suffering and their measurements do not provide good information about the degree of suffering the animal is undergoing.

Another justification for using tests of HPA function to assess chronic suffering in animals comes from findings that similar tests can detect human depression. A large body of research has found altered functioning of the HPA axis in people suffering from depression or anxiety (Risbrough and Stein, 2006). However, the nature of the changes in HPA functioning appears to vary greatly according to which type of depressive disorders the people suffer from (Watson et al., 2002). Thus, while it is clear that changed activity in HPA axis activity is associated with psychological disorders leading to intense human suffering, the details differ from disorder to disorder, making it very difficult for us to infer what sort of changes in HPA axis activity we would expect to see in animals suffering in similar ways. The research done to date does not provide sufficient justification for using these tests to assess the degree of chronic suffering.

We conclude that changes in HPA activity cannot be used with much confidence to infer that cattle are suffering.

7 Validating Physiological Measures of Welfare: Biological Consequences of Physiological Changes

A second way of validating physiological measures as indicators of animal welfare is to show that the physiological changes have biological effects that can themselves cause suffering, for example, through stress-induced immunosuppression leading to disease.

In reviewing the biological consequences of HPA axis activity, we must bear in mind the following caveats. First, the biological effects of stress are not always due to increased HPA axis activity, but may reflect stress-induced changes in other endocrine systems. Second, even if the HPA axis is involved, the effects may not always be due to circulating corticosteroids. Third, different components of the HPA response, i.e. CRH, ACTH, and corticosteroids, can have quite different biological effects. Fourth, the biological effects differ between acute and chronic HPA axis activation and will also depend on the magnitude of the changes in HPA axis activity. In some cases, the temporary activation of the HPA axis following acute stress may have beneficial effects on the welfare of the animals.

Of course, increased HPA axis activity per se is not necessarily detrimental for the animal – many of the effects of increased HPA activity following a stressor are beneficial to the animal, either by helping the animal cope with the stressor or by limiting the damage caused by other defensive mechanisms evoked by the stressor (Sapolsky et al., 2000). Thus, evidence of increased HPA axis activity is not necessarily a sign of reduced welfare. Whether an increase in HPA activity is detrimental or beneficial for the animal will depend on the magnitude and the duration of the change. Negative effects of increased HPA activity on welfare come most often when the increase is large or prolonged in time. More subtly, however, the effects of increased HPA activity on overall welfare will be judged differently according to which particular consequence of HPA (e.g. metabolic or immune) we are examining. While it may be possible to identify if a particular consequence of HPA activity is negative or positive for animal welfare, it is much more difficult to judge whether the sum total of all the consequences is positive or negative. For this reason, we believe that considerable caution should be exercised in using measures of HPA activity to assess animal welfare.

Although activation of the HPA axis has a wide range of effects on animals, including various pathological histological changes in a number of organs, research on cattle has concentrated mainly on documenting metabolic effects and effects on the immune system.

7.1 Metabolic Consequences of Stress-Induced HPA Activity

Evidence that increases in HPA activity may signal a reduction in animal welfare comes from the metabolic consequences of increased HPA axis activity. Animals subject to chronic stress generally suffer from marked disturbances of metabolism associated with reduced feed intake, negative energy balance, an increased metabolic rate, and subsequently loss of body weight or reduced growth (Elsasser et al., 2000; Dallman, 2001). In rodents, a decrease in body weight has been proposed as a reliable measure of the response to chronic stressors (Dallman, 2001). This loss of body weight following stress is not due solely to the reduced feed intake (Smagin et al., 1999) but results from a variety of tissue-specific metabolic changes (Zhou et al., 1999). These metabolic effects can also be seen in changes in the structure of a number of organs (Tsigos and Chrousos, 2002).

Activation of the HPA axis is partly responsible for these effects; intracerebroventricular (icv) infusions of CRH can mimic the stress-induced reductions in feed intake and loss of body weight (e.g. Linthorst et al., 1997), and icv injections of a CRH receptor antagonist can block these metabolic consequences of stress (Smagin et al., 1999). However, repeated stress results in long-lasting changes in metabolism associated with a sustained loss of body weight that outlasts the initial activation of the HPA axis (Smagin et al., 1999) and which may persist for some time after the stressor itself has been removed (Zhou et al., 1999). Corticosteroids are major catabolic hormones leading to a mobilization of energy for dealing with emergencies rather than long-term growth or development (Wingfield and Ramenofsky, 1999; Sapolsky et al., 2000) and the principle metabolic effect of corticosteroids is to increase glucose concentrations in the circulation via a variety of mechanisms (Sapolsky et al., 2000). Nevertheless, the release of CRH during stress may have metabolic effects independently of the peripheral secretion of corticosteroids (Linthorst et al., 1997). Furthermore, the different components of the HPA axis can influence metabolism in different ways: infusions of CRH tend to reduce feeding behaviour and feed intake (e.g. Linthorst et al., 1997) whereas slight increases in corticosteroids may increase feed intake (Wingfield and Ramenofsky, 1999; Sapolsky et al., 2000). Finally, the metabolic consequences of elevated corticosteroids can vary greatly according to the presence of other hormones, principally insulin (Sapolsky et al., 2000). These complexities need to be borne in mind in reviewing the effects of increased HPA activity on metabolism.

In cattle, the effects of increased HPA activity on metabolism have been shown mainly by the changes in plasma concentrations of metabolites following injections of either ACTH or cortisol. However, the importance of these changes for animal welfare is difficult to determine. The effects on animal welfare of the metabolic consequences of stress-induced increases in HPA axis activity would be more obvious if these were apparent in reduced growth rates of young animals, or drops in body weight or body condition of adults.

7.1.1 Effects of HPA Activity on Growth in Young Cattle

Evidence that stress-induced increases in circulating cortisol can reduce the growth of cattle comes from studies of castration of beef bulls. Castration generally leads to immediate increases in cortisol and reductions in growth rates (see Chapter 5). Chase et al. (1995) noted a small but significant negative correlation across animals between the size of the increase in circulating cortisol and the drop in growth rate that followed castration, while Fisher et al. (1996) noted that use of a local anaesthetic reduced the effect of castration on both the magnitude of the cortisol increase and the growth rate. By injecting metyrapone, which blocks the synthesis of cortisol in the adrenal glands, Fisher et al. (1997b) were able to reduce the acute increase in circulating cortisol that followed castration and noted an improvement in weight gains over 7 days. Increases in circulating ACTH were noted (because of the reduced negative feedback from cortisol) suggesting that the effect was due to a reduction specifically in circulating cortisol. However, feed intake was also increased so it is not clear if the effect was due to changes in feed consumption or because of a direct effect of cortisol on metabolism. In contrast to these findings, injections of cortisol, which mimicked the rise in cortisol that followed castration did not reduce growth rates (Fisher et al., 1997a), suggesting that the increases in circulating cortisol is not the only factor involved. Together these studies provide some evidence that the increased activation of the HPA axis that follows castration may have effects on the weight gains of the animals. However, the mechanisms behind this remain unclear.

When cortisol concentrations are elevated to the point where growth is reduced, and this elevation is sustained, we may conclude that some threats to long-term welfare are likely. However, again we must be careful not to conclude that all changes in metabolism of young animals due to increases in circulating cortisol will have an adverse effect on welfare. Bellows and Lammoglia (2000) noted lower body temperature in newly born calves after calving difficulties, which reduced their chances of survival in a cold temperature. This was associated with lower circulating cortisol than in normally born calves, which may have contributed to their reduced thermogenic response. In this case, elevated cortisol concentrations were associated with increased survival.

7.1.2 Effects of HPA Activity on Metabolism of Adult Animals: Transition Cows

Modern dairy cows suffer from a variety of disorders, such as milk fever, ketosis, displaced abomasum, and laminitis which become apparent or have their origin during the first few weeks after parturition, a period at which the cow additionally suffers from an increased risk of infectious diseases (Goff and Horst, 1997; Drackley, 1999; Chapters 2 and 9). During this period, the energetic demands for tissue maintenance and lactation exceed the cow's ability to obtain energy from feed, and the resulting negative energy balance that is often held partly responsible for the health problems that occur (Collard et al., 2000; Goff, 2006). This negative energy balance and many of the health problems that have their origin at parturition are partly due to the reduction in feed intake that occurs during the weeks before and after parturition (Ingvartsen and Andersen, 2000). Goff and Horst (1997) have reviewed much of the evidence of the endocrine changes that occur around parturition that may impact energy balance. Plasma cortisol concentrations increase during the day of parturition and the day after (Goff and Hurst, 1997; Ingvartsen and Andersen, 2000). Given the generally catabolic effects of increased HPA activity, it seems likely that increased HPA activity at or soon after parturition would increase the negative energy balance and thus exacerbate these problems.

In a number of species, central injections of CFH can reduce feed intake via a central neural mechanism (Ingvartsen and Andersen, 2000) suggesting that the reduction in feed intake around parturition may be due to activation of the HPA axis at that time. We have little information to judge the role of increased HPA axis activity in influencing feed intake in cattle, but Ingvartsen and Andersen (2000) suggest that the reduction in feed intake tends to last longer than does the increased activation of the HPA axis indicating that other factors may also be important.

Unfortunately, there is very little information to judge whether or not increases in circulating corticosteroids do produce metabolic changes that affect the transition cow. Horst and Jorgensen (1982) noted that cows suffering from milk fever (hypocalcaemia) had higher plasma cortisol concentrations on the day of calving and that across cows there was a negative correlation between plasma cortisol concentrations and plasma calcium concentrations. However, injections of cortisol did not reduce calcium concentrations, and Horst and Jorgensen (1982) conclude that elevated cortisol is a consequence rather than a cause of the reduced calcium concentrations. In fact, the elevated cortisol may help the animals cope with hypocalcaemia through a reduction in plasma phosphate concentrations (Horst and Jorgensen, 1982) again reminding us that elevations in circulating corticosteroids are not necessarily harmful to the animals.

Despite the important metabolic role of corticosteroids and their likely effect on energy balance, we have little information to judge the seriousness of the metabolic consequences of increased circulating cortisol in the periparturient cow. More evidence is available for the likely effect on the immune system, as discussed below.

7.1.3 Conclusions

In conclusion we have as yet no good reason to believe that the magnitude of the HPA responses to ordinary husbandry stressors in cattle is sufficient to reliably induce metabolic disturbances of a magnitude that affects welfare.

7.2 HPA Axis Activity and the Immune System

7.2.1 Overview

Early research emphasized the suppressive effects of stress on the immune system, and the idea that stress can render animals and people more susceptible to disease has entered the public consciousness. Activation of the HPA axis, especially the elevation of circulating glucocorticoids, can have wide-ranging suppressive effects on many parameters of the immune system (Munck et al., 1984). It is beyond the scope of this book to document the relationship between HPA activity and the immune system; fortunately a number of excellent reviews are available (e.g. Elenkov and Chrousos, 1999; Maule and Vanderkooi, 1999; Blecha, 2000; Sapolsky et al., 2000), and we shall rely on these. The mechanisms underlying the suppressive effects of HPA axis activity on the immune system have been extensively explored. Although CRH and ACTH may have direct effects upon immune system, most research has concentrated on the role of the glucocorticoids. The most general effect of these is to inhibit the action of cytokines and other mediators that promote immune and inflammatory responses.

However, recent research has now shown the complexity of the relation between the HPA axis and the immune system, and there is now ample evidence that activation of the HPA axis can also have either no effect on the immune system, or can have immunoenhancing effects (see reviews by Elenkov and Chrousos, 1999; Maule and Vanderkooi, 1999; Blecha, 2000; Sapolsky et al., 2000; Dhabhar, 2002). Whether stress will suppress or enhance immune responses appears to depend upon the state of the animal, which particular component of the immune system is being studied, and the relative timing of the stress and of the immune challenge (Maule and Vanderkooi, 1999). Sapolsky et al. (2000) have tried to explain the complex nature of the effects of HPA activity on immune function by suggesting that the immunosuppressive and immunoenhancing effects complement each other, first by initiating defensive responses to a stress and then preventing an overreaction of these defensive responses, which may prove damaging itself. The relative timing of the immune challenge in relation to the HPA axis response to stress may be particularly important. Sapolsky et al. (2000) argue that the immediate response to stress (which generally occurs at lower circulating concentrations of glucocorticoids) is primarily immunoenhancing. However, the latter effects of HPA axis activity (involving higher levels of glucocorticoids) are immunosuppressive, acting to limit the effects of the immune response that was initially elicited by the stressor. The immunosuppressive effects are functional primarily by reducing the risk of autoimmune responses. Thus the immunosuppressive effects of HPA axis activation tend to be more likely when circulating glucocorticoid concentrations are substantially elevated for a longer time after a stressor.

The immune system can also be affected by many other endocrine and neuroendocrine responses to stress besides the HPA axis (Maule and Vanderkooi, 1999; Sapolsky et al., 2000). The catecholamines of the SNS appear to have both suppressive and enhancing effects on the immune system (Madden, 2003), and there is interest in the immunoenhancing properties of growth hormone (Burvenich et al., 1999). Thus the overall consequences of stress on immune function will depend upon a number of physiological changes elicited by the stressor (Elenkov and Chrousos, 1999).

Although most research showing the complex effects of HPA axis activity on immune function has been done using laboratory animals, we have some evidence showing similar complexity for farm animals (Minton, 1994; Blecha, 2000). Rhind et al. (1999) have shown in sheep that suppression of lymphocyte proliferation in response to an antigen occurs only when large, occasional (every 6 h) transient increases in cortisol concentrations (achieved through infusions of cortisol) occur repeatedly over a number of days. The transient nature of the increase seems important; when cortisol concentrations were continuously elevated over several days or when more frequent infusions were given (every 1 h), no suppression of lymphocyte proliferation was observed.

In conclusion, it should be clear that an increase in HPA axis activity alone should not be considered as evidence of immune suppression.

7.2.2 HPA Activity and Immune Function in Growing Cattle

Injection of the synthetic glucocorticoid, dexamethasone, has wide-ranging effects on the immune system of 1-year-old steers (Anderson et al., 1999), including a reduction in circulating eosinophils, some classes of lymphocytes, immunoglobulin M, and a slight reduction in mitogen-induced lymphocyte proliferation. However, some enhancing effects were also noted such as an increase in circulating neutrophils, monocytes, and some classes of lymphocytes. Other aspects of the immune system were not affected, such as NK cell activity, interferon-y production or concentrations of immunoglobulin A and G. Thus the effects of increased HPA activity are likely to be complex and are not necessarily limited to immunosuppressive effects. However, in this study it was not documented how the circulating concentrations achieved related to concentrations of corticosteroids following stress.

Most of the research investigating the relation between HPA axis activity and immune function in growing cattle has focused on shipping fever or bovine respiratory disease, which often follows the transport of calves to feed lots (Blecha, 2000; see Chapter 2). Reviews of research related to this question conclude that activation of the HPA axis alone cannot account for the degree of immunosupression found in transported calves (Minton, 1994; Blecha, 2000). For example, transport of calves for several hours produced evident changes in circulating neutrophils and reductions in lymphocyte proliferation; however, these effects occurred without any apparent increase in cortisol concentrations (Blecha et al., 1984).

Castration of growing bulls results in temporary increases in circulating cortisol and a number of changes in immune function, but there is no relationship in terms of breed or treatment differences in the magnitude of the cortisol response and of the changes in white blood cell numbers (Chase et al., 1995). Administration of cortisol to mimic the rise in circulating cortisol seen after castration does not induce the reduction in interferon-y production in response to mitogens or the increase in acute phase proteins that occurs in response to actual castration (Fisher et al., 1997a). Suppression of cortisol synthesis by use of metyrapon did reduce the effects of castration on circulating cortisol concentrations but did not suppress the effects of castration on interferon production, white blood cells numbers, acute phase protein concentrations or neutrophil: lymphocyte ratios (Fisher et al., 1997b).

Few studies appear to have looked at the relationship between stress-induced HPA axis activity and immune function in young calves. Van Reenen et al. (2000) experimentally infected calves with bovine herpes 1 virus, and noted that calves with lower basal cortisol and a reduced cortisol response to ACTH showed less severe symptoms of respiratory disease even though no differences were found in antibody titres.

7.2.3 HPA Activity and Immune Function in Periparturient and Lactating Cattle

It has often been proposed that the increased HPA activity and the increased secretion of glucocorticoids that occur at parturition may be important factors leading to the immunosuppression noted at parturition and to the metabolic and other diseases of dairy cows that have their origin at parturition (Goff and Horst, 1997).

The most common approach to test this link has been to experimentally increase concentrations of glucocorticoids, either by in vivo injection of ACTH or of dexamethasone, or in vitro incubation of blood with glucocorticoids, and examine the resulting changes in immune function. There is abundant evidence from in vivo and in vitro studies showing that experimental increases in glucocorticoid concentrations in periparturient or lactating cows (usually achieved by injections of dexamethasone) can have marked effects on several parameters of immune function (e.g. Gwazdauskas et al., 1980; Burton and Kehrli, 1995; Roets et al., 1999; Diez-Fraile et al., 2000). These experimental studies provide good evidence that administration of glucocorticoids can certainly result in immunosupression and most likely increase the incidence of various diseases associated with parturition. The problem with such an approach is that it ignores the possible effects of other components of the HPA axis such as CRH or ACTH, which may independently influence immune function (Maule and Vanderkooi, 1999). Furthermore, it is never entirely clear whether the dose of glucocorticoids used corresponds to the increase in concentrations that occur following normal stressors. An alternative way to approach the issue, which overcomes these problems, is to see whether individual cows that have high cortisol concentrations at parturition are those that show the largest degree of immunosuppression.

Mallard et al. (1997) immunized dairy cows against ovalbumin during the weeks prior to parturition and noted that not all cows showed the reduced immune response typically associated with parturition. The magnitude of the antibody response of cows to ovalbumin could be used to predict later health problems: cows that responded highly to ovalbumin also responded highly to *E.coli* immunization and showed a lower incidence of health problems, especially mastitis. Although some correlations were found between the immune response to ovalbumin and concentrations of GH and IGF, no consistent correlations with concentrations of circulating cortisol were noted. Thus, a high concentration of circulating cortisol at the time of parturition is not associated with low-immune responsiveness.

Preisler et al. (2000a, b) noted a reduction in glucocorticoid receptor expression in bovine neutrophils, monocytes, and lymphocytes during parturition, and that the degree of this reduction was positively correlated across cows with circulating cortisol concentrations. Although the authors state that this finding has implications for the susceptibility of cattle to disease, it is not clear what these implications are: the reduced expression of glucocorticoids receptors may be a defensive mechanism to protect the cells against the increased cortisol concentrations that occur at parturition, and may explain the lack of correlation between disease incidence and parturient cortisol concentrations.

Hopster et al. (1998) examined lactating dairy cattle that differed in their HPA response to an acute stressor. Following intramammary endotoxin injections, they noted a marked increase in circulating corticosteroids and a reduction in the number of white blood cells. This reduction was most marked in the cows that showed a low cortisol response to stress. There was also a marked decrease in the number of lymphocytes, which was most evident in the cows that showed a high cortisol response to stress. They suggested that cows with low HPA responses to stress were better able to recruit leucocytes from the circulation to enter into the inflamed intramammary tissue. However, the peak cortisol response to the endotoxin treatment did not differ between the two groups of cows. Furthermore, there were no significant differences between the two groups in the disease incidence.

7.2.4 Conclusions

It would be foolish to argue that HPA axis activity does not have suppressive effects on the immune system or that changes in immune function and in HPA axis activity are unrelated. However, for our purposes, the question is whether measures of HPA axis activity that occur during stress can predict suppressed immune function. Clearly, elevations of circulating corticosteroids through administration of synthetic glucocorticoids such as dexamethasone can have marked suppressive effects on some aspects of the immune system of cattle, and may increase the incidence of disease. Unfortunately, it is difficult to determine whether the rises in circulating corticosteroids mimic those normally found in stress responses. Where equivalence has been established (e.g. Fisher et al., 1997a), the administration of glucocorticoids has not been found to mimic the effects of actual stress on the immune system. The few studies that have examined differences between individual animals have not found convincing evidence that cattle with high concentrations of circulating glucocorticoids show signs of a more impaired immune response or are more likely to suffer from increased diseases around parturition. Unfortunately, there are too few studies that examined correlations between immune function and other components of the HPA response such as ACTH concentrations.

In general then, there is as yet little convincing evidence that assessment of HPA axis activity will predict immune function. In view of the difficulties in predicting how any given change in HPA function will affect the immune system, researchers have increasingly tried to directly assess aspects of immune function.

8 Conclusion

An impressive body of research has documented the physiological changes that cattle and other animals undergo when they are subjected to various stressors. It is clear that the HPA axis is a central part of the response to stressors and that the HPA response is associated with the emotional state of the animal. Furthermore, there is little doubt that changes in HPA activity and in other physiological systems can have marked deleterious effects on the animals metabolism and immune system, and ultimately on their welfare. Thus understanding the physiological responses to stress can play an important role in research on animal welfare.

That said, measures of these physiological changes provide only an unreliable indicator of the animal's well-being. Our review has revealed continuing uncertainty about which are the most appropriate means of assessing HPA activity, particularly in response to chronic, prolonged stress. Despite the many uses of physiological data to assess animal welfare, there are surprisingly few attempts to demonstrate the validity of the measures used. We have argued that such validation can be demonstrated in two ways. The physiological responses to a stress may reflect the emotional response that the animal makes when exposed to the stressor, and so measurement of these physiological changes may be used to infer the nature of the animal's emotional state. Measurement of plasma cortisol and to a lesser extent of heart rate have been used with great success to examine the responses of animals to acute challenges such dehorning, branding, etc. as well as the effectiveness of pain prevention methods (see Chapter 5). It is in such situations that physiological measures have proved most useful in assessing animal welfare. However, in many cases, such responses can be measured equally as well and probably with less time and cost, by examining the behavioural responses of the animals. It is in the cases where behavioural responses cannot be measured, or where our confidence in the behavioural measures is limited, that such physiological measures are most useful. The ceiling effect, noted when plasma cortisol concentrations reach a maximum with no further increase, place limits on the ability of this measure in particular to detect variation in the intensity of the stressor. Such problems may be overcome by the inclusion of other measures such as ACTH concentrations. However, plasma cortisol concentrations also increase following sexual encounters, feeding, and milking, all which would appear to be positive and rewarding for the animal, or at least not highly aversive. Thus changes in plasma cortisol concentrations alone cannot be used to determine that the experience was aversive for the animal.

In the case of prolonged or chronic stress the situation is even less clear. The changes that occur in HPA function in response to prolonged stress mean that the usual measures of plasma concentrations of hormones are no longer sufficient. Unfortunately, none of the proposed challenge tests have been adequately validated. For example, the ACTH test is the measure used most often, but there is still continuing uncertainty about even the direction of the change expected following chronic stress, with some studies suggesting an increased sensitivity and others finding the opposite. Thus these tests would seem to provide little solid information about animal welfare.

The second way that physiological measures can relate to welfare is if the responses to stress lead to a pre-pathological state, either through metabolic changes or changes in the immune system. Although changes in HPA activity as a result of chronic stress can produce changes in metabolism and immune function, the complexity of the relationship makes it difficult to use changes in HPA activity to predict the nature or magnitude of the metabolic and immune consequences. Direct assessment of metabolism and immune function would seem preferable, although these too need to be validated as predictors of altered susceptibility to disease. We also lack knowledge of the other long-term effects of increased HPA activity in cattle.

Our review supports the traditional view of the role of the HPA axis as being an essential and major component of the response to stress and playing a central role in mediating the effect of stress on metabolism, the immune system, and reproduction. A greater understanding of the nature of the HPA axis, how it responds to stress and allows animals to cope with stress, is essential to our further understanding of animal welfare. However, we recommend considerable caution when we attempt to use physiological data to assess the welfare of cattle and other animals.

Chapter 4
Animal Behaviour

1 Introduction

Some of the most difficult issues in animal welfare deal with the behaviour of the animals. Unfortunately, the traditional training of veterinarians and animal scientists has not involved training in the science of animal behaviour, or ethology as it is sometimes called. While animal behaviour and animal welfare are not the same, as is sometimes mistakenly assumed, the study of farm animal behaviour has made major contributions to identifying and helping to solve some of the key problems in the welfare of farm animals, including cattle. Knowledge and observations of animal behaviour can both help to establish input-based welfare criteria and also serve as outcome-based criteria for animal welfare (see Chapter1).

To establish input-based criteria, tests based on animal behaviour can help us identify the types of housing and handling routines that are most likely to affect animal welfare. For example, the degree that various handling procedures are painful or frightening to animals can be assessed most directly by examining the animal's aversion towards them. A simple way of assessing alternative designs for housing animals is to allow the animals to choose between them. Knowledge of farm animal behaviour can provide information on how to design housing environments so as to provide opportunities for the animal to behave in ways that are important to it. The study of animal behaviour can help determine animals' needs, and so serve as the basis for input-based criteria for animal welfare. In addition, the occurrence of some behaviour patterns can serve as outcome- or animal-based indicators of welfare. This is most obvious for behavioural signs of pain or fear. In addition, farm animals in modern housing systems occasionally behave in ways that appear to be abnormal, and the performance of such behaviours has been proposed to indicate poor welfare.

However, there can be difficulties in interpreting animal behaviour. As we discussed in Chapter 1, the role of animal behaviour in animal welfare has long been one of its more controversial aspects, particularly with regards to behavioural deprivation. The issue of behavioural deprivation is a crucial one for animal welfare (Dawkins, 1988, 1998, 2004) and the possibility that farm animals are suffering because they cannot perform behaviour that they normally would perform is one of the enduring concerns that the public has about the welfare of animals in modern husbandry. Housing systems and management practices for cattle have been criticized on the grounds that they prevent natural behaviours. For example, cattle housed indoors are no longer able to graze; the very early separation of the calf from its mother prevents a whole suite of nursing and parent–offspring related behaviours from occurring; individual housing for veal calves or milk-fed dairy replacement heifers limits opportunities for social contact between calves and may prevent calves from turning around or performing many normal body movements. The ability of animals to perform "natural" behaviour has been used as an input-based criterion in assessing the adequacy of animal housing systems, for example in the "Five Freedoms" (Chapter 1). Yet the claim that farm animals are suffering because modern housing systems do not allow these behaviours has proven one of the more difficult claims for science to deal with. Many of the issues concerning the causes of behaviour and the relationship between behaviour and animal welfare are at, or sometimes even beyond, the cutting edge of science. In this chapter, we consider how animal behaviour has been used as input- and outcome-based indicators of animal welfare.

2 Behaviour as Input-Based Criteria for Animal Welfare

This section focuses upon how knowledge of animal behaviour can be used to design management techniques or housing that improve animal welfare. First, we consider aversion learning procedures whereby, animals provide us with information about which types of handling procedures they dislike the most. We then discuss the use of choice tests to examine animals' preferences for certain design features that can be incorporated into their housing. We then focus upon some of the more controversial aspects of animal welfare, especially the concept of natural behaviour, and we discuss the different techniques available to discover which sorts of behaviour an animal "needs" to perform, and consequently, which sorts of behavioural deprivation have the greatest consequences for animal welfare, and which should be avoided when designing housing and rearing systems for cattle.

2.1 Aversion Learning

To assess welfare we need information on the extent that animals suffer as a result of the way they are housed, managed, or handled. Most obviously, animals may suffer when they are undergoing operations that are painful, for example, branding or dehorning, or when they are frightened of the situations in which they find themselves. One particularly powerful method of assessing the amount of suffering resulting from this pain or fear is aversion learning. Kirkden and Pajor (2006) provide a useful review of aversion learning and related techniques.

Aversion learning is based on the assumption that the main function of unpleasant feelings or emotions is to help animals avoid the situations that cause them. Modern scientific approaches to studying consciousness or mental states (including emotions) can be roughly divided into those that study the "hardware", where the emphasis is upon understanding the structural basis of consciousness by studying neural activity, and those which focus on understanding the function of consciousness, asking the question "what is it for?" and "what does it do?" This latter approach often involves trying to understand how different mental states influence an animal's behaviour. In people, suffering can arise from a number of different emotions or mental states, such as being in pain, being frightened, being hungry, or bored. From our own experience, it is apparent that these different emotions feel very different and make us behave differently. It is likely that the neural hardware underlying these different emotions is also different. What they have in common, however, is that they are all aversive; that is, we generally try to avoid or escape from situations that cause these states. The recognition that some situations are aversive is one of the most basic ways that we have of categorizing our experiences. By measuring the extent that we try to avoid such situations, we can begin to get some idea of the relative degree of suffering these situations cause, irrespective of the specific affective state involved.

A similar logic can be applied to measuring suffering in animals. When an animal is put into a situation that causes it distress, we cannot directly perceive the suffering it is experiencing. Furthermore, the immediate behavioural response will depend upon the nature of the situation. For example, a cow frightened by a dog will usually stare at the dog, try to run away, or perhaps attack it (e.g. Welp et al., 2004). A cow isolated in a small room will bellow and defecate (Rushen et al., 1999a). Thus, no one behavioural response will let us judge whether the presence of a threatening dog causes more or less distress than social isolation. However, cows can learn to use cues to predict the onset of these aversive situations, and their responses to these cues can be used to compare the degree of aversion they feel to the two situations.

Two articles by Pajor et al. (2000, 2003) demonstrate how aversion learning techniques can be applied to cattle. The purpose of the experiments was to determine which handling practices are most aversive to the animals. The types of handling examined were some commonly used to move cattle, such as shouting or yelling at the animal, or use of an electric cattle prod. These were compared with ways of handling animals that were thought to be more positive or rewarding for the animal such as providing a small food reward.

In the first experiment (Pajor et al., 2000), dairy cattle were moved individually down a short raceway. At the end of the race, the cattle were briefly restrained. Control cows were then simply released without being handled, but others were shouted at, or prodded with an electric cattle prod, or given a small amount of highly palatable food before they were released. This procedure was repeated a number of times, and each time the animal was handled in the same way. The logic was that the cattle would soon learn what type of handling to expect and, if the handling was aversive, would attempt to avoid it by refusing to move down the race. On the other hand, if the handling was rewarding they would move down the race more quickly.

The time taken by each cow to run down the race was measured and a score was given depending on the amount of pushing that had to be done whenever she stopped. Figure 4.1 shows the results.

For the control cows that were not handled in any particular way, the time taken to run down the race was highest during the first run and gradually decreased as the animals got used to the procedure. It is clear, however, that when the animals were shouted at, or prodded with the electric prod, they took longer to run down the race on each subsequent run, and required more pushing by the handler. This result shows that these handling procedures were aversive to the cattle and that the animals were trying to avoid them. When the cattle were given food at the end of the race, they tended to run down the race faster the next time, suggesting that the cattle found this to be rewarding.

A more simple and direct way of comparing handling practices is to let animals choose between the handling techniques. This is an example of the preference (or choice) tests described in Section 2.2, and shares many of the advantages and disadvantages discussed. Pajor et al. (2003) used this approach to compare similar handling procedures to those described earlier. The cattle were placed in a bifurcated race (a "Y" maze). The cows moved down the race and then had to choose between one of the two arms. When the animal chose the right arm, she was always handled in a particular way. When she chose the left arm, she was handled in

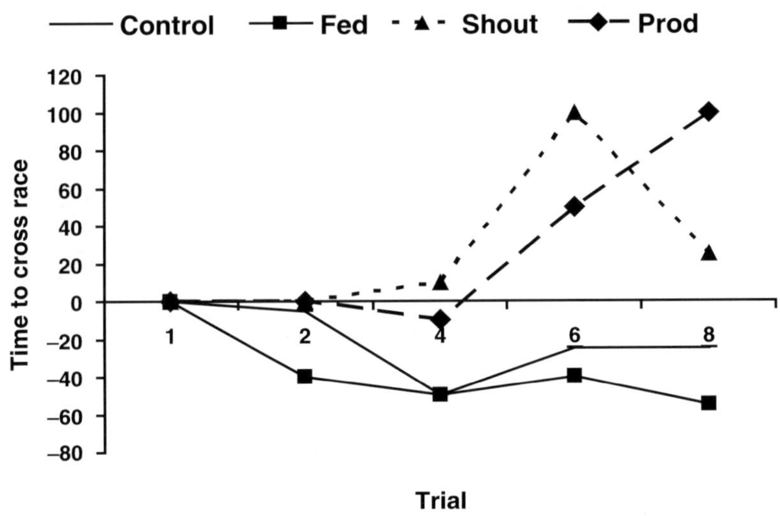

Figure 4.1 The time that cows take to run down a race when they are fed, shouted at, or touched with an electric prod each time they reach the end. Control cows were not handled in any way. The time is shown as the difference in seconds from the first trial (trial number 1) which occurred before the cows had experienced any of the treatments. The increased time taken by cows that were shouted at or touched with the prod indicates that they find these treatments aversive. The reduced time taken by cows that are fed show that they found this a positive or rewarding experience (Based on data presented in Pajor et al., 2000.)

another way. Figure 4.2 shows the relative per cent of animals choosing among different pairs of treatments.

The cattle preferred the arm in which they were fed rather than the arm where they were hit or shouted at, indicating that cattle are indeed able to choose treatments and that their relative preferences can be measured fairly easily. The results of this choice test confirmed the findings of the first experiment with the raceway. That is, cattle find being shouted at as aversive as being hit or prodded with a cattle prod. Although cattle preferred being spoken to gently over the more aversive treatments, they did not find gentle talking or brushing and patting to be rewarding. These experiments show that we can compare handling techniques using either the aversion or choice methods. However, while these techniques hold promise for assessing the extent that animals themselves perceive the treatments that we impose on them, there are limitations with aversion learning techniques that should be considered (Rushen, 1996).

First, the techniques rely on the animals learning to associate the end of the raceway or a particular arm of the Y maze with a particular treatment. Thus, factors that influence the animals' memory or learning will influence the results of aversion learning tests. Most obviously, animals will usually need to experience the treatments a number of times before learning that they will occur in a particular place. Figure 4.1 illustrates this phenomenon. As the experiments progressed, the differences between the different handling practices became more evident; the aversive handling treatments differed significantly from the control group after five applications of the

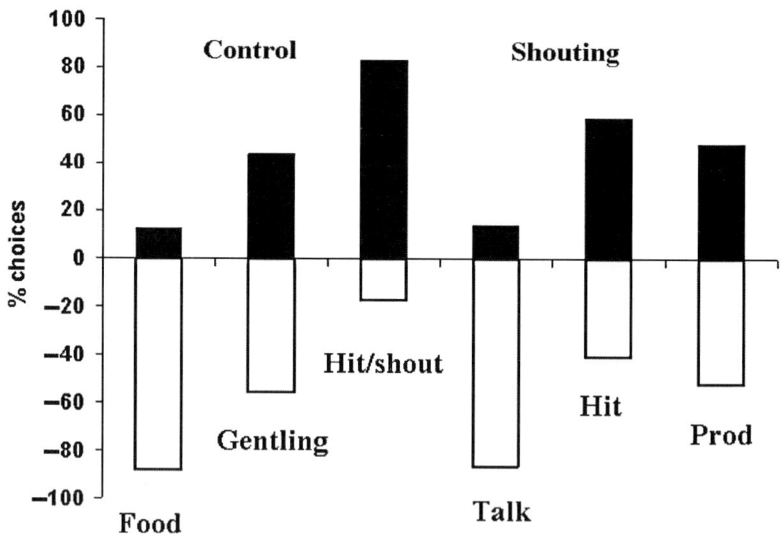

Figure 4.2 Cattle show clear preferences for being handled in certain ways. The results show the per cent of cows' preferences when they were able to choose between a control condition (which involved no handling) and being given food, gentled, or being hit and shouted at, and when they could choose between being shouted at and talked to in a gentle voice, being hit or being touched with an electric cattle prod (From results of the experiment described in Pajor et al., 2003.)

handling treatment. Thus, aversion learning techniques work best with handling treatments that can be applied a number of times within a short period. This is much more difficult in the case of surgical operations such as castration or branding which for practical or ethical reasons can only be applied once or a few times. For example, Schwartzkopf-Genswein et al. (1997b) found that aversion learning techniques were not useful in estimating the aversion associated with hot-iron branding or freeze branding of cattle.

Individual animals differ in how well or quickly they can learn such associations, so it is difficult to use aversion learning techniques to examine differences between different animals in how aversive they find treatments. For example, if we find that cows vary in the strength of their preference for being spoken to gently compared to being shouted at, this may be because they genuinely find shouting to be less aversive, or because they have not learned as well to expect how they will be handled when they make a given choice. When cattle have learnt to associate a particular aversive treatment with a particular place, they find it difficult to subsequently learn to associate another treatment with that place (Grandin et al., 1994). Consequently, when different treatments are being compared, it is important to ensure that each treatment is consistently applied in the same place and preferably in a place that is unfamiliar to the cattle. Further discussion of difficulties with aversion learning techniques is given in Rushen (1996) and Kirkden and Pajor (2006). If such methodological issues are borne in mind when designing and interpreting these tests, the results can provide useful insights into which types of handling animals dislike the most.

2.2 *Preference Tests*

Designing appropriate environments for animals is much aided by knowing what the animal prefers. Choice or preference tests have been used often in studies of housing and management for cattle. For example, the technique has been used to determine the response of dairy cattle to various aspects of stall design (e.g. Tucker et al., 2003; Manninen et al., 2002) and to various types of food (e.g. Rutter, 2006). Chapters 6 and 7 discuss in more detail the way that preference techniques have been used to assess housing for dairy cattle. In this section, we discuss some of the more general limitations with this technique.

Testing animals' preferences would seem one of the most obvious and direct ways of asking what it is that an animal wants in its environment. However, there can be problems with interpretation. A number of excellent discussions of the pros and cons of preference testing are available (e.g. Fraser and Matthews, 1997; Lawrence and Illius, 1997; Kirkden and Pajor, 2006). We summarize the most important concerns later, but we encourage readers to consult these articles if they are considering using these methods in their own research.

One important limitation in preference tests is that the animals' choices are restricted to the options provided, so a strong preference for one option only means that it is better than the alternative and cannot be taken to mean that it is an ideal or even a reasonable solution for the animal. In addition, the tests detect how animals rank various options, but provide little information on the strength of their preferences. One way to evaluate the relative value of the various alternatives is to measure how much each is used when animals are limited to a single option at any one time. For example, Tucker et al. (2003) compared dairy cows' preferences for stalls containing three types of bedding and found that most cows showed a preference for deep sawdust bedding. When the cows were given access to only one of the options at a time, they also spent more time lying down in stalls with the deep sawdust, indicating that access to the preferred option was of some importance to the cows. Another way of assessing whether access to a preferred option is really important to the animal is to impose some cost on the animal's choice. This issue is explored in depth by Kirkden and Pajor (2006). Unfortunately, there are few good examples of this approach with cattle, and the best illustrations of this approach involve other species. Marian Dawkins (1983) took advantage of this principle in a well-known experiment on laying hens. She found that recently fed hens showed a strong preference to enter a cage with litter (in which they could dust-bathe) but no food, over another cage with food and a plain wire floor. In this case, the loss of access to food was the cost that the hens were willing to pay to gain access to litter. However, if hens were deprived of food for 3 h before the test they showed no clear preference, and after 12 h of deprivation the preference was reversed.

Some experiments on animal preferences sum the choices of a number of animals. This technique may provide us with information on the "average" animal but it overlooks differences between animals. Although the majority of the choices may be for one alternative, certain animals may not share the preference, or indeed may have a strong preference for the alternative treatment (e.g. Tucker et al., 2003). Another important methodological consideration in choice tests is that preferences may change according to age, physiological state, previous experience, and other factors. For example, animals may initially prefer the option that is familiar, yet if encouraged to try the alternative or if given sufficient time, the animals may switch preferences. Tucker et al. (2003) showed that dairy cows that were reared on stalls with sawdust bedding preferred such stalls to those bedded with sand. However, cows that had been reared on sand choose sand-bedded stalls as often as sawdust-bedded ones. Preference tests need to ensure that animals have been given sufficient experience with all the options available. The results of preference tests can also be affected by how the tests were performed. For example, cows' preference for types of bedding that differ in thermal conductivity can differ according to whether the tests are performed in summer or winter (Manninen et al., 2002). The results of preference tests must be interpreted bearing in mind the conditions under which the tests are performed.

The results of choice tests are particularly difficult to interpret when the options presented to an animal differ in a number of properties. For example, the types of flooring in stalls for cattle may differ in both the degree of softness and the degree of thermal insulation provided, and careful experiments would be required to determine which properties are the most important to the animal. The general conclusion of reviews is that preference tests are most successful when there are only relatively small differences between the options.

The social context in which the test is made can also affect preferences. For example, herd animals like cattle are often tested in groups. In this situation, it is essential that a sufficient number of each option be presented so that all animals can exercise their choice, but even then the animals' choices may not be independent. For example, two cows may be close companions and choose to lie in neighbouring stalls, even if these did not reflect their individual preferences for stall type. In such cases it may be preferable to test animals individually.

Finally, and most importantly, choice tests reflect the animals' immediate response to the alternatives, but we do not know to what extent animals are capable of judging the long-term effect of a housing treatment on their overall welfare (Lawrence and Illius, 1997). For example, as we discuss in Chapter 6, dairy cows tend to choose wider stalls even though wider stalls can become dirtier potentially increasing the risk of mastitis. Cattle simply may not have the capacity (or the experience) to judge the long-term consequences of their choices. We should not assume that animals are perfect judges of their own welfare any more than people are perfect judges of theirs. The underlying assumption behind choice tests is that an animal's welfare will be improved if it is housed under the conditions it prefers, but this assumption has never adequately been explored. Furthermore, we do not really understand how animals weigh the different options before choosing. Lawrence and Illius (1997) conclude that the strong impact that "procedural" variables have upon the results of preference tests means that we have not yet developed an adequate conceptual model of how animals make choices. They further conclude that in using preference tests, "animal welfare research has placed too much emphasis on measurement and too little on understanding the rules that govern short-term behavioural decisions" (p. 24).

In conclusion, preference experiments need to be carefully planned and carefully interpreted to avoid some of the problems outlined earlier. With these cautions in mind, the technique can tell us which options the animal prefers – information that will always be an important one in improving living environments for animals.

2.3 *Natural Behaviour*

Aversion learning and preference tests provide two examples of how we can use observations of animal behaviour to determine what animals like and dislike. In the following sections, we discuss how a greater fundamental knowledge of the causes of animal behaviour can help avoid the welfare problems that arise from behavioural deprivation.

Thorpe (1965) provided one of the earliest and clearest scientific arguments that the ability to perform natural behaviour was essential if an animal was to have an acceptable level of welfare. His arguments were rooted in the ethological theory developed and popularized primarily by Lorenz and Tinbergen arguing that each species has a distinct repertoire of behaviour patterns that could be used to characterize the species in much the same way as morphological features. It was further argued that domestic animals had retained much of the repertoire of their wild ancestors, but that modern farming conditions sometimes prevented animals from performing these behaviours, leading to suffering.

The problems with this concept have been discussed many times (e.g. Dawkins, 1998, 2004; Spinka, 2006), and do not need to be repeated in detail. Briefly, there is no reason to think that an animal will inevitably suffer simply because it does not perform all the behaviour patterns shown by its wild ancestors. Indeed, allowing animals to perform some natural behaviour, such as aggressive behaviour or infanticide, may lead to reductions in animal welfare. Furthermore, even within a population of animals in natural environments, there are differences between individual animals in the behavioural strategies adopted, raising the question of what should be considered as "the" natural behaviour (Spinka, 2006). Finally, a multitude of detailed studies on different species has revealed how much artificial selection has altered the behaviour of domestic animals (e.g. Price, 2003; Jensen, 2006).

Despite these criticisms, the longevity and the ubiquity of this concept suggests that it captures some of the disquiet that modern farming systems provoke in many people. Furthermore, comparisons between farm animals and their wild ancestors or feral counterparts may give some clues as to where to look for welfare problems that can arise from behavioural deprivation (Spinka, 2006).

2.3.1 Inability to Perform Natural Behaviour as a Welfare Concern in Cattle

Applying the concept of natural behaviour is troublesome for cattle since the ancestor of domestic cattle, the wild ox or aurochs (*Bos primigenius*) of the near east (Clutton-Brock, 1999; Troy et al., 2001), has been extinct for several centuries (Clutton-Brock, 1999). Furthermore, genetic analysis shows that domestic cattle (*Bos taurus* and *Bos indicus*) diverged from the wild species of Bovini (e.g. bison and buffalo) hundreds of thousands of years ago (Ritz et al., 2000; Figure 4.3).

This makes it difficult to determine how much domestication has influenced cattle behaviour and whether or not modern breeds of cattle have retained the behavioural repertoire of their ancestors. In contrast, for both pigs and poultry, populations of the wild ancestors still exist in relatively natural environments.

Evidence of human uses of cattle comes from over 8,000 years ago (Clutton-Brock, 1999; Figure 4.4), and during this time there has been considerable artificial selection. This may be the reason for the genetic divergence found both between European and African populations of *Bos taurus* (Troy et al., 2001) and between the two main domesticated species of cattle, *Bos taurus* (e.g. "European" breeds of cattle) and *Bos indicus* (Zebu or Brahman cattle) (Ritz et al., 2000; Troy et al., 2001). Mitochondrial DNA analysis suggests that the aurochs fall well outside of the genetic range found within populations of modern day *Bos taurus*, and are even

Figure 4.3 *Bos indicus* cattle can be recognized by the hump on the back and the large droopy ears. They are commonly found in warmer climates since they are better adapted to dealing with heat than the more common *Bos taurus* cattle that are found as beef and dairy cattle in more temperate climates. The wide phenotypic variation found within domestic cattle is the result of many thousands of years of artificial selection for characteristics that are appropriate for how the cattle are used by people. This, and the absence of any of the original ancestors of cattle, the aurochs, makes it difficult to describe the natural behaviour of cattle

Figure 4.4 Cattle have been domesticated and used by people for various ends for over 8,000 years. The modern breeds of animals that are today used in dairy and beef production show considerable differences from the cattle in use early in domestication. These phenotypic differences probably are evidence of marked genetic differences, making it difficult to infer what their "natural behaviour" is or was

further from present day *Bos indicus* (Troy et al., 2001). Together, these results show the extent to which domestication has influenced the genetic basis of cattle populations, and most likely their behaviour. For all of these reasons it is difficult to use information about the behaviour of wild ancestors of cattle to assess which behaviours are "natural" in domestic cattle.

2.3.2 Behaviour of "Feral" Cattle

Despite the absence of the wild ancestor of cattle, there are available a number of populations of feral cattle and cattle that have had minimal human interference, often for several decades or centuries. These populations provide us with a rough glimpse at the "natural" behaviour in cattle, although the following caveats must be borne in mind. First, in no way can these animals be considered equivalent to the ancestral wild cattle. At most, the populations have been without substantial human interference for only a few centuries (maximum 800 years), which contrasts with the thousands of years of domestication that had occurred before. Even from the beginning of domestication of cattle, there is evidence of genetic selection for animals to perform particular roles (e.g. to provide milk or meat or as work animals in a particular environment) that resulted in phenotypically distinct, and quite possibly behaviourally distinct animals (Clutton-Brock, 1999). Second, these cattle are different from modern breeds of cattle (Giovambattista et al., 2001), usually being far smaller, producing far less milk, and with distinct colouration. It is questionable whether modern breeds of cattle would behave similarly if placed in similar environments.

Generally these unmanaged cattle live in herds, although there are marked differences between studies in the size of the groups and the areas over which they range. Adult and immature females and males tend to live together in relatively small herds. These herds occupy non-overlapping home ranges reported to vary between 1,243 and 2,635 ha for feral cattle in Spain (Lazo, 1994) and between 4,500 and 5,000 ha in Mexico (Hernandez et al., 1999). Cow-herd size was found to range between 13–32 animals by Lazo (1994), although Hernandez et al. (1999) found that over 80% of cows were in groups containing fewer than 10 animals. Lazo (1994) suggested that female calves born within a herd remained in that herd throughout their lives: virtually no female migration between neighbouring herds was observed. Within female herds, Lazo (1994) and Reinhardt and Reinhardt (1981) found affinities between particular individuals that were long lasting, although this was not reported in the Chillingham cattle studied by Hall (1986). Reinhardt and Reinhardt (1981) report complex patterns of social preferences between cows within herds.

The closest associations occur between mothers and their offspring (Reinhardt et al., 1977; Reinhardt and Reinhardt, 1981; Lazo, 1994; Figure 4.5). Studies of a herd of Massai *Bos indicus* cattle in semi-natural conditions (Reinhardt et al., 1977) showed that just prior to parturition, the cow separated from the herd and gave birth in a concealed place. The cow remained on the periphery of the herd for up to 2 weeks, remaining in close proximity to her offspring, and often threatening other cows that approach too closely. Between 1 and 3 weeks after calving, the cow and her calf join the rest of the herd; the cow grazing with others, while the calf tends to associate with other calves in the creche (Reinhardt et al., 1977). Calves suck an average of 38 min per day and are weaned between 7 and 9 months for female calves and between 9 and 14 months for male calves (Reinhardt and Reinhardt, 1981). Free-living *Bos taurus* cattle also wean their calves around 10 months (Reinhardt et al., 1986). However, cows continue to associate with their offspring long after weaning and choose them as grooming or grazing partners for many years (Reinhardt and Reinhardt, 1981). When not with their mothers, calves usually associate with each other (Hall, 1986; Lazo, 1994). Social interactions involving mock fighting are quite high between older calves, as are "mock" sexual interactions between male and female calves (Reinhardt et al., 1978). Reinhardt et al. (1978) and Reinhardt and Reinhardt (1981) concluded that calves did not interact at random but formed clear preferences that were responsible for the formation of small social units. They speculated that these preferences would persist into adulthood.

Within female herds, relatively stable, near-linear dominance hierarchies occur, with social rank strongly related to the age of the animals (Hall, 1986; Reinhardt et al., 1986). However, the incidence of aggression appears quite low (Hall, 1989) with little evidence of aggressive defence of feeding or grazing areas (Hall, 1986, 1989).

Adult bulls are typically found in groups of 2–3 individuals occupying non-overlapping home ranges (Hall, 1986; Lazo, 1994), with stable dominance relationships reflecting the animals' age (Reinhardt et al., 1986). Bulls generally live separately from the cows, although Lazo's (1994) data suggest that each group of bulls tended to associate with one herd of cows. Hall and Moore (1986) report that the bulls of the feral cattle in the Orkneys tended to herd with the cows although one mature bull was consistently found alone. Hall (1989) reported year-round breeding, while Reinhardt et al. (1986) report highly seasonal breeding with 90% of calvings occurring in early spring. Lazo (1994) found that calvings were highly synchronized with herds, with 75% of calvings occurring during 2–4 months.

Hall (1989) provides some data on the time budgeting of the cattle. Grazing occupied 10–11 h per day, a figure that corresponds to that found for feral cattle in Ireland (Linnane et al., 2001). This latter study found marked circadian rhythm in grazing, with peaks of grazing occurring at dawn and dusk. Some grazing was spread throughout the day, with low levels occurring during the early morning. Although there were marked seasonal effects on grazing time, Linnane et al. (2001) noticed some consistency across days and across seasons when grazing occurred, and suggested some internal control of grazing time. According to Hall (1989), Chillingham cattle spend about 7–8 h lying down, 5 h ruminating (of which 75% was done while lying down), 40 min sleeping, 70 min in social behaviour, and 30 min walking. Feral cattle in Mexico were reported to walk up to 20 km per day (Hernandez et al., 1999).

There are a number of discrepancies between different studies likely due to genetic or environmental differences, illustrating the danger of considering "the" natural behaviour of cattle as a single standard. Nevertheless, there are a number of consistencies that throw into relief the behavioural differences between these unmanaged animals and managed cattle (Table 4.1).

While we cannot necessarily conclude anything about the welfare of the animals solely on the basis of these differences, they may help us to understand the welfare problems that arise in modern production systems (Spinka, 2006). We will discuss evidence of the extent to which such behavioural differences may indicate welfare problems later.

Table 4.1 A contrast between what is known of the natural behaviour of cattle, based on studies of feral cattle, and the way that cattle are kept in intensive management systems. The contrast is not meant to imply that the welfare of feral cattle is better. However, these differences may help us pinpoint some of the possible threats to animal welfare in intensively housed animals

Feral cattle	Cattle in modern, intensive systems
Close association between mothers and grown-up daughters	Mothers and daughters often separated
Close social associations between calves	Calves often reared individually
Complex age structure in matrilineal herds with long-lasting associations between individual animals	Herds often consist of a few generations of unrelated individuals. Frequent culling. New animals often bought into a herd. Cows and calves often housed separately
Small herds	Herds can be very large (>100 individuals is common)
Spend many hours a day grazing	Often no access to pasture

2.4 *Behavioural Needs and Motivational Analysis*

The obvious difficulties in deciding what is "natural", and the uncertainty about the link between natural behaviour and animal welfare led scientists to develop alternative conceptual frameworks for trying to decide what sorts of behaviour are important for animals. Dawkins (2004) and Spinka (2006) provide useful reviews of these developments. Many approaches have been adopted in an effort to document the behavioural "needs" of animals; that is, the behaviours that animals need to perform in order to ensure good welfare.

One approach involves trying to understand what causes the animal to behave in a particular way. The principle arguments of this approach can be stated simply: the inability to perform a behaviour will cause greater suffering when the factors that stimulate the behaviour are internal to the animal (rather than being a feature of the environment) and where the performance of the behaviour itself is necessary to reduce the underlying motivation. Dawkins (1983, 1998, 2004), Hughes and Duncan (1988), Jensen and Toates (1993), and Spinka (2006) provide some particularly good discussions of the concepts and issues underlying this approach. The importance of external (i.e. triggered by stimuli in the environment of the animal) versus internal (i.e. triggered by events occurring within the animal) causes of behaviour can be understood with some simple examples. The natural behaviour of most species includes anti-predator responses that are performed when potential predators are detected. Many species of animals have alarm calls that serve to alert conspecifics that predators are present. Anti-predator behaviour in cattle is less obvious, consisting of increased vigilance, bunching up, and fleeing or sometimes attacking the predator. An obvious question is whether farm animals "need" to express this anti-predator behaviour. Should we expose cattle to wolves or dogs so that they can do this? Most would say no (e.g. Hughes and Duncan, 1988; Dawkins, 1998), and the most common reason is that such anti-predator behaviour is stimulated by the sight of the predator itself. In the absence of any predators, which is generally the case in most modern farming systems, the animals are simply not motivated to perform anti-predator behaviour, and so the fact that they do not perform the behaviour in modern husbandry settings is not a problem for their welfare. This situation is often captured by the phrase "out of sight, out of mind".

However, other behaviours are not caused in such an apparently simple fashion. For example, a few days before calving, cows in late pregnancy show changes in behaviour that lead them to separate from the herd and seek an appropriate place to calve (Figure 4.5). Such behaviours are far more elaborate in other ungulates such as pigs that show complex nest building behaviour (described in detail by Spinka, 2006). Little is known about the causes of these behaviours in cattle, but they are probably similar to other ungulates. Briefly, the pre-parturient behaviour is triggered by a combination of the hormonal or neuro-endocrine changes that occur during the last stages of pregnancy. Cattle in many modern environments are unable to fully perform these behaviours because the environment in which they are kept does not provide the appropriate resources. For example, cattle may not be able separate themselves from the herd when kept in indoor housing. The argument above would indicate that since these behaviours are triggered by internal changes (rather than by environmental features that are absent in captivity), then the cows are still likely motivated to perform the behaviours even in environments that prevent them from doing so. The inability to perform a behaviour that animals are motivated to perform is a welfare concern, both directly (as this is what the animal prefers) and indirectly (as the motivation is likely linked to either positive or negative affective states).

In terms of the factors responsible for stopping the behaviour or turning off the motivation, the question is whether these result from the animal achieving the functional goals of the behaviour or whether it is necessary to perform the behaviour itself. As a simple example, the functional goal of eating is the consumption of various nutrients. This is usually achieved by the animal eating, but it need not be: the nutrients could be supplied by, for example, a stomach tube. The question in this case is whether feeding motivation is turned off simply by the ingestion of the nutrients, or whether the animal also needs to perform the behaviour of eating. It is generally thought that the diets of farm animals are formulated to provide the necessary nutrients (but see Chapter 8 for some counter-examples). Certainly most farm animals absorb these nutrients by eating, but there are many components of feeding behaviour that modern housing systems do not permit animals to perform. For example, many milk-fed calves do not suck to ingest their milk, and many dairy cows no longer graze. How much is the performance of these behaviours necessary to reduce feeding motivation, even when nutrient intake is adequate?

A large body of research has focused on the motivation of different behaviours, with a particular focus on whether the main factors are internal or external to the animal and whether or not performance of the behaviour itself is necessary to fulfil the underlying motivation. In cattle, this research has been most complete when examining the causes of sucking behaviour in calves, which we discuss later. However, one of the main findings is that the theories of how behaviour is caused are far too simple. Control of behaviour is complex and often involves a mix of internal and external factors. Mason et al. (1997) present a good discussion of different models of motivation and how changes in our understanding of motivational systems have an impact on how we conceptualize the

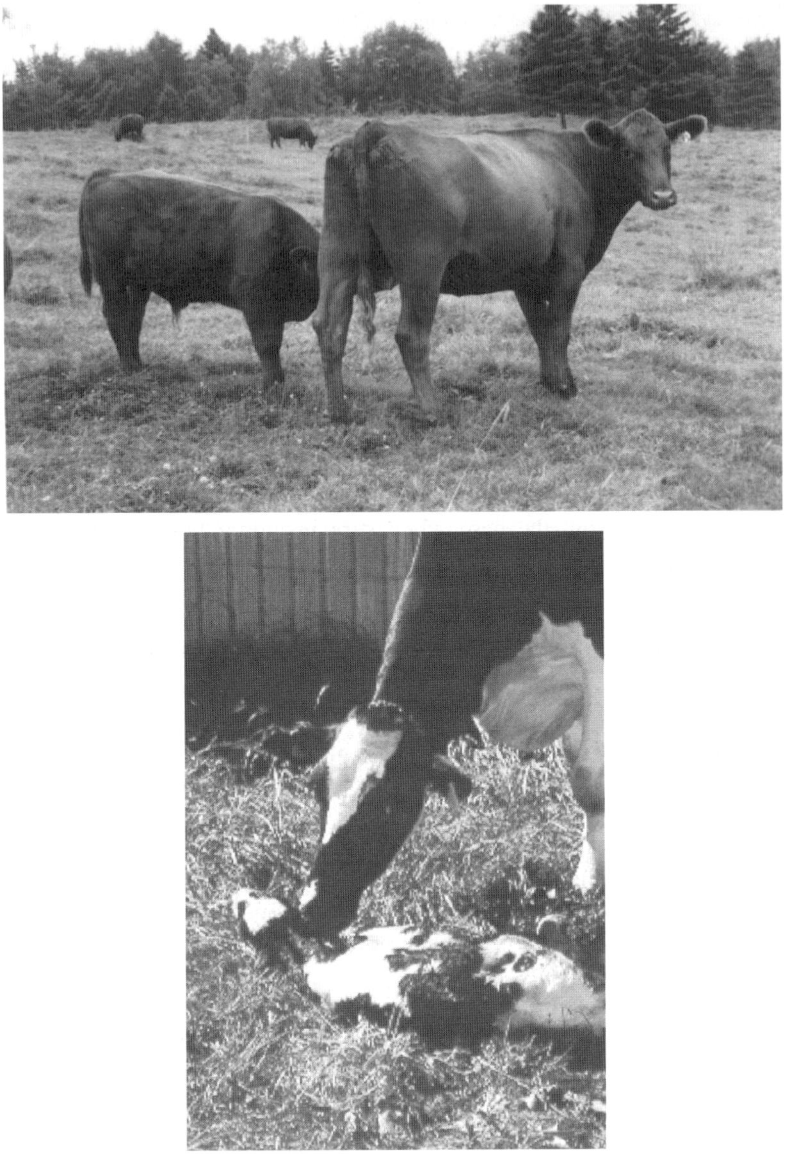

Figure 4.5 One of the most important social relationships among feral cattle is that established between a mother and her calf at the moment of birth. Usually this relationship is established while the mother is separated from the rest of the herd and occurs as a result of the intense social contact between mother and calf in the hours immediately after birth. In the case of female calves, the relationship is maintained long after weaning and social relationships between related females of different generations form the basis of the matrilineal herds found among feral cattle. Modern cattle still show much of the mother–offspring behaviour typical of feral cattle, such as the prolonged licking of the calf after birth. In beef cattle, the mother–offspring relationship is maintained for a few months in cow-calf herds but in modern dairy production, this relationship is terminated early or never allowed to develop. Usually, the calves are removed from the mothers at birth. Nowadays, herds of dairy and beef cattle are likely to consist of large numbers of animals of a similar age with no bulls present

issue of behavioural deprivation for animal welfare. As a result of the complexity of behaviour control, interest in the concept of behavioural needs appears to have waned. However, the underlying issues posed by the concept of behavioural needs have not been resolved. While we agree that motivational analysis will not produce clear and simple answers to the question of behavioural needs, we believe that an understanding of the factors controlling behaviour is essential in understanding the importance of behavioural deprivation for animal welfare. We illustrate these issues by discussing the work on the causes of sucking behaviour in calves.

2.4.1 Calves' Motivation to Suck

Young mammals typically obtain their milk by sucking on their mothers' teats but most dairy and veal calves are reared separately from their mother and fed by buckets and so cannot perform much of their normal sucking behaviour. This has raised concerns for the calves' welfare, and stimulated research to understand the motivation of sucking behaviour.

The main function of sucking seems obvious: to obtain milk. However, when calves and other young ruminants are raised separately from their mothers, they suck each other and at parts of their pens despite apparently adequate nutrition. Furthermore, non-nutritive sucking is common in infant mammals (Wolff, 1968), and occurs during normal nursing in cattle (Lidfors et al., 1994; de Passillé and Rushen, 2006b). This suggests that young calves may be motivated to perform sucking behaviour independently of hunger and that something other than the ingestion of milk is responsible for controlling this motivation. de Passillé and Rushen (2006c) and de Passillé (2001) have reviewed research on factors that cause or inhibit non-nutritive sucking by calves and Section 2.4.2 below is a summary of these findings.

Much of the research in this area was done by examining calves sucking on a dry rubber teat that provides no milk (Figure 4.6). Non-nutritive sucking on such artificial teats can be readily elicited in young calves and the fact that the calves continue to suck even though they never receive milk through the teat itself suggests that ingestion of milk is not the only factor controlling sucking behaviour.

What factors stimulate the calves to suck and what factors reduce the sucking motivation? Non-nutritive sucking is slightly higher in calves receiving a lower ration of milk (Jung and Lidfors, 2001) indicating that non-nutritive sucking is dependent on food intake in the longer term. Nevertheless, halving the amount of milk the calves drink during a meal (Rushen and de Passillé, 1995) does not increase the amount of non-nutritive sucking that occurs after the meal, suggesting that neither the amount of milk in the stomach nor the oral sensations from ingesting milk inhibit non-nutritive sucking in the short term.

Furthermore, non-nutritive sucking is far more common immediately after the calves have drunk their milk than before (de Passillé et al., 1992) and disappears once calves have been weaned off milk (Lidfors, 1993; Krohn et al., 1999). These results suggest that non-nutritive sucking is elicited, rather than reduced by the ingestion of milk. Simply injecting small quantities of milk into the mouth of the calf

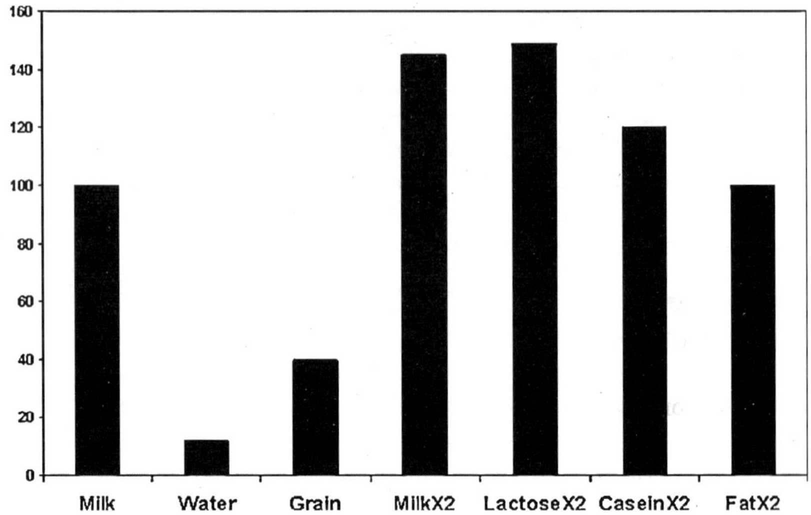

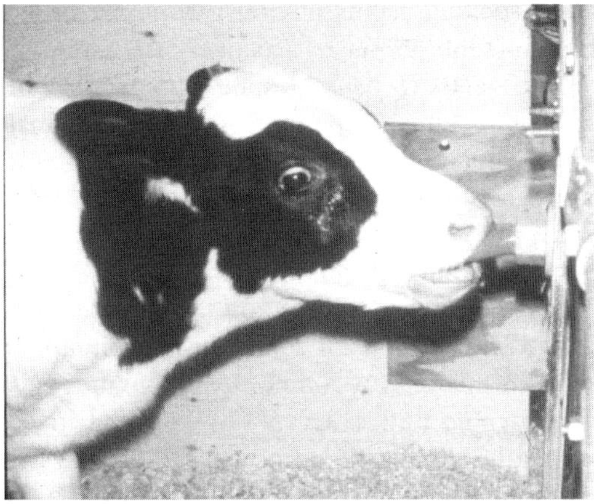

Figure 4.6 Milk-fed calves will readily suck a rubber teat after drinking milk even if they never receive milk through the teat. Generally, this non-nutritive sucking occurs after a milk meal and is stimulated by the taste of milk. Injecting small quantities of milk or milk replacer into the calf's mouth stimulates it to suck vigourously. Increasing the concentration of milk replacer or just of lactose increases the effect, suggesting that it is the taste of lactose which is mainly responsible. Other conditions tested include water, milk, a grain solution, milk replacer with twice the normal amount of (caseinX2) or fat (fatX2). The amount of sucking is expressed as a percentage of the amount shown when normal milk was injected (Adapted from de Passillé et al., 1997, 2006b.)

is sufficient to stimulate considerable sucking (Rushen and de Passillé, 1995; de Passillé and Rushen, 2006a; Figure 4.6). Much less sucking was found when the calves taste either water or a suspension of grain and the amount of non-nutritive sucking increases as the concentration of milk replacer increases (de Passillé et al., 1997; Figure 4.6). These findings indicate that it is specifically the taste of milk that elicits sucking. Changes in the concentrations of butter fat, casein, and lactoserum proteins in reconstituted milk were found to have little effect on the amount of non-nutritive sucking, but increases in lactose concentrations increased non-nutritive sucking (de Passillé and Rushen, 2006a).

If the ingestion of milk does not reduce sucking motivation, what does? Rushen and de Passillé (1995) examined whether the performance of sucking behaviour itself reduces the motivation to suck. Calves were given a small portion of milk in a bucket after which only half of them were given a dry teat to suck. All calves were then given a second portion of milk and allowed to suck the teat. Those calves that sucked the dry teat after drinking the first portion of milk had a lower duration of sucking on the dry teat following the second portion of milk. This result indicates that the performance of sucking behaviour itself is more effective in reducing the underlying motivation than is the ingestion of milk. This conclusion was supported by experiments of Haley et al. (2001) and Loberg and Lidfors (2001) who experimentally altered the flow rate of the milk delivered to calves by a nipple feeder. At a lower flow rate, the same quantity of milk would require longer to drink so that the amount of time sucking on the teat was increased. Even though the same volume of milk was drunk in all cases, the amount of non-nutritive sucking that occurred after the meal was lower with slower flow rates.

Even though sucking motivation is reduced when the calves can suck on an object, the motivation appears to decline spontaneously after the meal is finished. For example, de Passillé et al. (1992) found that sucking motivation elicited by the ingestion of the milk waned during the 10 min following the meal and Rushen and de Passillé (1995) found only low levels of sucking motivation 40 min after a meal. These results suggest that the motivation to suck declines quickly even in the absence of opportunities to suck.

Figure 4.7 illustrates some of the factors that control calves' motivation to suck.

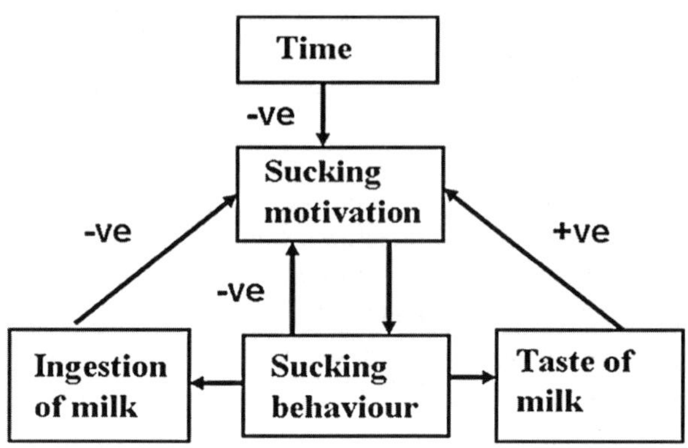

Figure 4.7 The motivation of even an apparently simple behaviour like sucking by young calves can be complex and is affected by a variety of negative and positive feedback from both the consequences and by the performance of the behaviour itself. Research has shown that sucking motivation is turned off by a combination of negative feedback from performing sucking behaviour and, to a lesser extent, by ingesting milk. However, the taste of milk, especially of the lactose in the milk, provides positive feedback, stimulating the calf to suck. This stimulatory effect wanes with time, such that non-nutritive sucking is most common immediately after calves have drunk milk

These results give some support to the idea that allowing the calves to perform sucking behaviour is a more effective way of reducing sucking motivation than providing milk. It seems likely that until calves drink milk they are not strongly motivated to suck, regardless of sucking deprivation, and that sucking motivation elicited by ingesting milk soon decays regardless of whether any sucking occurs. In general, the widespread non-nutritive sucking by calves reflects a relatively strong motivation to suck during or after a milk meal. As we discuss in Chapter 8, the model of the motivation of non-nutritive sucking corresponds well to what we know of the factors leading to cross-sucking by group-housed calves.

2.4.2 Motivation for Locomotion

The research into the motivational basis of sucking found no evidence that the animals' motivation to suck continued to increase the longer the animals were deprived of the opportunity to suck. One type of behaviour where such a motivational "build up" seems to occur is calves' motivation for locomotion.

One long-standing concern about intensive husbandry systems for cattle is that they limit animal movement. For example, veal calves kept in a typical veal "crate" or dairy cows in tie-stalls are virtually immobilized and can usually take only a few steps backwards or forwards. But are animals motivated to move if all the resources they need are directly in front of them? Unfortunately, animals' motivation for movement is one of the least understood aspects of animal behaviour. The limited research that has been done into calves' motivation for locomotion suggests that this motivation does increase if calves have restricted opportunities for movement.

Early studies noted that when calves kept in small crates were placed in a larger room, providing them more opportunity to move around, they would often run and jump far more than calves reared in larger pens (Dellmeier et al., 1985). This was interpreted as the result of a "rebound" effect due to the increased motivation that resulted from the period of deprivation. Subsequent studies have given some support to this interpretation. For example, Jensen (1999) and Jensen and Kyhn (2000) confirmed that calves kept in a small pen galloped and jumped more when allowed access to a larger pen than did calves kept in larger pens. Jensen (2001) reported that calves housed in small pens showed less running and jumping when placed in a larger pen if they had access to an exercise area immediately prior to the test. This effect was interpreted as showing that the initial confinement increased the motivation for locomotion, while the movement that occurred in the exercise area reduced this motivation.

2.4.3 Conclusions

The motivation underling apparently simple behaviours such as sucking by young calves can be complex and models of motivation that stress dichotomies, such as that between internal and external sources of motivation, are inadequate in capturing this complexity. This is even more so for more complex behaviours such as locomotion that are likely to be affected by a very wide range of motivational systems. Research on both sucking behaviour and locomotion support the idea that the performance of the behaviour itself, rather than just achieving the normal functional consequences of the behaviour, is necessary to fully satisfy the animals' motivation, and this research provides grounds for concern that welfare is compromised when animals are prevented from performing normal behaviour.

2.5 Demand Analysis

Research on sucking motivation by calves illustrates the complexity in the motivational systems that underlie even relatively simple behaviours. This approach was based on trying to understand the factors that stimulate and reduce motivation, but the complexity of the motivational systems complicates the task. An alternative approach pays less attention to the causal factors underlying why animals perform the behaviour, and instead attempts to directly measure the strength of the motivation to perform a behaviour. This approach, based on theory developed by economists to analyse consumer demand for resources, was largely pioneered by Marion Dawkins and her own writings remain some of the best explanations of the underlying principles (Dawkins, 1983, 1988, 1990, 1998). Although consumer demand theory is often used to assess behavioural "needs", demand theory addresses different issues in that the actual motivational causes of the behaviour need not be specified when one tests the animal's demand. All that we need to know is how much the animal is willing to work or "pay" in order to perform the behaviour by imposing a measurable cost on the performance of the behaviour.

There are a number of ways of imposing costs. For example, in a choice situation, animals may have to forgo access to a valued resource, such as food, in order to access another resource. As another option, animals can be trained to perform a task such as pressing a bar or pecking a key, in order to access a resource. Once animals have learnt to do this, the "price" of the item can be increased by, for example, making it press the bar more often in order to gain the reward. The extent that the animal values the item can then be assessed by determining how much work it is willing to expend. An alternative approach is to restrict the amount of time that animals have available to spend in performing different behaviours. As the time available decreases, only the most valued behaviours should continue to be performed. In these different ways, we can estimate the value of different resources to animals.

2.5.1 Calves' Demand for Social Contact

As an example of this approach, Holm et al. (2002) carried out an experiment based on demand functions to determine how much priority calves placed on different types of social contact. Dairy and veal calves are often kept in individual housing and the lack of social contact in such housing systems has been criticized as having a detrimental effect on the calves' welfare. The experiment aimed to determine how much social contact calves "want". Calves were housed individually but when they pressed a panel with their heads, a gate opened allowing them to enter another pen in which there was another calf. The calves were allowed to interact with the other calf for 3 min before being returned to their own pen. Some calves could interact fully with the calf, whereas other calves could only have some head-to-head contact through metal bars. The "price" the calf had to pay in order to gain access to the other calf was increased by increasing the number of times the calf had to push on the panel before the gate opened (from 6 to 30 times).

The calves were clearly prepared to work for social contact. When the calves had to press the panel only six times to gain access to the calf, they still opened the door almost 10 times in sessions that lasted between 20 and 50 min. Thus, the calves were motivated to seek social contact. As the number of pushes required for the door to open increased, the calves did open the door less often. When they had to push the panel 30 times to open the door, they did so less than five times in each session. Nevertheless, as the cost of making social contact increased, the calves were prepared to work harder, in that they pushed on the panel more times. They also worked harder for full access to another calf than for the head-to-head contact through the metal bar, indicating that calves valued the full social contact. The authors concluded that "calves are willing to work to get access to a conspecific, suggesting that the calves' welfare may be threatened if they are not allowed to perform social behaviours" (Holm et al., 2002, p. 189).

This experiment illustrates how demand theory can be used to improve our understanding of how much animals "value" certain resources. Since the introduction of techniques based on demand theory, many researchers have documented the potential pitfalls with these techniques (e.g. Mason et al., 1997; Kirkden and Pajor, 2006). The results of the test can depend greatly on how the test was done. In order to assess the value of a resource to an animal, the animal must be given sufficient time to interact fully with the resource. For example, adult cattle usually rest in bouts of at least 15 min. If a test of demand for rest allowed cows to rest for only 5 min, then the cattle may not be prepared to work for the resting period. Furthermore, the motivation to perform behaviours often varies with time. For example, calves spend most time sucking during the 5–10 min following a meal of milk (de Passillé et al., 1992). Tests of calves' motivation to suck on a teat therefore are best done during this relatively brief time. Tests of demand work best when the animal is kept in a "closed economy", that is, where the animals' only access to a resource is during the test itself. For example, if a calf is normally housed with conspecifics, then it is free to engage in social interactions whenever it wishes. If such an animal is tested in an experiment like that of Holm et al. (2002), it may well show less motivation to get access to the conspecifics during the test period.

Other factors have been shown to affect the results of demand analysis with other species, which may also be true for cattle. For example, the presence of social companions during the test can affect pigs' motivation to work for both food and straw (Pedersen et al., 2002). The presence of cues or incentives that may trigger the animals to behave in certain ways can also influence the tests (Warburton and Mason, 2003).

Clearly, the results of demand testing can vary greatly depending on how and when the testing is done. Thus there is danger in misunderstanding effects on animal welfare from poorly thought out or executed experiments. Mason et al. (1997) conclude: "One way of producing results that are useful for making welfare recommendations is to use subjects who are in a state and environment as similar as possible to the captive animals whose welfare we are aiming to improve" (p. 16).

2.5.2 Motivation to Rest

Demand theory has been applied to cattle most often in the attempt to establish the importance of rest. Cattle have been forced to "pay" for the opportunity to rest by making them choose between resting and some activity, usually feeding. This method was pioneered by Ruckebusch (1974) who deprived cattle of both lying (by use of a rope harness) and feeding except for a period of 4 h per day when the cattle could do both. Initially, feeding appeared to be the priority behaviour with the cattle spending most of the 4-h period feeding and very little time lying down. As the treatments continued however, the speed of eating was increased so that feeding time decreased, allowing the cattle to spend more time lying down. In a later study, Metz (1985) prevented cattle lying down for 3 h each morning by placing them on slatted floors. Generally it was found that the amount of time resting in the subsequent period was increased, indicating that the cattle did attempt to compensate for the reduced rest time, and a number of other studies have found evidence of this compensation (Munksgaard and Simonsen, 1996; Munksgaard et al., 1999; Fisher et al., 2002). In a second experiment by Metz (1985), the animals were also prevented from feeding for the same 3-h period. When the cows had not been prevented from lying down, the feeding deprivation resulted in a significant reduction in resting time during the subsequent 3-h period, because the animals were eating. However, when the cattle had been prevented from resting, the feeding deprivation did not result in a reduction in resting time during the subsequent period. The results indicate that cattle protect lying times and compensate for reduced resting times. They also show that resting can take precedence over feeding.

From these findings we can infer that cattle value rest and will work to obtain it, indicating that insufficient rest will reduce the animals' welfare. Further evidence of the welfare consequences can be derived from studying how animals respond when they are unable to perform behaviours that are important to them.

2.6 Examining the Consequences of Behavioural Deprivation

Motivational testing involves understanding the nature and strength of the causes of the behaviour and provides one way of examining the importance of different behaviours for animal welfare. An alternative approach is to assess the consequences of preventing animals from performing certain behaviours. This approach can be used to confirm results from other testing methods. For example, Mason et al. (2001) conducted a series of experiments using demand analysis to understand captive mink's requirements for certain resources. A swimming pool was found to be high on the list. When the mink were systematically deprived of each resource, urinary cortisol (taken as a measure of stress; see Chapter 3) was highest when mink were unable to access the swimming pool. For cattle, the previous discussions have suggested that sucking for calves and rest for cows is a high priority. Thus, we would expect that deprivation of these behaviours would have deleterious consequences for the animals.

2.6.1 Consequences of Inability to Perform Sucking Behaviour

The performance of non-nutritive sucking by calves following a milk meal appears to have important physiological consequences. de Passillé et al. (1993) found that insulin and cholecystokinin (CCK) in the hepatic portal vein after the meal were higher when the calves sucked a dry teat after the meal. The increase in CCK and insulin concentrations was positively correlated with the duration of sucking, but not with the duration of other oral behaviour directed at the teat. This result suggests that actual sucking behaviour is important rather than just the extra sensory stimulation from having the teat in the mouth. The physiological mechanism underlying the effect is unclear, although increased vagal stimulation may be involved (Veissier et al., 2002). Veissier et al. (2002) provided some evidence that sucking for milk may increase postprandial satiety and reduce heart rates, while increasing heart-rate variability, which they interpreted as an indicator of greater calmness in the calves. The possible satiety effects and the widespread metabolic effects of insulin and CCK mean that deprivation of sucking behaviour cannot be assumed to be inconsequential for animal well-being and growth even if this does not affect nutrient intake. Furthermore, as we discuss further in Chapter 8, not allowing calves to suck can increase the incidence of cross-sucking between the calves.

2.6.2 Examining the Consequences of Altered Resting Behaviour

Preventing the performance of certain behaviour has been also used in trying to understand the importance of resting time for cattle. Cattle have been prevented from lying down by the use of a specially designed harness. Use of such harnesses to prevent cattle from resting for 2 periods of 7 h each day reduces growth hormone concentrations in lactating cows (Munksgaard and Løvendahl, 1993). Munksgaard and Simonsen (1996) noted that deprivation of resting behaviour by cows resulted in an increase in grooming, and tended to increase the incidence of licking the pen equipment. Although the duration of eating and ruminating was not affected, feeding tended to occur in more but shorter bouts, and more rumination occurred while the animal was standing rather than lying down. Neither overall basal concentrations of cortisol and ACTH, nor the cortisol response to ACTH were affected. However, the authors did note a slight increase in plasma cortisol concentrations at the beginning of the period when the harness was attached, and a markedly increased cortisol response to being placed in a novel area (some of the problems in interpreting these physiological responses are described in Chapter 3).

In a similar experiment with growing bulls, deprivation of resting time did not affect basal cortisol or ACTH concentrations but tended to reduce cortisol and ACTH responses to CRF injections 3 days after treatment and increase cortisol responses to ACTH injections after 53 days of treatment (Munksgaard et al., 1999). Behavioural changes were also noted: use of the harnesses did not affect the total time spent eating or ruminating, but ruminating occurred in more frequent and presumably shorter bouts, mainly while the animal was standing. There was also an increase in the occurrence of licking and chewing at the pen fixtures and an increase in the frequency of grooming. Overall, there was an increased frequency of transitions between behaviours, which the authors interpreted as due to frustration (Munksgaard et al., 1999).

Fisher et al. (2002) prevented cows from lying down for a single 15-h period each day by using a harness that delivered a slight electrical shock each time the cow attempted to lie down. This reduced the time that the cows lay down from 8.1 h per day to less than 4 h per day. After 5 days, basal cortisol concentrations were increased and ACTH and cortisol responses to CRH were reduced.

2.6.3 Interpreting the Consequences of Behavioural Deprivation

Together these studies show the effects of preventing cows from lying down or preventing calves from sucking on their behaviour and physiology. However, our ability to draw conclusions about these effects in terms of animal welfare is limited for the following reasons. The link between the behavioural and physiological responses that were seen and the overall welfare of the animals is not clear. Less ambiguous indicators of reduced welfare, such as reduced body condition, loss of weight, or drops in feed intake were not seen. While the changes in HPA activity noted (such as increased basal cortisol concentrations or increased cortisol responses to ACTH) are often considered as evidence of "stress", the relationship with the level of animal welfare is not easy to establish (see Chapter 3), and the findings are somewhat inconsistent from study to study. What these studies show is that depriving animals of resting time or of sucking behaviour is not without consequence for the animals. Also the procedures used to reduce the behaviour (e.g. the harness used to prevent lying) may themselves have been stressful to the animals.

One of the problems with the studies that have tried to assess the consequences of deprivation of rest on the welfare of cattle is that they tend to ignore what the cattle do when lying down. When lying down, cattle tend to either ruminate, idle (i.e. doing nothing that is obvious to an observer) or sleep. Sleep itself can generally be divided into the two categories: (1) paradoxical or rapid eye movement sleep (REM) and (2) slow wave (SW), quiet or non-rapid eye movement (NREM) sleep (Hänninen, 2007). A reduction in lying time may have very different effects on the animal depending on which of the behaviours are prevented. Rumination usually occurs when cattle are lying down, but when prevented from lying, cattle will ruminate while standing, with little change in the total duration of rumination (Munksgaard and Simonsen, 1996; Munksgaard et al., 1999). It is not known whether the rumination that cattle do when standing is different from the rumination they perform when lying down.

If the reduction in lying down is sufficient to interfere with sleeping, especially with REM sleep, then the effects on animal welfare are more likely. Unfortunately, apart from the pioneering studies of Ruckebusch (1972, 1974, 1975) we still know relatively little about sleep in ruminants. Based on polygraph recordings and behavioural observations, Ruckebusch (1972, 1974, 1975) calculated that cattle spend somewhere between 200 and 300 min in SW sleep and between 20 and 45 min per day in REM sleep. Under normal conditions, both REM and SW sleep occur when the animals are lying down, although Ruckebush (1974) presents some evidence that SW sleep can occur while the animal is standing. Generally, REM sleep requires that the animals are able to adopt a relaxed posture (Hänninen et al. 2007) (Figure 4.8).

Figure 4.8 Adequate rest and sleep is an important behaviour, especially for young animals. Cows spend over half the day lying down. Tests based on consumer demand theory have shown that cattle are prepared to pay a price in order to lie down. In these experiments cows were prevented from lying down by a harness, but could release the harness by pressing their noses against a panel. The price of release could be increased by increasing the number of times the cows must push on the panel. As the price increased cows would push more often, showing that they are prepared to work in order to rest. Rapid eye movement (REM) sleep generally requires that the muscles are relaxed and so probably occurs when the animals are resting in relaxed postures, such as shown above (Hänninen et al. 2007)

Although the different estimates of sleep time vary, they all suggest that a reduction in resting time would need to be fairly severe to limit sleep. Ruckebush (1974) noted a reduction in time spent in REM and SW sleep only when cattle were kept standing for 20–22 h per day and could only eat during the time that they were able to lie down. The fact that Ruckebush's experiments were done with a very small number of animals must be borne in mind. Unfortunately, we still know very little about sleep (as opposed to rest) in cattle, although new techniques being developed for non-invasive electrophysiological data collection and for using resting postures to estimate the time spent in different phases of sleep (e.g. Hänninen, 2007).

In conclusion, while these experiments show that depriving animals of the ability to perform certain behaviours can have marked behavioural and physiological consequences, we cannot yet unambiguously interpret these consequences in terms of their effect on animal welfare.

3 Behaviour as an Outcome-Based Criterion for Animal Welfare

In this section we discuss the use of animal behaviour as potential outcome- or animal-based criteria of animal welfare, that is, where the performance of certain behaviours by the animal is used to make some inferences about the actual state of welfare of the animal. We discuss three types of behaviour. The first is behaviours that directly reduce the welfare of the animals. These include injurious behaviours, such as fighting. The second category includes behaviours that do not themselves reduce the animal's welfare but that may be an indirect sign that the animal's welfare is threatened. These include vocal signals, behaviours indicating fear or anxiety, and so-called abnormal behaviours, such as stereotypic behaviour. The third type of behaviour includes those like play that may indicate that the animals' state of welfare is good.

A central theme of this book is that proposed indicators of animal welfare must be validated before they can be used with confidence. That is, we must have some confidence that these welfare indicators are reflecting the welfare of the animals. Validating behavioural (or any other) indicators of welfare is far from simple and much of the following section revolves around how this can be achieved.

3.1 Injurious Behaviours

The least controversial behavioural indicator is one that results in injury. In cattle, the most obvious example of this is aggressive behaviour, or a behaviour known as "bulling". Bulling is the term used to refer to cases when one or more animals repeatedly mount other animals. "Bullers", which, ungrammatically, are the animals being mounted, can suffer from obvious injury and bruising and appear more likely to become sick and die from respiratory disease (Taylor et al., 1997b). Such behaviours are a clear threat to the welfare of the recipient animals, so measures of their occurrence are useful in assessing animal welfare.

Although fighting can be frequent in some situations, for example, during initial encounters between adult bulls (e.g. Mounier et al., 2005), aggression between cattle is fairly rare (e.g. Menke et al., 1999; Veissier et al., 2001) and what does occur tends to be concentrated at certain times of the day (Menke et al., 1999). This means that forming accurate estimates of how much aggression is occurring requires intensive observations. It is often easier to observe the consequences of the behaviour, for example, by counting wounds. However, the correlation between the levels of aggression and the incidence of injuries can be variable: in one study, the incidence of skin injuries was found to be uncorrelated with the incidence of horning and butting in a herd of dairy cows (e.g. Menke et al., 1999). Furthermore, although the most obvious results of aggression are injury to the animals, the consequences of the dominance relationships that are formed as a result of aggression may be more evident in difficulties in competing over food (e.g. Phillips and Rind, 2001, 2002; Val-Laillet et al., 2007). A lack of injuries or even overt fighting should not be taken as evidence of a lack of social tension.

3.2 Behavioural Indicators of Fear

Injurious behaviours are fairly straightforward signs of poor welfare. Other behavioural indicators of poor welfare are more indirect in that the behaviour can only be taken as a sign of some problem. For example, cattle in pain or frightened may express this in some way in their behaviour, and these behaviours would be a sign of poor welfare. We do not intend to review all uses of such behavioural indicators of fear or pain. In this section we focus on one fundamental issue: how can we validate these behavioural changes as indicators of poor welfare. We focus this discussion specifically on measures of fear in cattle and we ask to what extent does the performance of the behaviour actually show that the animal is frightened? In the following section, we extend this discussion to signalling behaviour, specifically vocalizations.

Unfortunately, it is often assumed that behavioural responses to pain or fear can be easily recorded, and often purported behavioural signs of pain or fear are not validated. The main points that we wish to make can be summarized as follows:

1. The control of the behaviour of animals in response to frightening situations is complex and we cannot interpret behavioural responses, nor use them as indicators of fear, until we understand the underlying causes of the behaviour. This understanding can come from investigating the motivational controls on the behaviour.
2. Behavioural responses to frightening situations are derived in part from predator avoidance behaviours, and studies on the functional significance of these behaviours can help us understand how animals respond under fear.
3. Because the behavioural responses that animals make when frightened are done to help them deal with the situation, the types of responses are often specific to

the particular source of the fear. Thus, there may not be "general" behavioural responses that animals will perform in response to fear, regardless of the specific cause of the fear.

While we focus on fear responses, similar points could be raised regarding behavioural responses to pain or other behavioural indicators of poor welfare. To validate behavioural responses as a measure of fear, it is important to understand the motivational system underlying the behaviour. The interpretation of behavioural indicators of fear is and will remain difficult primarily because of the complexity of the motivational systems underlying behaviour. The considerable research into the motivational basis of behaviour has shown that the behaviour of an animal at any one time is the result of an interaction between different motivational systems that appear to "compete" for control of the animal's behaviour (see McFarland, 1989). The important point for this discussion is that the behavioural responses that animals make in any fearful situation will reflect the relative strength of a mix of different motivations. The difficulty interpreting behavioural responses can be illustrated by the work that has been done on the very popular open-field test.

3.2.1 The Open-Field Test

The open-field test was originally developed to measure rather poorly defined characteristics of rats, such as "fear" or "emotionality" (Hall, 1936). A single animal was placed in a large, novel area and the rats' responses, which usually involved defecation or increased activity, were interpreted as reflecting the degree of fear, either in response to novelty or to the "openness" of the enclosure. However, detailed analysis of the test showed problems with this simple interpretation. Careful experimental work suggested that the responses probably reflect a mix of motivations such as freezing in response to fear, exploration, escape attempts, and specific responses to social isolation (Archer, 1973). More recently it has been shown that open-field activity of rats can be affected by factors apparently unrelated to the degree of fear, such as aerobic capacity (Friedman et al., 1992).

Despite these problems, the open-field continues to be used as a measure of animals' responses to stress. The open-field test has long been used as a way of measuring cattle's responses to fear-provoking situations (Kilgour, 1975; Kilgour et al., 2006; Van Reenen et al., 2004, 2005). Often this involves placing a cow or calf in a novel area for a few minutes and then recording some aspect of its behaviour. In many cases, the open-field area is divided into a number of squares and the researchers count the number of squares that the animal enters. The research, however, shows the same problems with interpretation as the rodent research. For example, some authors interpret the degree of activity that cattle show in an open-field as the degree of nervousness (Warnick et al., 1977), general agitation (Kilgour et al., 2006), or vigilance (Müller and von Keyserlingk, 2006) while others interpret the same degree of activity in terms of the level of "locomotory" motivation (e.g. Dantzer et al., 1983; Dellmeier et al., 1985).

It is most likely, as is the case with rodents, that the behaviour of cattle in an "open-field" reflects a changing mix of motivations, rather than activity on a single dimension such as fearfulness. Cattle's responses to a variety of tests thought to measure fearfulness in different situations often are not strongly correlated (e.g. Kilgour et al., 2006; Van Reenen et al., 2004, 2005) suggesting that the causes of the behaviour are multidimensional. de Passillé et al. (1995) used factor analysis as a means of analysing the mix of motivations that underlie the behaviour of calves in an open-field. Even when only seven different behaviours were observed (sniffing/licking, walking, running, jumping, vocalization, defecation, and standing immobile), it was still necessary to use three factors to account for more than half of the variance in these behaviours. Based on the behaviours that had the highest correlations, these three factors were tentatively labelled as "fear" (vocalization, defecation), exploration (sniffing and licking), and locomotion (running and jumping). The correlations between these three factors were small, suggesting that they were effectively independent sources of motivation. Furthermore, the total activity of the calf was correlated with all three factors. Thus, a calf could show a lot of activity either because it had a high level of fearfulness, or a high level of exploration, or a high level of locomotion, or a mixture of all three. Thus it would be dangerous simply to measure total activity and to interpret the calves' activity in this test as reflecting only or primarily the level of fearfulness. Van Reenen et al. (2004, 2005) also report a multidimensional analysis of calves' responses to an open-field test.

While factor analysis can be a useful way of examining the complex motivational structure underlying behaviour, the result can depend on the assumptions made in carrying out the test, so results are best thought of as hypotheses about the motivational structure of behaviour. These hypotheses still need to be tested before we can be confident in the interpretation. One approach to testing these hypotheses is to correlate the behavioural and physiological responses that are thought to be a sign of stress (e.g. Müller and Schrader, 2005; Van Reenen et al., 2005). However, the physiological responses are themselves not easy to interpret (see Chapter 3) and so de Passillé et al. (1995) attempted to test their interpretation of calves' open-field behaviour by experimentally varying the degree of motivation of each calf. For example, the factor analysis suggested that the amount of defecations and vocalizations might indicate the degree of fear the calves showed in response to the novelty of the enclosure. This was tested by examining how these aspects of the behaviour of the calf changed in response to three factors that were likely to alter the degree of novelty of the enclosure and hence the degree of fear the calves showed: prior experience of the enclosure, the presence of a novel object, and the age of the calves. As expected, the amount of defecations and vocalizations was higher for younger calves and was increased by adding a novel object, while allowing the calves to become familiar with the arena reduced the occurrence of these behaviours. Thus these behaviours seem to be useful in assessing the degree of fear that calves show in response to novelty.

Although older calves also showed more running and jumping, these behaviours were not affected by the degree of familiarity of the enclosure or by the presence of a novel object, suggesting that measures of these types of activities could not be used to measure calves' degree of fearfulness. Van Reenen et al. (2005) similarly found that measures of locomotion in the open-field did not correlate well with the other behaviours shown. In contrast, there is evidence that the occurrence of these specific behaviours may well reflect locomotor motivation, as proposed by de Passillé et al. (1995) and Van Reenen et al. (2005). For example, Jensen et al. (1999) found that calves that had been confined in a smaller space before the open-field test showed more galloping and bucking during the test than calves that had access to more space. The social context of the test may have an influence: Jensen (2001) found that calves showed less locomotor activity in an open-field when tested in pairs compared to when tested alone.

Generally, it is likely that any behavioural response to stress will reflect a mix of motivations. Multivariate statistical analysis can help unravel the nature of the motivations underlying these responses (Müller and Schrader, 2005; Van Reenen et al., 2005; Kilgour et al., 2006) but interpretation of the results often appears subjective. However, the study of de Passillé et al. (1995) shows that use of multivariate statistical analysis, combined with experimental manipulations, can help us interpret the behaviour that animals make in an open-field and help choose the most appropriate measures of fear. Alternative ways of measuring the degree of fearfulness of cattle extend the use of statistical techniques described earlier to combine results of multiple tests, and try to find common measures of fear that are apparent in a range of tests (Müller and Schrader, 2005; Van Reenen et al., 2005; Kilgour et al., 2006).

3.2.2 Alternative Approaches to Fear Assessment

One criticism of the open-field test is that it places animals in an artificial situation. Blanchard and Blanchard (1988) have criticized this use of artificial stressors, arguing that stress responses shown by animals have evolved to deal with the specific stressors that the animal is likely to encounter in its natural life. For example, many fear responses are specific to predation threats. Anti-predator behaviour is likely to involve a series of graded responses, and the first line of defence is to detect the predator before it is too close. For most prey animals this is achieved by routine scanning of the environment, a phenomenon that has become known as "vigilance" (Figure 4.9).

To improve detection of predators, animals need to spend time being vigilant, but this scanning can interfere with other behaviours like grazing. Thus animals are facing competing selective pressures on time spent scanning; reducing the risk of predation but allowing time to perform other important activities like foraging. The degree of vigilance may be affected both by the animals' own assessment of the risk of predation and the strength of motivation to perform competing behaviours. Thus, studies of vigilance may provide a useful paradigm for assessing perceived risk under different conditions of motivational conflict.

Considerable research has been carried out on vigilance behaviour in free-living animals. One robust finding is that vigilance by individual animals in a group is inversely related to the size of the group (Elgar, 1989; Roberts, 1996). A plausible

Figure 4.9 Vigilance is an important part of anti-predator behaviour of many prey species, increasing the chance that predators will be detected. Cattle are more vigilant in the presence of a dog than in the presence of a person. Cattle are also more vigilant in the presence of a handler who has handled them aversively compared to an unfamiliar person, or a person who has handled them gently. These results support the idea that measures of vigilance can be used to measure the degree of fearfulness of cattle (From results presented in Welp et al., 2004.)

functional explanation for this relationship is that the additional group members increase the chance of predator detection so that, in a larger group, any one individual can reduce vigilance time but still have the same chance of detecting a predator. This effect of group size on vigilance shows that anti-predator behaviour can be sensitive to environmental factors that affect the risk of predation. Other factors that influence the risk of predation have been shown to affect vigilance. For example, vigilance has been shown to be higher at places where predators are more likely to be encountered such as at waterholes (e.g. Rose and Fedigan, 1995; Burger and Gochfield, 1992). Vigilance is also higher when animals are further from cover or a refuge (Frid, 1997). Studies on vigilance also provide information on how different sources of fear or risk interact. For example, work on Dall's sheep showed that vigilance decreased as group size increased and as distance to cliffs (which the sheep use to escape from predators) decreased, and the effect of changes in group size was much smaller when the animals were close to cliffs (Frid, 1997). These results indicate that animals rate the risk of predation lower when they are close to a refuge, reducing the relative importance of rapid predator detection.

A few studies have used vigilance as an indicator of fear in cattle. Welp et al. (2004) investigated fear in dairy cattle by using a dog as a potential predator-like stimulus. The results showed that cows were more vigilant in the presence of a dog (Figure 4.8). This vigilance increased when cows were tested in a novel environment, and there is considerable evidence that cows are fearful of new environments as shown by increased defecation, heart rate, and cortisol levels (e.g. Munksgaard and Simonsen, 1995; Rushen et al., 1999a).

These results support the idea that measures of vigilance can be used as a measure of fearfulness in cattle. Although research on such behaviours in farm animals is rudimentary, these types of measures have the advantage over the use of more artificial stressors like the open-field. Vigilance functions to improve predator detection – other individuals might be able to detect these behaviours and use them for their own benefit, but this is not necessarily the intention of the animal performing the behaviour. Other behaviours do function specifically as signals to other animals, and it is to these behaviours that we now turn.

3.3 Signalling Behaviour

Signals, such as vocalizations and displays, can provide information about the signaller's state that is useful in welfare assessment. Since these behaviours are directed towards others, they are susceptible to eavesdropping, allowing us to listen in to animal conversations and potentially learn the animal's view of its own condition. Griffin (1981) suggested that the study of communicative behaviour could provide a "window on the minds of animals" (p. 149), and animal vocalizations have been promoted as a means of assessing animal welfare (e.g. Manteuffel et al., 2004). Watts and Stookey (2000) have reviewed much of the work on the vocalizations of cattle, and argue that these behaviours can provide information about the animals' emotional or affective state.

3.3.1 Measuring and Describing Cattle Vocalizations

Many tools are available that allow for the quantitative and qualitative analyses of animal sounds. Two of the graphical displays commonly used for this analysis are the time waveform, that allows us to assess changes in amplitude (loudness) with time, and the frequency spectrogram, that also allows us to assess how call frequency (pitch) changes with time (see Bradbury and Vehrencamp, 2000, for more information on the analysis of animal vocalizations). Figure 4.10 shows a frequency spectrogram of a call produced by a dairy calf after separation from the cow.

Like all calls produced by cattle, this one consists of a stack of harmonically related frequencies, much like those in vowel sounds in human speech. Some of the measures that can be taken from this type of analysis include call duration, the fundamental frequency and the number of the loudest harmonic. In Figure 4.10, the loudest harmonic appears as the darkest band, and the fundamental frequency is equal to the frequency difference between adjacent bands. Because these measures vary little within each separation call, they can provide a reasonable description of the call and have been used to assess how calves respond to separation (Weary and Chua, 2000; Flower and Weary, 2001).

These types of call features, and especially changes in such features over the duration of the call, have been used by some scientists to divide calls into types. For example, Dellmeier et al. (1985), distinguished between two types of vocalizations: "moo" and "baaocks". The "baaock" tended to occur when calves were jumping and running and were interpreted as part of playful behaviour. In her classic work on the vocalizations of ungulates, Kiley (1972) identified many different call types in cattle on the basis of such differences in spectral characteristics and her impressions of how and when the calls were used. Correctly identifying call types can be very important, especially when using vocal measures to assess welfare. For example, Taylor et al. (2001) examined the vocal responses of piglets to castration. In some calls, the loudest frequencies were well below 1 kHz. Other calls emphasized frequencies much above this level. Moreover, there was a clear bimodality in the distribution of the frequencies, providing an objective basis for classifying calls into two types. When calls were classified in this way, the different call types showed very different responses. Piglets produced a similar number of calls less than 1 kHz regardless if they were castrated or simply held. However, the castrated piglets produced many more calls that were greater than 1 kHz. New work is now required to determine if the calls of cattle can be objectively identified in this way, and if such categorization improves our ability to use vocal behaviour in welfare assessment.

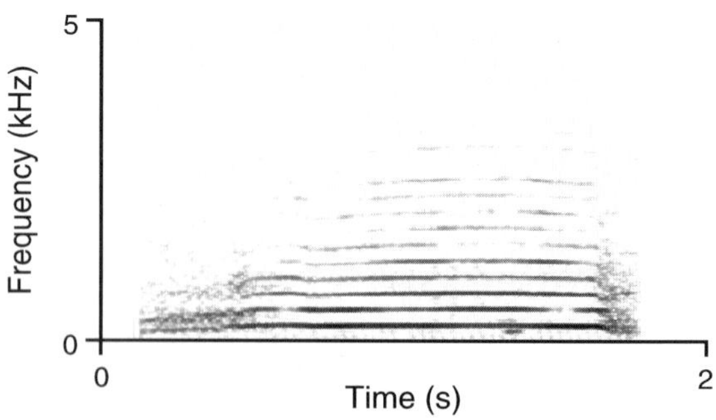

Figure 4.10 Sound spectrogram of a calf's call

3.3.2 Using Vocal Behaviour to Assess Welfare: Theory and Evidence

As with other behaviours, we must approach the use of signals with some caution. Weary and Fraser (1995) review some of the issues involved in using signals as a means of assessing animals' needs, and the reader is referred to this article. In order to use signalling to assess animals' internal states, there must be some variability in the vocalization given by the animals, either in terms of the likelihood of vocalization occurring, or the rate, amplitude or some aspect of the acoustic structure of the calls, and some aspect of this variability must provide reliable information about the state of the animal.

The view that communicative behaviour is essentially honest, in providing reliable information about an animal's internal state, especially its motivational state, was implicit in the classical, ethological studies of animal displays (e.g. Tinbergen, 1969). The underlying assumption was that natural selection should favour accurate signals, that is, signals that reliably communicate information about an animal's intentions or likely behaviour. However, more detailed analysis of the evolutionary pressures on animals, particularly those that focused upon understanding the costs and benefits to individuals rather than species or groups of animals, showed that in some cases natural selection could be expected to favour "dishonest" communication, that is, where animals conveyed inaccurate information about their state in order to manipulate the behaviour of other animals to their advantage. Indeed, there are a number of instances (reviewed in Weary and Fraser, 1995) where animal signals appear to convey inaccurate information about an animal's needs, intentions, or internal state. Experiments have also shown that signalling behaviour can be affected by the presence of an audience: roosters will sometimes give "food calls" to attract hens, even when no food is present (Gyger and Marler, 1988). Also, roosters will sometimes fail to give alarm calls after seeing a predator, depending on the sex and status of other birds in their "audience" (Evans and Marler, 1992). If communication is performed with the intent of influencing the behaviour of others, rather than just expressing an animal's internal state, it is not surprising that the absence or presence of an audience would have an effect.

Under some circumstances natural selection will favour "honest" signals that correctly reflect some attribute of the signaller. Some signals can serve to indicate the quality of the signaller, for example, as a potential mate. Such signalling systems will likely be honest if the higher-quality individuals are capable of signalling more (i.e. they pay a lower cost to produce a given signal) with the result that condition and signalling are positively correlated (Grafen, 1990). For example, dominant roosters crow more than their more subordinate flock-mates, perhaps because subordinates pay the cost of being attacked by a dominant when they do attempt to crow. Other signals indicate an animal's degree of need for certain resources, such as the young calf's need for milk. Honest signalling of need requires that individuals in greater need signal more. Evolutionary models (Johnstone, 1999) show that this type of signalling is most likely to occur if: (1) listeners derive a fitness benefit from providing the resource to the signaller, (2) signallers vary in their need for the resource, and (3) that performing the signal entails a fitness cost. Rushen (2000) discusses some of the ways that have been used to test the reliability or "honesty" of animal communication.

Thus under the right conditions animal signals can be a useful tool in welfare assessment. Particularly useful are those signals that provide information about the signaller's need for certain resources (Weary and Fraser, 1995). Domestic cattle are known to vocalize in response to food deprivation. For example, dairy calves are normally silent when fed milk *ad libitum*, but become highly vocal when the milk is withdrawn at weaning (Thomas et al., 2001). Moreover, calves often vocalize when separated from the cow soon after birth (Weary and Chua, 2000; Flower and Weary, 2001) and this response can be much diminished simply by feeding the calves more milk (Figure 4.11; Thomas et al., 2001).

Subjective states, like fear and pain, are obviously important considerations in animal welfare and the development of reliable methods of assessing these states is a priority for research. Unfortunately, the theoretical background described earlier provides little guidance for when and how we might expect vocal signals to relate to these subjective states. Any assessment of subjective states rests on J.S. Mills' "argument by analogy". Briefly stated, this argument specifies that: (1) subjective states in another individual cannot be assessed directly, but (2) inferences regarding these states can be drawn on the basis that you experience this state and the other individual is similar to you. The strength of this argument rests on the evidence of the similarity. For example, if considering a pain response, we might look for evidence of similarities in neuroanat-

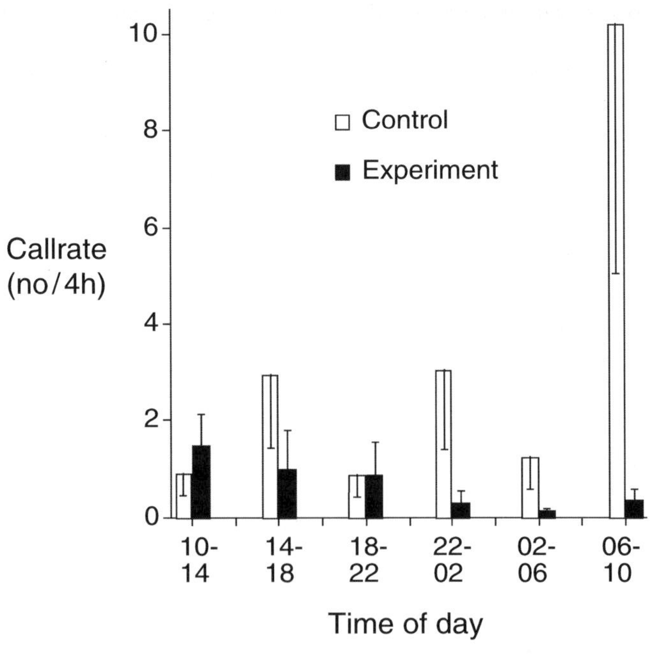

Figure 4.11 Rate of calling by calves at various times relative to separation from the mother. Experimental calves received extra milk after weaning (Adapted from Thomas et al., 2001.)

omy, neurophysiology, and responses to pharmacological treatments, such as nerve blocks. Vocal responses, such as that produced by piglets during castration, have been used with some success in pain assessment. A number of studies have also used vocal responses to assess pain in domestic cattle, during processes such as branding (e.g. Lay, 1992a, b; Schwartzkopf-Genswein et al., 1997b; Watts and Stookey, 1999; see Chapter 5). These studies on cattle have typically found that only some individuals respond vocally to the pain, meaning that the vocal measures were relatively insensitive to treatment effects. This may be because contextual factors can play an important role in influencing vocalization rates: Rushen et al. (2001b) showed that adult cows vocalize repeatedly when placed in social isolation but the cows did not vocalize when a person was present. This may reflect that vocalizations in cattle are primarily a response to the social environment (Van Reenen et al., 2005). The propensity of cattle to vocalize is also be affected by genetic factors (Watts and Stookey, 2001; Watts et al., 2001). Regardless, it would seem that vocal measures are less useful for pain detection in cattle. In particular, observers should not conclude that a procedure is not painful for an individual cow, simply because she failed to vocalize during the procedure. This problem of inconsistent vocal responses is much reduced if considering large samples. For example, Grandin (1998a, 2001a) has found that many beef cattle vocalize during pre-slaughter handling associated with painful events, such as use of cattle prods, etc. When considering the hundreds of animals that pass through the slaughter line, she found that changes in the incidence of vocalizations were useful in tracking welfare improvements that occurred following changes in operating procedures at the plant.

Cattle also vocalize in situations that appear to be fear-provoking. For example, both adult cattle (e.g. Boissy and LeNeindre, 1997; Watts and Stookey, 2001; Rushen et al., 2001b) and younger calves (de Passillé et al., 1995; Watts et al., 2001) vocalize particularly when socially isolated. Cows also vocalize when separated from their calves (Hopster et al., 1995; Weary and Chua, 2000; Flower and Weary, 2001; Haley et al., 2005). Calves also vocalize when placed in novel environments, even where this does not involve social isolation (Carson and Wood-Gush, 1984; Dellmeier et al., 1985).

In other species, the link between vocalization and fear has been supported by studies showing a relationship between vocalization and physiological systems involved in fear. For example, in pigs, Schrader and Todt (1998) demonstrated a correlation between the occurrence of vocalization following social isolation and changes in stress hormones such as epinephrine and cortisol. In cattle, the link between vocalization and physiological responses is less clear. For example, placing adult cattle in social isolation increases the incidence of vocalization and results in increased cortisol and suppression of oxytocin. However, the presence of a person eliminates the vocalization but does not alter the cortisol or oxytocin responses, suggesting that different mechanisms underlie these behavioural and physiological responses (Rushen et al., 2001b).

Thus vocal signals can be of use in welfare assessment, including assessments of subjective states such as fear and pain. However, we must remember that the functional bases and mechanisms controlling vocal responses in cattle are still poorly understood, such that inferences regarding these responses must be made with caution. New work is also required on if and how cattle vocalizations can be objectively divided into categories and if so, whether these different call types have different meaning for the listeners. More generally, we know very little about other types of signalling in cattle (e.g. visual and chemical signals), and how these signals might be used in understanding and improving their welfare. Recent work has suggested that the amount of eye-white shown by cattle may signal emotional state (Sandem et al., 2006). The types of signals used by cattle clearly needs further research.

3.4 Abnormal Behaviour

Cattle, especially calves, show a variety of sucking, licking, nibbling, and tongue-related behaviours that are often considered "abnormal" and have figured prominently in attempts to assess welfare (Figure 4.12). The use of the word "abnormal" to describe such behaviours in some ways suggests that they are associated with poor welfare, but the word can have many meanings. First, there is the purely statistical meaning of rare or not usual. This meaning is of little relevance in the present case since such behaviours can often be performed by a large majority of animals in some housing systems (Mason and Latham, 2004). Second is the meaning of unnatural, in the sense that performance of these behaviours is not a part of the natural behaviour of the species. However, this meaning is of little help in discussing animal welfare largely because of the problems with the concept of natural behaviour discussed in Section 2.3. In many cases, such behaviours are labelled abnormal because the function of the behaviour is not clear to the observer. This is more a recognition that we do not understand the behaviour, rather than a property of the behaviour itself. Our own point of view is that the labelling of such behaviours as "abnormal" is not helpful. Instead, we urge readers to think carefully about why animals are performing the behaviours. A mistaken understanding of the motivation behind other abnormal behaviours, such as stereotypic behaviour in swine, has led to incorrect interpretations of what these behaviours show about animal welfare (Rushen, 2003). We must also be able to validate these behaviours as an indicator of welfare by linking performance to reduced welfare.

In Chapters 6, 7, and 8, we provide some examples of how housing and feeding methods can influence the occurrence of a variety of abnormal behaviours. In Section 3.4.1 we illustrate how the motivation underlying one abnormal behaviour, tongue rolling, has been studied and related to animal welfare. This behaviour is an example of a stereotypy: behaviours that are highly repetitive, performed for long periods, and that are without obvious function (see papers in Mason and Rushen, 2006). Mason and Latham (2004) provide a discussion of how apparently abnormal behaviours are related to animal welfare, and we encourage the reader to consult this article. Although we use tongue rolling as an example, we believe that the same approach as described here can be applied to the analysis of other abnormal behaviours in cattle, including cross-sucking or non-nutritive sucking (discussed earlier in this chapter and in Chapter 8).

Figure 4.12 An example of non-nutritive sucking by a milk-fed calf (above) and tongue rolling by a lactating cow (below). These are two examples of behaviour that are commonly thought of as "abnormal", the occurrence of which may indicate a welfare problem. However, in order to use the occurrence of such behaviours as a means of assessing the welfare of the animals, we need to better understand why the animals perform them

3.4.1 Stereotypic Tongue Rolling

The most common stereotypic behaviour among adult cattle is tongue rolling, occurring either inside or outside the mouth as described in detail by Redbo (1990; Figure 4.12). This behaviour occurs in association with bar biting and licking parts of the pen or stall equipment such as metal bars or chains. In a well-integrated series of experiments, Ingrid Redbo has explored the causal basis of these behaviours and has documented the role that feeding motivation plays in their occurrence. A considerable body of evidence has shown that stereotypies in other farm animals are also closely related to feeding problems or high levels of feeding motivation (Bergeron et al., 2006). Tongue rolling was found to be far more frequent among tethered cows than among cows on pasture or in loose housing, where it was almost absent (Redbo, 1990, 1992, 1993). Initially, this suggested that physical restraint was the cause of these behaviours. However, tongue rolling also tended to be preceded or succeeded by feed-searching behaviours such as licking and sniffing at the empty food trough, behaviours that suggested frustrated feeding motivation as an underlying cause (Redbo, 1990, 1992). Redbo (1992) suggested that tethering might increase the incidence of these behaviours by limiting the amount of time that the cows could manipulate feed.

Compared to *ad libitum* fed tied cows, those fed a limited amount show more tongue rolling and feed-searching behaviours, and spend less time feeding and ruminating (Redbo et al., 1996). Adding long straw to cows' diets markedly reduced the incidence of tongue rolling even though the energy content of the diet was not altered (Redbo and Nordblad, 1997). Using cows fitted with rumen fistulas, Lindström and Redbo (2000) were able to transfer rumen content between cows and thus were able separate out the effects of the amount of feed ingested (including the duration of feeding) from the effects of rumen fill and the metabolic effects associated with digestion. Cows provided reduced amounts of feed showed high levels of tongue rolling, even if rumen content was increased via the rumen fistula. Moreover, cows that were allowed to eat a large meal, but that had some of their rumen content removed, showed only a low level of stereotypic tongue rolling. It is not yet clear whether these effects were due to the reduced time eating or to the reduced amount of feed intake. However, the results show that the ingestion of feed plays an important role in the occurrence of these behaviours, and likely in feeding motivation.

Together, these results indicate that tethering increases the incidence of tongue rolling, primarily because it reduces the time that the cows take to eat. However, it should be noted that the herd studied had a very high incidence of tongue rolling, with over 70% of cows showing tongue rolling (Redbo et al., 1996). This may have been due to cows being fed restricted amounts and that the feed was placed in sight of the cows but out of their reach (Redbo, 1990, 1993). The extent that the occurrence of tongue rolling is influenced by the method of food delivery or the type of food provided (e.g. the amount of forage and concentrate provided in the diet) has not yet been examined. Furthermore, there is some evidence that salt deprivation increases oral stereotypies (Phillips et al., 1999).

Tongue rolling is also seen in young calves (Redbo, 1998) and fattening bulls, where the incidence also can be high (Sato et al., 1994). The incidence does not differ between tethered and untethered, individually housed animals (Wilson et al., 1999). Providing calves with objects that they can manipulate orally, such as a piece of rubber tire or a chain, reduces tongue rolling (Veissier et al., 1997). Tongue playing in older beef cattle has been less studied but Sato et al. (1994) reported that such behaviours were performed most often at the same time as feeding behaviours and were less common at the end of the fattening period, when the feeding of concentrates was *ad libitum*. This again suggests the importance of feeding motivation in the occurrence of these behaviours.

Although tongue rolling is related to feeding motivation, other factors also influence its occurrence. Redbo (1992) noted a large difference between individual cows in the extent that they showed tongue rolling. Furthermore, the distributions of the frequency of this behaviour tended to be positively skewed, suggesting that most individuals show little or no tongue rolling, while a small number of cows perform the behaviour very frequently. These differences appear to remain as the animals' age. Redbo (1998) noted that 4- to 7-month-old calves that showed frequent tongue playing, also showed more tongue rolling when they were 17–20 months old.

What evidence is there that the occurrence of such behaviours reflects a reduced level of animal welfare? At a herd level the evidence seems clear, as the incidence relates to underfeeding and hunger is clearly a welfare problem. It is, however, less clear that variation among animals within a herd is indicative of variation in how well these animals are coping with the underfeeding. No one has yet reported that cows that show more tongue rolling are at greater risk for health problems, and no clear correlations were found between the incidence of tongue rolling and basal urinary cortisol concentrations when 16- to 20-month-old heifers were tethered (Redbo, 1993). However, high levels of tongue rolling were associated with lower basal concentrations of ACTH in 4–7-month-old calves and lower cortisol responses to ACTH in 17- to 20-month-old heifers (Redbo, 1998). The degree to which an animal showed tongue rolling was also related to some aspects of its behaviour in an open-field. Although the pattern of these results is difficult to interpret, Redbo (1998) suggests that differences between individual animals in the extent of tongue rolling may be related to differences in their response to stressful events. The major problem, however, lies in the difficulty interpreting HPA activity (Chapter 3) and open-field activity (Section 3.2.1 of this chapter) as animal welfare indicators.

Despite the widespread use of stereotyped behaviours in assessment of animal welfare, we still lack clear evidence that they are associated with poor welfare at an individual level in cattle and in other species (Mason and Latham, 2004). Indeed, in some cases the performance of stereotypies may actually help animals. For example, horses increase saliva production through the performance of stereotyped oral behaviours, and this saliva can help horses cope with the acidosis produced by diets rich in concentrates (Nicol et al., 2002). Mason and Latham's (2004) review of the evidence linking the stereotypies and animal welfare concluded that "environments that induce or increase stereotypy are indeed typically worse than those that do not, but within a stereotypy-inducing environment, the most stereotypic animals are likely to be the least welfare-compromised individuals. However, even this distinction is clearly not the whole story. Some treatments or housing conditions that cause good welfare also enhance stereotypy and stereotyping animals do not always have good welfare" (p. S60).

3.5 Play Behaviour

Many of the behaviours reviewed earlier are useful in detecting negative emotions or states in animals. Unfortunately, little research on animal welfare has focused on ways of assessing and improving positive emotions in animals. One class of behaviours that are relevant in this respect is play, which animals seem to enjoy (Spinka et al., 2001). Play sequences often consists of interactions that imitate the process, if not the end point, of more clearly functional behaviours like fighting (Figure 4.13).

Figure 4.13 The occurrence of play has been proposed as one behaviour that may indicate "positive" animal welfare. Animals give the impression that they enjoy playing (top). The types of play behaviour seen among cattle have not been well described. Young cattle are often seen fighting in a playful manner (bottom). The difficulty is knowing whether this really is play or a form of aggression

How animals benefit from play is not well known, but play could help animals improve social skills, especially for species that need practice to develop effective courtship, appeasement, or competitive behaviour. For example, pre-weaned dairy calves reared in groups (as compared to the more conventional individual housing) spend time playing (Jensen et al., 1998), and these calves are more likely to become dominant when mixed with animals that have been individually reared (Broom and Leaver, 1978). Play may also prepare animals for coping with unusual situations, such as maintaining balance on a slippery surface. Indeed, play sequences often involve some aspect of self-handicap, such as an ungainly body posture that make animals particularly likely to fall or move in unusual ways (Spinka et al., 2001). Little is known about the motivations underlying play behaviour, although in some animals, preventing young from playing results in an increased time spent playing when they are given the opportunity, suggesting some internal sources of motivation (e.g. Holloway and Sutter, 2004).

Play in cattle has not been well studied, and it is not clear which behaviours should be described as playful. Young calves show mainly locomotor play, such as the jumping described by Dellmeier et al. (1985). This behaviour does decrease with age and is reduced when calves are kept at higher densities (Jensen et al., 1998), but at present we do not know enough about the causes of this behaviour in calves to use this with much confidence to assess their welfare.

4 Conclusions

In this chapter, we have focused upon the issues that arise when we use animal behaviour to assess welfare. Behavioural deprivation remains one of the crucial issues in animal welfare, and developing scientific techniques for deciding how any given instance of behavioural deprivation affects animal welfare remains an ongoing need. Although the concept of natural behaviour captures some of the popular disquiet with modern housing systems, it presents scientific difficulties. We have discussed a number of techniques that have been used to try to assess the importance to the animal of being able to perform a particular behaviour. However, these techniques have so far only been used on a small number of behavioural systems, limiting our ability to use behavioural indices of animal welfare with much confidence. Demonstrating the validity of such indices remains an urgent task.

Part II
Challenges to Animal Welfare

Chapter 5
Acute or Short-term Challenges to Animal Welfare

1 Introduction

The various challenges to animal welfare can be roughly divided into acute challenges (such as branding, dehorning) that last for only a short period of time (seconds, minutes, or hours) and which consequently affect the animals' welfare mostly for a relatively short period of time, and chronic challenges (such as the method of housing) that last considerably longer (weeks, months, or years) and that consequently affect the animals for much longer. In this chapter, we discuss short-term procedures.

2 Painful Procedures

Procedures that cause pain and distress to animals are likely the most contentious of all animal welfare issues. Intentionally causing pain to another human is considered repulsive in most societies, and causing pain to animals normally generates a similar response. Despite these concerns, painful procedures like branding, castration, and dehorning are routinely performed on cattle without benefit of pain relief. In this section, we address how pain in cattle can be assessed and how painful but necessary procedures can be modified to reduce or prevent the pain that cattle experience.

2.1 Assessing Pain

Pain, like other affective or emotional states in animals (fear, anxiety, pleasure, etc.), is difficult to assess. However, a substantial body of literature is now available that discusses the best methods of assessing pain. These general ideas and methods also apply to the assessment of the other affective or emotional states. For reviews of pain and pain assessment, see Weary et al. (2006) and Rutherford (2002). Here we summarize some of the important ideas and approaches.

The scientific assessment of pain in animals normally relies on one of three approaches: measures of general body functioning such as food and water intake or weight gain, measures of physiological responses such as plasma cortisol concentrations, and measures of behaviour such as vocalizations (Rutherford, 2002). Changes in body weight, food and water intake are often easiest to record when animals are singly housed, but these measures need to be taken over a period of time and thus tell us more about what the animal was experiencing in the past, that is during the interval between observations (which are typically hours for food and water consumption and days for changes in body weight), rather than what the animal is experiencing at present. Behavioural responses give the most immediate information on the animal's present emotional state, but there are obvious difficulties in interpretation (Chapter 4). In relatively stoic prey species, such as cattle, that rarely show pronounced behavioural responses to pain, physiological measures of pain and distress (see Gregory, 2004) can be useful. However, these measures require techniques and equipment that make them impractical for on-farm assessments, and typically require restraint and blood sampling that are also distressing for the animal. Furthermore, there are many difficulties in interpreting these responses in terms of animal welfare (Chapter 3). Regardless of the method used, the assessment techniques need to be validated to establish that variation in the response provides meaningful information about the pain that the animal is experiencing.

The most convincing evidence that a measure of pain is valid comes from experiments that examine animals' responses in four different situations: those with and without a painful treatment and with and without analgesics known to be effective at treating this pain. The most useful measures will be able to identify the painful treatment (i.e. the painful procedure without the analgesic), and will show no difference among the other three situations. For example, in one experiment, Faulkner and Weary (2000) measured pain response of calves in the hours after hot-iron dehorning (Figure 5.1). Calves were either dehorned or put through a sham procedure. These two treatments were done either with or without the non-steroidal anti-inflammatory drug (NSAID) ketoprofen. Three behavioural responses (number of ear flicks, head rubs, and head shakes) were recorded for 24 h after the procedure. Calves that were dehorned without the NSAID showed high frequencies of the three behaviours throughout the 24-h period after the procedure. However, calves that underwent the sham procedure with or without the NSAID showed almost none of these behaviours. Calves that were dehorned with the NSAID also showed a much lower occurrence of these behaviours, demonstrating that these responses are valid indicators of post-procedural pain (Figure 5.2).

2.2. *Assessing Pain Through Behaviour*

The types of behavioural observations used to assess pain can be roughly divided into subjective measures, which require the observer to make some subjective rating based on the occurrence of certain types of behaviours, or objective measures,

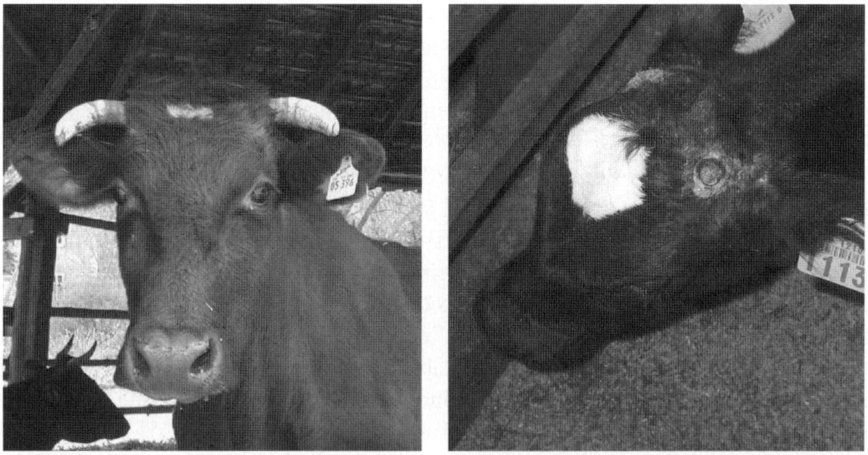

Figure 5.1 The horns of cattle (left) can be dangerous to the people who handle them and to other cattle when they are kept together in enclosed areas. For this reason, horns are often removed on both beef and dairy cattle during the first weeks or months of life (right). Many different procedures are used. Research has shown that most of these procedures cause pain to the animals and that a combination of local anaesthetics, analgesics, and tranquillizers substantially reduce the pain and stress

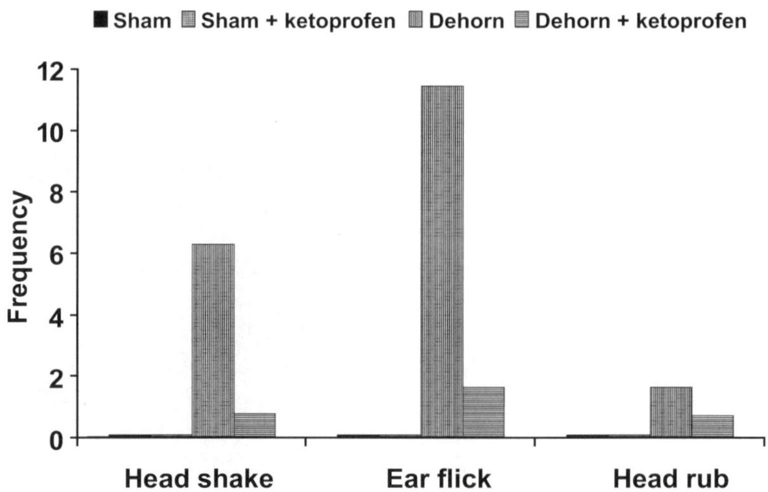

Figure 5.2 The frequency of three behaviours (head shaking, ear flicking, and head rubbing) that are thought to indicate the calf's response to the pain of dehorning. These behaviours hardly occurred when the calf was subjected only to sham dehorning but the frequency was significantly higher when the calf was dehorned. Use of an analgesic (ketoprofen) reduced the frequency of these behaviours almost to the level that was found when the animals were only subject to sham dehorning. This shows that the frequency of these behaviours is a valid measure of how painful the calves find the procedure (From data presented in Faulkner and Weary, 2000.)

which require a measurement of some parameters of the behaviour itself (e.g. frequency of occurrence, durations, etc.). Subjective scoring systems have been particularly popular in the veterinary literature, perhaps because these are considered relatively easy to apply in a clinical setting. A well-known example for farm animals is gait scoring to assess lameness in cattle (discussed in Chapter 2). Unfortunately, there is often little standardization or justifications of which behaviours are included in the scoring system, and subjective schemes often suffer from poor reliability or repeatability. The reliability of a measure refers to its potential for obtaining the same results when scoring is repeated. This can be evaluated by having the same observer rescore animals on multiple occasions (intraobserver reliability), or by having different observers independently score the animals (interobserver reliability). The degree of repeatability will vary depending on the actual scoring technique used. For example, Winckler and Willen (2001) found that lameness scores for three observers were in agreement for 63–74% of observations, while Flower and Weary (2006) found intraobserver consistency of 76–85% and interobserver consistency of 69%. Lower levels of inter and intraobserver reliability obviously limit the usefulness of a measure, and place an upper limit on the extent to which the measure can be validated.

Three main classes of behaviours can be useful in pain assessment. The most obvious of these are pain-specific behaviours, like head rubbing following dehorning, as described above, or licking the tail area during tail docking (Eicher et al., 2000; Figure 5.3). Defensive behaviours can sometimes also be seen when the animal or site of injury is manipulated, such as bucking in lambs upon palpation of the scrotum following castration (Thornton and Waterman-Pearson, 1999).

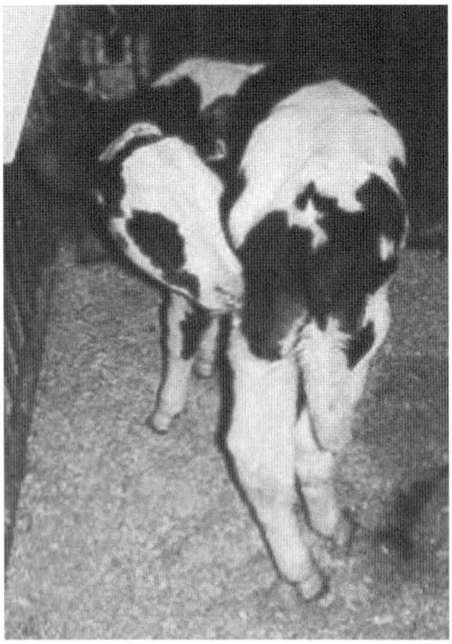

Figure 5.3 One useful indicator of pain is the occurrence of pain-specific behaviours. For example, calves that have been dehorned rub their heads, flick their ears, and shake their heads. Calves that have had their tails banded as part of tail docking (above) frequently lick their tails or their posterior. Although such behaviours occasionally occur spontaneously, the frequency of their occurrence is much higher when animals are in pain

A painful injury will sometimes increase the animal's sensitivity to other sources of pain, and such hyperalgesia is typically assessed by exposing the animals to a painful stimulus (e.g. heat) and measuring the withdrawal response (e.g. Whay et al., 1998). The site and intensity of pain and its duration will influence the severity of these responses, and the sorts of behaviours that are observed. For example, calves respond to application of the hot iron by performing vigorous escape behaviours such as tipping forward onto the front legs and rearing, but respond to the post-operative pain with more subtle injury-directed behaviours such as ear flicking and head shaking (e.g. Grøndahl-Nielsen et al., 1999). The fact that different types of behavioural responses occur at different times following a painful event adds to the complexity of assessing pain in animals.

A second class of pain response involves declines in the frequency or magnitude of certain behaviours. General lethargy has long been regarded as a sign of pain in animals (Morton and Griffiths, 1985), and pain studies often include measures of both reduced activity and reactivity. Particularly informative are those behaviours that animals would otherwise be highly motivated to perform. For example, cows with painful hoof lesions may spend less time standing and feeding at the feed bunk. Such a decline comes at a clear cost to the animal (reduced food intake), indicating that the pain is important to the cow.

A third class of pain measure are those of choice or preference. Measures of choice were among the first behaviours to be used in the field of animal welfare science (e.g. Hughes and Black, 1973), and can be used to evaluate how animals perceive the relative value or aversiveness of different treatments (Chapter 4). In what is arguably the most convincing form of choice study from the perspective of pain assessment, animals can be trained to self-medicate with analgesics, and researchers can directly assess the frequency and amounts administered. For example, Danbury et al. (2000) trained lame and sound broilers to discriminate between two feeds, one containing an analgesic, and found that lame birds consumed more of the food containing analgesic. To our knowledge, this approach has yet to be attempted with cattle.

2.3 *Pain Prevention and Mitigation*

Recognizing pain only takes us part way – we also need to find ways of reducing or preventing its occurrence. Obvious approaches include preventing injury and disease that cause pain, and minimizing the effects of these ailments through improved diagnosis and treatment. Many cases of painful injury and disease can be avoided by refinements in animal care. For example, mastitis is a painful infection of the udder in lactating cows that can be largely prevented through proper management practices. The issue of animal disease and its impact on animal welfare is discussed at length in Chapter 2.

In other cases we may also be able to dispense with the procedures that cause pain, but the practicality of this option will depend upon the purpose of the procedure and the availability of feasible alternatives. Painful procedures performed on animals are normally assumed to provide some concrete benefit to either the animal or its caretaker, but research can sometimes allow us to reconsider whether the procedure is really needed, at least in its current form. For example, in the past few decades, dairy farmers began tail docking (see Section 3.1 for more details) their animals in an attempt to reduce the risk of mastitis on their farms. Reducing the risk of mastitis is a laudable aim, but a series of experiments – involving thousands of animals with docked and intact tails – have found no effect on udder health (Eicher et al., 2001; Tucker et al., 2001; Schreiner and Ruegg, 2002). This evidence is helping producers make better-informed decisions, and today fewer dairy producers are using this procedure on their farms. In other cases, a painful but necessary procedure can be eliminated through the use of selective breeding for desired traits. For example, dehorning of calves is normally considered necessary to prevent injury to other cattle and to animal handlers, but the development of polled lines (i.e. animals that are genetically hornless) through selective breeding has negated the need for this procedure in some breeds of beef cattle.

One of the most obvious refinements for injury and disease and for painful procedures is the provision of anaesthetics and analgesics. Although there are many excellent examples where such treatments are effective and appropriate (see Benson, 2004), it is also important to consider that any restraint required to administer the drugs may be distressing for the animal, and methods of administration (such as repeated injections necessary for a ring block) can themselves be painful. It is perhaps equally important to consider the practicality of measures used to mitigate pain. In the calf-dehorning example presented above (Faulkner and Weary, 2000), the pain of hot-iron dehorning could be controlled using a sedative, a local block consisting of a series of injections, and NSAIDs to control post-operative pain (Figure 5.2). However, it is difficult to convince dairy producers to adopt such an elaborate approach. One way of increasing adoption is to find methods that are inexpensive and easy to apply, or provide benefits to the producers, or ideally both. For example, Vickers et al. (2005) showed that pain and distress due to caustic paste dehorning could be controlled using only one injection of the drug xylazine (an inexpensive sedative and mild analgesic). Producers also benefit because the sedative makes the chore easier to perform.

There are a number of practical constraints on adopting refinements to painful procedures. Although better scientific and technical solutions play an important role in encouraging animal users to adopt procedures that prevent and mitigate pain, it is also important to recognize some of the cultural and economic constraints on adoption. For example, some effective drugs are not certified for use in all countries, and in other cases can only be dispensed by a veterinarian. In some countries, the drug xylazine (mentioned above) can only be used under the supervision of a licensed veterinarian. In some cases this may not be a problem – large dairy producers well serviced by local practitioners can include this treatment as part of regular herd health visits. However, for isolated ranchers living sometimes hundreds of kilometres from the nearest veterinarian this becomes an important constraint.

3 Painful Procedures for Cattle

Having reviewed above the more general issues of pain assessment and prevention, we now turn in more detail to common procedures that are generally considered painful.

3.1 Tail Docking

Cows use their tails as a natural fly swat, and with each swat, the tail comes into contact with the rest of the body. When the tail becomes contaminated with faeces containing pathogens, it can contaminate other areas of the cow's body, perhaps increasing the risk of udder infections, and the tail becomes more of a threat to milkers and others who work with the cows. For these reasons the practice of tail docking dairy cattle (Figure 5.4) gained popularity in the 1980s and 1990s. Dairy farmers vary in when and how they perform this procedure (Barnett et al., 1999; Stull et al., 2002). For example, docking is sometimes done using elastic rings that restrict blood flow and kill the distal portion of the tail and sometimes using a docking iron that both cuts the tail and cauterizes the stump (see Tom et al., 2002b for comparison).

A number of studies have examined the effect of tail docking on the welfare of dairy cattle. In general, there seems to be little evidence that the procedure causes

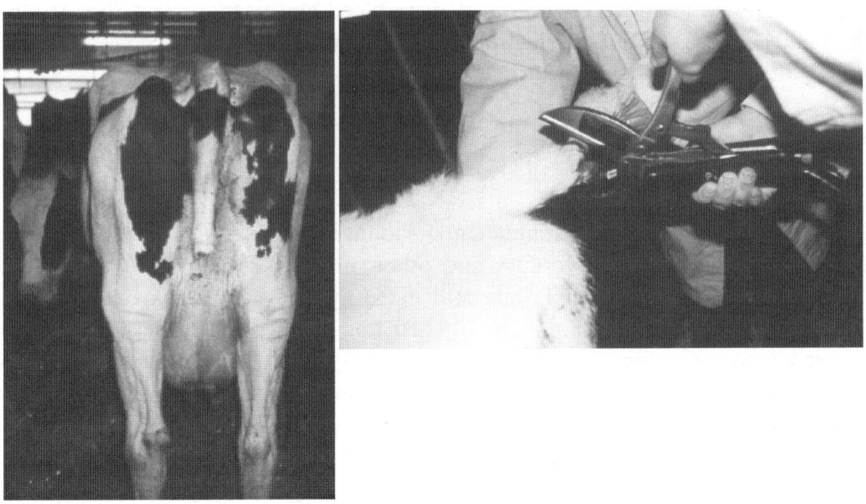

Figure 5.4 In the belief that the procedure keeps cows clean, dairy farmers in some countries will sometimes dock the tail of their cows, either when they are adult or as calves. Research has shown that the procedure is not particularly painful but may make the cows more susceptible to flies and can cause neuromata that may lead to chronic pain. Research has also shown that tail docking has little effect on the cleanliness of the cows

much acute pain. This has been shown by a variety of behavioural, physiological, and immune measures of both calves and adult cows (e.g. Petrie et al., 1996b; Eicher et al., 2000, 2001; Tom et al., 2002a, b). There is no clear evidence that anaesthetics are necessary or that tail docking calves is preferable to docking adult cows (Tom et al., 2002a, b). In calves, evidence of acute pain is sometimes apparent when rubber rings are used rather than a hot docking iron (Petrie et al., 1995; Tom et al., 2002a), but even here the effect is small.

Despite the absence of evidence of acute pain following the procedure, tail docking may have longer-lasting effects on the animals' welfare. Sectioning the nerves in the tails of both young calves and adult cattle results in neuroma formation (Lunam et al., 2002), which could result in chronic pain, similar to the phantom pain felt following limb amputation (Eicher et al., 2006). In addition, docked cows have more flies on them and show more fly avoidance behaviours (Eicher et al., 2001), both of which could reduce their welfare.

The issue of improved milker comfort from tail docking (Petrie et al., 1996b) is now becoming less relevant and most modern milking parlours prevent contact with the tail (Stull et al., 2002). Moreover, in contrast to the beliefs of many dairy farmers (Barnett et al., 1999), multiple large-scale, controlled experiments have now shown that docking tails provides no systematic advantage in terms of cow cleanliness or udder health (Schreiner and Ruegg, 2002; Tucker et al., 2001), although one smaller-scale study did report an increased cleanliness of docked cows (Eicher et al., 2001). The study by Schreiner and Ruegg, for example, found no differences in cleanliness, somatic cell counts or bacterial cultures of mastitis causing pathogens from docked and undocked cattle on nine commercial dairies. Given the obvious disadvantages to the cow, especially their reduced ability to control flies (Eicher et al., 2001), there seems little justification for continuing this procedure.

3.2 *Dehorning*

Few disagree that intensively reared cattle should be kept without horns: the horns of cattle are a danger to workers and other animals if they are not removed (Figure 5.1). Hornless animals can still cause injuries, but the extent of these injuries is reduced (Meischke et al., 1974). For this reason, horns are typically removed using a number of methods described below. All dehorning methods cause pain, but research is showing that certain methods and interventions can be used to reduce this pain (see Stafford and Mellor, 2005a for a review).

The developing horns of cattle 3 months of age or older are normally removed surgically using a number of techniques (e.g. scooping, shearing, and sawing), and physiological responses indicate that these procedures are painful (Sylvester et al., 1998). Dehorning of older animals can lead to a setback in weight gain that can be detected more than 100 days after dehorning (Goonewardene and Hand, 1991). It is therefore generally recommended that dehorning be done when animals are less than 3 months of age. Horn buds of calves are typically removed using a caustic paste or a hot iron. There is again good evidence that both methods are painful (Morisse et al., 1995), but most of the studies on the pain of dehorning and how it can be reduced have focused on hot-iron dehorning. This procedure is known to cause an immediate behavioural response, including tail wagging, head movements, tripping, and rearing (e.g. Graf and Senn, 1999), as well as post-operative pain indicated by head rubbing, head shaking, and ear flicking (e.g. McMeekan et al., 1999) and increased levels of circulating corticosteroids in the hours following the procedure (e.g. Petrie et al., 1996a). It is well known that local analgesics can reduce the pain caused by the burn injury, but it is now becoming clear that use of local anaesthetic alone is not completely satisfactory (Stafford and Mellor, 2005a; Figure 5.5).

One concern is that local anaesthetic does not provide adequate post-operative pain relief. The most popular local anaesthetic, lidocaine, is effective for 2–3 h after administration (McMeekan et al., 1998), and calves treated with local anaesthetic actually experience higher plasma cortisol levels than untreated animals after the local anaesthetic loses its effectiveness (Graf and Senn, 1999; McMeekan et al., 1998; Petrie et al., 1996 a). However, the use of NSAIDs (such as

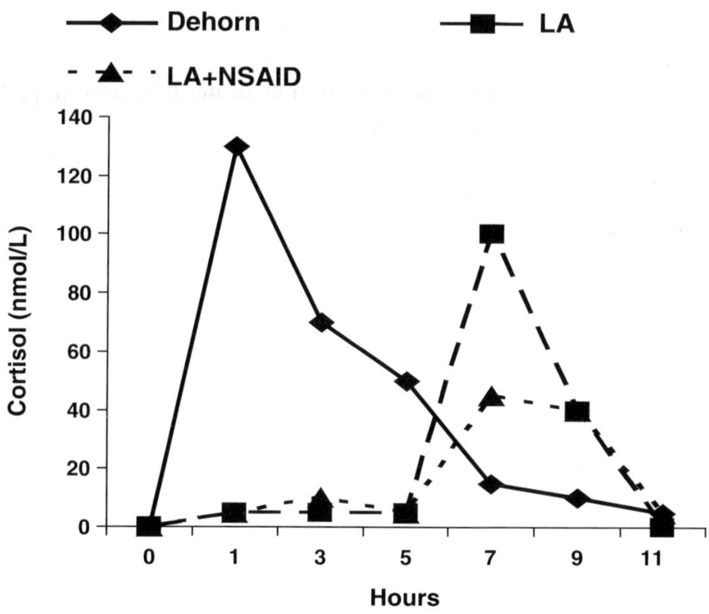

Figure 5.5 Increases in plasma concentrations of cortisol following dehorning show that the procedure is painful (Dehorn– solid line). Local anaesthetics (LA– dashed line with squares) prevent the cortisol increase while the drug is active. However, once the drug wears off, the cortisol concentrations rise, indicating post-operative pain. The most effective pain relief, involves the combination of a local anaesthetic and a non-steroidal analgesic, such as ketoprofen (LA + NSAID– dashed line with triangles), which reduces the post-operative pain (From data presented in Stafford and Mellor, 2005a.)

ketoprofen), in addition to a local anaesthetic, can keep plasma cortisol and behavioural responses close to baseline levels in the hours that follow dehorning (Stafford and Mellor, 2005a).

A second consideration is that animals respond to both the pain of the procedure and to the physical restraint. Calves dehorned using a local anaesthetic still require restraint, and calves must also be restrained while the local anaesthetic is administered. The use of a sedative (such as xylazine) can essentially eliminate calf response to the administration of the local anaesthetic and the need for physical restraint during the administration of the local anaesthetic and during dehorning (Grøndahl-Nielsen et al., 1999). Thus a combination of sedative, local anaesthetic, and an NSAID reduces the response to the pain both during dehorning and in the hours that follow. Unfortunately, such a combination of treatments is unwieldy for farmers and may itself have drawbacks for the animal. For example, an effective local block requires repeated injections (into the cornual nerve within the occipital groove of each eye and a ring block around each horn bud) that are themselves painful.

One common alternative to hot-iron dehorning is using caustic paste to cause a chemical burn. This method of dehorning is still painful for the calves (Morisse et al., 1995), but as described above Vickers et al. (2005) found that this pain is easier to control. This research shows how methods of pain treatment can be developed that are both effective and practical for use on farm.

One practical alternative to dehorning for many breeds of cattle is to breed cows to polled (i.e. genetically hornless) sires (Prayaga, 2007). Horns are inherited as an autosomal recessive gene with polled as the dominant condition (Long and Gregory, 1978), making it easy to reliably produce polled calves from horned cows. Recent molecular biological research has begun to identify the genes involved (Prayaga, 2007). The quality of many polled beef sires is similar to that for horned animals and there is no evidence that polled cattle have lower productivity (Prayaga, 2007). For example, Goonewardene et al. (1999a, b) found no differences between horned and polled cattle in birth weight, weaning weight, carcass weight, carcass characteristics, pregnancy rates, dystocia scores, cow weights, and cow condition scores. Unfortunately, dairy producers still have only a relatively small selection of polled sires available. This is an obvious area for continued development by companies that sell cattle genetics, as the use of polled sires save both a chore for the producer, and provides an easy way of avoiding what is obviously a painful procedure for the calf.

3.3 Branding

Like dehorning, branding involves at least three distinct welfare issues: stress due to restraining the animal before and during the procedure, the immediate pain during branding, and post-operative pain that can occur in the hours following the procedure. Research to date has focused on the second of these three issues, although some data is also available to assess post-operative pain.

Cattle are typically branded using a hot iron (heated electrically or over fire) that burns the skin and creates scar tissue on which no hair will grow (Figure 5.6).

Figure 5.6 In some parts of the world, cattle are branded with a hot iron for purposes of identification. Research has shown that this procedure is painful but there has been little research aimed at finding practical ways of mitigating this pain. The development of alternative methods for identifying animals may eventually eliminate the need for this practice

One alternative method is freeze branding with an iron that has been cooled in liquid nitrogen or a combination of dry ice and alcohol. The freeze brand works by killing the cells that pigment the hair, such that white hair grows from the area that has been branded. It seems clear from all the research that has been completed to date that both methods are painful, but freeze banding consistently results in a lower pain response than hot-iron branding (Lay et al., 1992a, b; Schwartzkopf-Genswein et al., 1997a, b, 1998; Watts and Stookey, 1999; Figure 5.7).

During branding, cattle respond by vocalizing, kicking, flicking their tail, falling in the chute, and making avoidance or escape movements that have been characterized using subjective scores and objective methods. In an elegant example, Schwartzkopf-Genswein et al. (1998) quantified the head movements of cattle during branding, and found that the number of head movements, the distance the head was moved, and the speed of movement were all greater for the cattle that were hot-iron branded compared to freeze branded. In another study, Schwartzkopf-Genswein et al. (1997a) evaluated the post-operative inflammatory responses to branding using infrared thermography. Both freeze and hot-iron branding resulted in a pronounced inflammatory response, with skin temperature almost 2°C higher than baseline and differences persisting throughout the 7 days of post-procedural monitoring. However, this inflammatory response was greater and persisted longer for cattle that were hot-iron branded compared to their freeze-branded counterparts.

One obvious gap in the research on branding is lack of attention paid to finding practical methods of pain mitigation. Given that the work on hot-iron dehorning has resulted in successful treatment methods, this seems like a useful

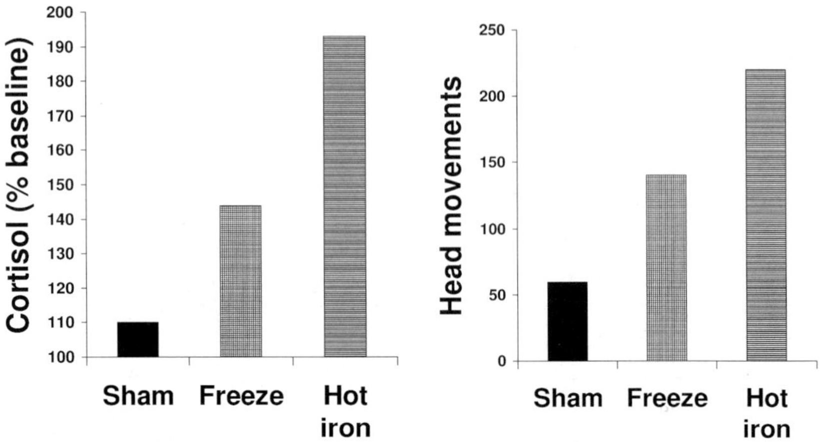

Figure 5.7 Research using a variety of measures of pain responses, such as increases in plasma cortisol concentrations (left) and increases in head flicking (right) has shown that both hot-iron and freeze branding are painful but that freeze branding is slightly less so (From data presented in Schwartkopf-Genswein et al., 1997a, 1998.)

area to pursue. More generally, researchers and the cattle industry need to work to develop and adopt modern methods of identifying cattle that do not involve injuring the animal.

3.4 *Castration*

Of all the routine surgical procedures performed on cattle, one of the most ancient and best researched is castration. The effects of castration on the welfare of cattle have been reviewed by Stafford and Mellor (2005b) and Bretschneider (2005), and we encourage readers to consult these articles. Below we provide a brief review of some of the key issues and research in this area.

As with other procedures that can be accomplished using several techniques, research on cattle castration has tended to focus on comparisons of alternative procedures. The most common methods are those in which the testicles are either removed (surgery), or killed by crushing (Burdizzo) or constricting (rubber rings or latex bands) the tissues that supply blood to the testes. There are several variations on the surgical method, including: (1) whether the scrotum is simply incised to allow extraction of the testes versus removal of the bottom of the scrotum and (2) cutting the spermatic cord versus tearing it by pulling on the testicle. All methods of castration are known to cause pain, but the evidence reviewed below indicates that the constriction methods (rubber ring and latex band) are most problematic.

Early scientific assessments of castration centred on production effects. Production measures will, at best, be indirectly related to the pain, as discussed in Chapter 1. The assessment of production parameters is further complicated for castration, because of the role of testosterone in mediating growth. However, production measures are still worthy of attention, in part because they can allow us to identify win-win solutions that provide economic benefits to producers and welfare benefits to their animals. One major problem with studies of weight gain is that cattle can vary greatly in body weight depending upon the last time they last ate, drank, defecated, or urinated. Studies using weight changes as a response measure should weigh subjects multiple times each day. Studies based on infrequent measures of body weight will be prone to Type II error (i.e. fail to find differences even when these exist) and negative results should always be treated with a grain of salt. That said, previous work has indicated that all methods of castration can cause reductions in weight gains with this being most pronounced when older animals are castrated (e.g. Bretschneider, 2005). Some studies have reported differences among castration methods in the magnitude of growth check that results. For example, Knight et al. (2000) found that cattle castrated with latex bands showed a greater setback in growth than did those that were castrated surgically. However, Bretschneider's (2005) review showed that most studies did not find such a difference.

A more directly relevant class of measure is that related to wound healing and complications associated with the procedure. All else being equal, methods producing wounds that heal quickly, and that are less prone to post-surgical infections and complications should be preferred by veterinarians, producers, and the cattle. Intuitively the "bloodless" methods of Burdizzo and ring castration might be considered superior in this regard, but the scientific evidence suggests that tissue trauma actually heals most quickly with surgical castration. For example, Stafford et al. (2002) reported that wounds from surgery were completely healed within 28 days, while healing continued more than 7 weeks after the rubber ring procedure.

Studies attempting to directly assess pain have used both physiological and behavioural measures, with physiological studies focusing on plasma cortisol. In some situations, plasma cortisol has proven to be a valid indicator of pain; levels are lower following castration with the local anaesthetic lidocaine than following castration without a local block (Fisher et al., 1996). However, cortisol will also respond to other stressors potentially masking pain effects (Chapter 3). For example, separation from the cow results in a pronounced cortisol response, such that calves that are separated and castrated cannot be distinguished from those that are simply separated and not castrated (King et al., 1991). For this reason many of the well-designed studies using physiological responses to assess pain have been on individually housed cattle. All methods of castration appear to cause a pronounced cortisol response, with this being greatest for older animals (Bretschneider, 2005). However, the relative ranking of measures varies depending upon the way cortisol is interpreted (e.g. peak response, duration above baseline). One way of integrating these measures is to consider the area under the response curve (Chapter 3), and by this measure the Burdizzo method seems slightly better than surgical or ring methods (Stafford et al.,

2002). Interestingly, the cortisol response to the surgical methods in which the spermatic cord is cut (rather than torn by traction) is highly variable, suggesting that some ways of performing this procedure may be less painful than others.

Because of the time course of the response, behavioural responses may be better able to distinguish between the distress due to restraint and separation from herd-mates versus pain due to castration, and also distinguish between the immediate effects of the procedure, and longer-term post-operative pain. However, the physical restraint can also make it difficult for calves to express certain behaviours, and for observers to properly quantify these responses. Unfortunately, relatively little behavioural data is available to address pain due to castration in cattle. Fell et al. (1986) showed that during castration calves struggle and kick with the hind legs, but this response is more evident during surgical castration than during the placement of a rubber ring. In the hours that follow castration all methods cause behavioural change although the nature of these changes can vary with method. As Stafford and Mellor (2005b) argue, there are no clear differences that can allow us to conclude which procedures cause more or less pain.

A number of studies show that this pain is likely reduced when the procedure is performed at younger ages. For example, Ting et al. (2005) reported that the cortisol response to the Burdizzo procedure is greater when applied at 5 months of age than when applied at younger ages. Restraint is also easier with younger animals, so all procedures requiring restraint are best performed at younger ages. However, castration results in a pronounced physiological and behavioural pain response at all ages, so performing the procedure at a young age should not be considered as sufficient to eliminate pain – methods are still needed to control or prevent the pain.

As with other procedures, pain mitigation strategies should consider distress due to restraint, the immediate pain associated with the procedure, and post-operative pain. In our view, restraint stress is the most difficult to address, and for range cattle the distress associated with capture and restraint may be a greater welfare problem than any pain the animal experiences. As described for dehorning (see above), drugs such as xylazine can be used to sedate calves, facilitating the procedure and removing the need for any physical restraint. However, innovative solutions are still required to ease the administration of the drugs, avoiding the distress due to capture and separation from the dam and other herdmates. Xylazine has the added advantage of providing some analgesic effect. Experiments on cattle have shown that this can effectively prevent the immediate pain due to the procedure (e.g. Ting et al., 2003), but this study used epidural application, a procedure that is unlikely to be practical on many commercial farms. New work is needed to determine if intramuscular injections of xylazine or other drugs of this class (alpha-2 agonists) can be used to both sedate the calf and provide adequate analgesia for the procedure, as we saw above for caustic paste dehorning of young calves.

As described earlier in the current section, the local anaesthetic lidocaine is clearly effective at preventing the immediate pain due to castration. Unfortunately, administering the drug requires extra restraint for the animal. Lidocaine should be administered several minutes before the procedure is performed, meaning either a prolonged period of restraint, or capturing and restraining the animal twice: first to administer the local and again to perform the procedure.

Regardless of how we control the immediate pain, castration will cause pain that extends for hours and sometimes days that follow. This pain can also be treated, although practicality becomes more difficult, the longer the pain endures. As described for other procedures like dehorning, NSAIDs are effective for treating pain following all of the castration methods described above (Stafford et al., 2002). A single treatment with the NSAID ketoprofen can be effective for up to a day, and if provided before the tissue damage occurs can also prevent the sensitization that otherwise contributes to post-operative pain. However, as with administration of the local anaesthetic, the largest difficulty is how to administer the NSAID prior to castration without imposing the distress due to additional restraint. Clearly, new approaches are needed to develop treatment protocols for pain due to castration that are both effective and practical for use on farm.

Unfortunately, there are few alternatives to castration currently available for commercial producers. Still at the research stage is the idea of immunocastration. In one study, Hernandez et al. (2005) immunized cattle against luteinizing hormone-releasing hormone (LHRH), a naturally occurring hormone important in reproductive development stimulating the release of other reproductive hormones and growth of the testicles. This and earlier studies have shown that this immunization can be highly effective at reducing testicle size and circulating levels of testosterone, even when immunization happens after the males are sexually mature. This technique shows great potential in that it avoids pain entirely, and allows farmers to capitalize on the improved growth characteristics of intact males.

3.5 Pain During Calving

Parturition is likely painful for all mammals, both during labour and in the hours that follow, but pain may be a particular problem during difficult calvings (dystocia) (Figure 5.8). Under extreme conditions, dystocia can result in stillbirth, and the incidence of stillbirth is increasing. For example, a study on primiparous Holsteins showed that frequency of stillbirth has now reached to 9% (Hansen et al., 2004). Even when the calf survives dystocia, both the cow and calf may continue to experience pain. One obvious area of concern is reluctance of the cow to eat and drink after calving. Cows are at high risk of metabolic and infectious disease in the days after calving, partially due to a negative energy balance associated with low intakes (see Chapter 2). To date no research has examined the role of pain due to calving on subsequent intakes and health, but we view this as an obvious and important area for future research.

Indeed, there has been almost no research at all on the pain associated with calving in cattle. The one exception that we are aware of is the work of Pinheiro Machado et al. (1997) on the analgesic effects of consuming amniotic fluid.

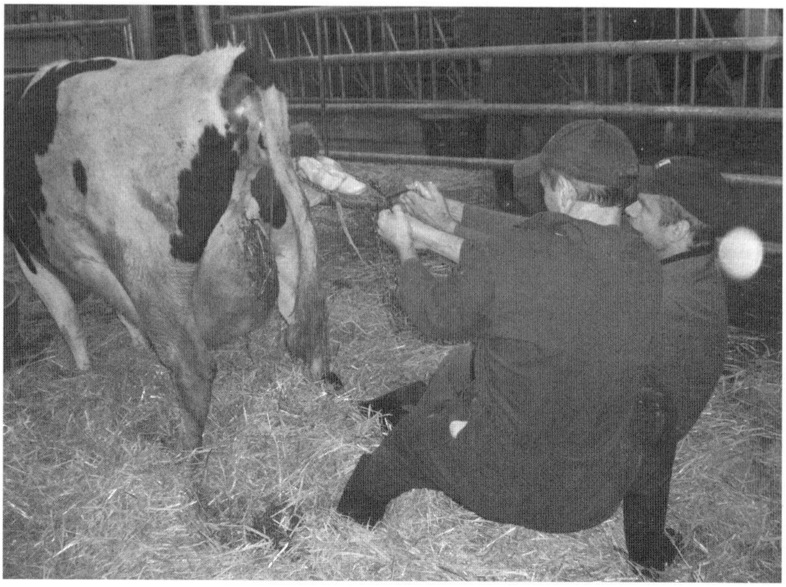

Figure 5.8 A difficult calving

Most terrestrial mammals lick and ingest at least some of the amniotic fluids, membranes, and placenta. Pinheiro Machado et al. (1997) showed that pain sensitivity was reduced at calving, and that cows that were allowed to consume amniotic fluid showed a further reduction in pain sensitivity, an effect that has also been documented in laboratory rats (Kristal et al., 1990). However, current recommendations for the management of calving for dairy cows encourage the rapid separation of cow and calf and limit opportunities for cows to ingest amniotic fluid. Also, with difficult calving and stillbirth, membranes typically rupture well before delivery of the calf limiting the role of amniotic fluid ingestion in pain control.

The effects of dystocia on the calf's welfare may also be important. In the case of stillbirth there is likely minimal opportunity for suffering; for neonates that never breath, low oxygen levels in the blood seem to keep animals unconscious, likely preventing any sensations of pain or distress (Mellor and Gregory, 2003). Difficult calving can, however, affect the behaviour of calves that survive parturition; these calves are less active and take longer to nurse (Metz and Metz, 1987). Low intakes after birth put the calf directly at risk from starvation, and indirectly due to an increased risk of hypothermia especially for outdoor calving in temperate climates. The extent of suffering due to hunger and hypothermia have not been studied, although some authors assume that the effects are likely to be modest (Mellor and Stafford, 2004), at least in comparison with the pain due to disease and the procedures discussed elsewhere in this chapter.

4 Stressful or Frightening Procedures

In the previous sections, we considered a number of procedures that cause physical pain to animals. Pain is an obvious concern in terms of animal welfare, but animals can also suffer other emotions, such as fear and anxiety. For example, just placing a cow alone in an unfamiliar place provokes strong behavioural and physiological responses associated with distress (e.g. Rushen et al., 1999a). For beef cattle that are handled rarely, contact with people can be a major source of stress. Such practices rarely receive the same attention as procedures that are painful. However, all of the painful procedures discussed above require that animals be handled, and this can be a welfare problem if poorly done. Intensively reared animals are generally easier to handle than extensively reared animals (Croney et al., 2000) and providing extra contact with people generally makes handling easier (Boivin et al., 1994; Lensink et al., 2000, 2001c; Chapter 9). Research has not been carried out to determine which types of handling practices are the most effective at moving animals, but has examined which are the least aversive. For cattle, use of electric prods and shouting appear to be the most stressful (Pajor et al., 2000, 2003), and should be replaced with other methods wherever possible. Use of electric prods may be more effective than other means at making animals move but also tend to make animals fall and bump into the chute walls (Croney et al., 2000). Loud noises, particularly the sound of people shouting, are disturbing to cattle (Waynert et al., 1999) and can easily be avoided. The need to handle animals roughly is reduced by having good facilities and equipment for moving and restraining animals. Grandin (1997) provides an excellent discussion of appropriate facilities for handling cattle, and a good discussion of principles to follow. Unfortunately, we know little about the extent that cattle are exposed to various frightening but non-painful procedures and this remains an important area for new research (Figure 5.9).

5 Weaning

In nature, weaning involves both a gradual reduction in milk intake and increasing social independence from the mother, but farmed cattle are often weaned abruptly by separation of the cow and calf. In addition, farmed animals are normally weaned at much younger ages than in the wild, and sometimes face additional stressors such as changes in the social and physical environments. Under these conditions, weaning can result in decreased nutrient intake combined with high levels of distress vocalizations and activity (for review, see Weary et al., 2007; Figure 5.10).

Weaning involves a dietary change, that is, from milk to a solid diet, and the effects of this on the welfare of the calf are discussed in Chapter 8. However, weaning also involves stress associated with the rupture of the social bond between the calf and the cow. Some research on beef cattle has attempted to separate these factors by preventing calves from accessing the dam's teats while allowing continued physical contact between the cow and calf (Price et al., 2003; Haley et al., 2005).

Figure 5.9 Hoof trimming. Cattle are subjected to a number of routine procedures, which they can find frightening or stressful. Many, such as hoof trimming, are done for the animals' own benefit. Regular hoof trimming has been shown to reduce the incidence of hoof lesions and is an essential component of hoof care for dairy cattle. Nevertheless, we know little about how stressful such procedures are for the cattle, or what are the best ways of reducing this stress

Dairy calves offer the opportunity to study events in the opposite order; calves are often separated from the cow within the first day of life but then fed milk artificially for several weeks before they are weaned onto solid food. Both examples provide models for understanding the effects of separation from the dam independently from the effects of the loss of the milk supply.

This work has shown that the presence of the emotional stressors associated with weaning increase the effects of the change of diet. For example, studies on beef cattle have demonstrated that calves show little distress when prevented from nursing if they have continued contact with the cow, but calves showed the typically strong behavioural response when simultaneously separated and prevented from nursing (e.g. Haley et al., 2005). Similarly, Jasper et al. (2007) found that dairy calves show little behavioural response to dilution of their milk ration, even when only warm water was provided from the feeding apparatus, so long as this change in diet was not combined with other stressors at weaning. The work on beef calves also shows that once the calf has had several days without access to the udder, cow and calf can be separated with little distress (Price, 2003; Haley et al., 2005).

Figure 5.10 Weaning the calf from the cow can be a significant source of stress. For beef calves, weaning involves both the physical separation of the calf from its dam and a change of diet from milk to solid foods. Separating the two sources of stress, so that the calf is allowed to stay close to its mother even after it is no longer allowed to nurse, can reduce the stress of separation. For dairy calves, the two events are normally separated in most dairy production systems. Calves are taken from the cow soon after birth, but continue to drink milk so that the change of diet occurs several weeks later. In such cases, the calf does not form much of an emotional bond to the cow. However, if the calf is kept with the cow for a few days, an emotional bond can form, which increases the calf's response to the separation, when it does occur

A fundamental question regarding the effect of early weaning on animal welfare concerns the importance of the bond between the calf and the dam and the effects of breaking this bond. Dairy calves are typically separated from the cow at birth and then fed milk artificially for several weeks before weaning from milk, and thus provide a model to examine the effects of maternal separation separately from the end of milk feeding (see review by Flower and Weary, 2003). Consistent with the results described above for dietary changes when no other stressors are imposed, evidence from the dairy calf suggests that early separation from the cow causes little distress response so long as calves are well fed. Thomas et al. (2001) measured the distress response of dairy calves following separation from the dam and found that calves showed little response during the first hours after separation. Instead, response peaked almost 18 h after calves were separated indicating that this response is due to the calf being hungry. To test this idea some calves were given almost double the colostrum normally provided; as expected, these calves showed almost no distress response to the separation.

It is important to note, however, that dairy calves are typically separated only hours after birth, with the intention of reducing the time available for cow and calf to form strong social bonds. If dam and young are really separated before they have been able to establish a social bond, then the response we see at this stage should not properly be considered a separation response. Consistent with this interpretation, a number of studies have shown that the strength of the calf's response to separation from the dam increases with the age of the calf (e.g. Lidfors, 1996; Flower and Weary, 2001). Age effects can be difficult to interpret, in part because older animals may also have more difficulty making the transition to the artificial feeding system, but these results indicate that the bond becomes increasingly strong, and the effects of breaking this bond become more evident, the longer the calf stays with the cow (Flower and Weary, 2001). This is an important issue for organic dairy farming where the calf is often kept with the cow for longer than is usual with traditional dairy farming.

In some cases, the availability of an alternative social partner may reduce the calf's response to weaning. In dairy calves, the reduction in growth at weaning from milk is mitigated if calves are weaned in pairs as compared to individually housed calves (Chua et al., 2002). The reasons for this difference are not clear, but our experience is that calves respond much less to weaning when group housed. More work on dairy calves may be of special interest, as these animals are typically housed singly before weaning and at weaning grouped with other calves for the first time.

In summary, separation from the cow, cessation of milk feeding, and changes in the physical and social environment can all be important stressors at weaning. The research results summarized above suggest that calves experience less of a distress response at weaning when these stressors are staggered (such as by first stopping nursing and later separating cow and calf).

6 Transport

Animals can be transported in a manner that is compatible with good welfare but, in many cases, transport is a highly stressful event and represents a major acute challenge to an animal's welfare. The transport of animals is the most frequently regulated aspect of animal production, partly because its effects on animal welfare are often highly visible. Most developed countries have regulations or codes that give relatively precise details about stocking densities, journey length, etc. Despite this, the actual amount of research upon the effects of transport on animal welfare is actually quite small.

Compared to other species of farm animals, cattle are transported for many reasons. Finished beef cattle, dairy bulls (young and old), culled dairy cows, and veal calves are transported to slaughter; weaned beef calves are transported to feedlots; unweaned dairy calves to veal operations and specialty heifer growers. Furthermore, the types of transport can vary greatly, including the long-distance mustering seen in Australia, and road, rail and sea transport. There is now more transport of pregnant dairy heifers and lactating cows (Eicher, 2001) as well as day-old calves (Moore et al., 2002). Since the purpose of this book is primarily a review of research, we will limit ourselves to discussing the findings of research rather than listing optimum conditions for transporting cattle. Fortunately, the Scientific Committee for Animal Health and Welfare of the European Commission (Scientific Committee on Animal Health and Animal Welfare, 2002) have reviewed this topic and discussed many of the practical issues. Moreover, the research on transport in cattle has been reviewed by, e.g. Hemsworth et al. (1995), Knowles (1999), Tarrant and Grandin (2000), Eicher (2001), and Swanson and Morrow-Tesch (2001). The Scientific Committee on Animal Health and Animal Welfare (2002) report describes animal transport within the European Union (EU), while Swanson and Morrow-Tesch (2001) provide a useful historical overview of how animal transport in North America has changed over the last century.

6.1 Effects of Transport

Transporting cattle even for short periods (less than an hour) can produce many of the physiological changes usually seen following a stressor. These include increases in sympathetic nervous system activity (Locatelli et al., 1989) and heart rate (Jacobson and Cook, 1996, 1998), suppression of LH secretion (Nanda et al., 1989, 1990), changes in acute phase proteins (Arthington et al., 2003) and most obviously, increases in hypothalmo-pituitary-adrenocortical activity (e.g. Warriss et al., 1995; Lay et al., 1996; Grigor et al., 2001; Morrow et al., 2002). The rise in cortisol can be detected in the faeces of transported cattle (Palme et al., 2000). Transport usually involves deprivation of feed and water, which leads to loss of body weight (Tarrant et al., 1992; Warriss et al., 1995; Knowles et al., 1997, 1999; Gallo et al., 2003). Warriss et al. (1995) found an average weight loss in steers of nearly 5% following a 5-h journey and 7% following a 15-h journey; the latter required 5 days to recover to pre-transport values. Biochemical signs of dehydration and mobilization of body reserves for energy include changes in blood osmolality and concentrations of urea, free fatty acids, glucose, and packed cell volumes (Tarrant et al., 1992; Warriss et al., 1995; Jarvis et al., 1996; Knowles et al., 1999). Cattle can remain standing for longer than normal (Kent and Ewbank, 1983, 1986; Warriss et al., 1995; Knowles et al., 1997, 1999), which leads to fatigue: many studies show increased concentrations of creatine kinase after transport (Tarrant et al., 1988, 1992; Warriss et al., 1995; Knowles et al., 1999), evidence of muscle exertion or injury. During transport rumination is suppressed at least for 6 h (Kent and Ewbank, 1983). Transport of cattle can also lead to reactivation of dormant viral and parasitic infections (Genchi et al., 1986; Thiry et al., 1987) as well as increased shedding of salmonella (Barham et al., 2002).

There are a number of studies examining the effects of transport on the immune system of cattle; largely provoked by the obvious problems of shipping fever, which is one of the most serious consequences of transport on animal welfare (Tarrant and Grandin, 2000). Shipping fever has attracted less attention in Europe because cattle tend to be transported directly for slaughter (Knowles, 1999). In North America, the use of feedlots as an intermediate fattening stage before slaughter highlights the economic, as well as the animal welfare costs of this disease. Studies of the effects of transport on the immune system were reviewed by Swanson and Morrow-Tesch (2001). The effects can be complex, leading to suppression and enhancement of the immune system depending upon the component measured (e.g. Mackenzie et al., 1997). This reflects the generally complex effects of stress on the immune system (Chapter 3). One difficulty is correlating the changes in the immune system with subsequent health problems. For example, Grigor et al. (2001) found no direct evidence of immunosuppression in month-old calves following two 9-h journeys, but did find increased respiratory disease after the transport.

Transporting animals involves a complex mix of stressors that can affect the welfare of the animals in quite different ways. This makes it difficult to determine the overall effect of transport on animal welfare, and limits the usefulness of attempts to simulate transport. Transport is preceded by collecting animals at the farm and loading them onto the truck. At this stage, the animals may be subjected to rough handling by people, unusual physical exercise in climbing ramps etc., as well as exposure to unfamiliar surroundings and possibly unfamiliar animals. During the journey, the animals may be crowded together, deprived of food and water for lengthy periods, be unable to lie down, subject to falls and loss of balance as the truck brakes and turns, and be subject to temperature extremes with inadequate ventilation. On arrival at the destination, the animals are subjected to further stresses associated with unloading and relocation. Finally, the stresses associated with the journey and the mixing with other animals increase the risks of disease that continue to affect welfare long after the transport is finished (Tarrant and Grandin, 2000; Scientific Committee on Animal Health and Animal Welfare, 2002). These different stressors will impact the animal in different ways, and so their effect on animal welfare will be apparent in different measures. Often, the effects of single components of transport are much easier to assess than the overall effects, but since the different components interact it is dangerous to assume that the effect of transport as a whole will simply be equal to the sum of its parts.

6.2 *Which Aspects of Transport Have Most Effect on Welfare?*

Collecting animals prior to transport and loading them onto a truck can be highly stressful: where the animals are subsequently transported for short distances in good conditions, the loading (and unloading) may be the most stressful part of the transport (Scientific Committee on Animal Health and Animal Welfare, 2002). This conclusion is supported by studies of the changes in plasma cortisol concentrations during transport, that appear to reach maximum values early during the journey, often gradually declining as the journey progresses (Kent and Ewbank, 1983, 1986; Warriss et al., 1995; Grigor et al., 2001). Loading and unloading also can be responsible for increased bruising of animals when they slip, fall, or come into contact with solid objects such as the walls of chutes (Blackshaw et al., 1987; Jarvis et al., 1995). In extensive beef production in countries such as Australia, cattle may be mustered for considerable distances before being loaded on trucks, which can be a major source of stress for cattle (Petherick, 2005).

No aspect of transport is more studied than the space allowance provided to animals. As well as having implications for animal welfare, stocking density is an important economic component of transport, as the fixed costs of the transport (driver, truck, etc.) are spread among more animals. Obviously, the minimum space provided can be no smaller than that taken up by an animal, and most recommendations on stocking density are based on this (Knowles, 1999). Of course, this must be considered as the minimum space allowance, since it does not allow the animals to lie down, which can be a problem for very young calves, or for pregnant or injured animals, or where the journey is long. There is some evidence that adult cattle will remain standing during journeys of up to 20 hours (Tarrant et al., 1992), although young calves (below 21 days) will tend to lie down during even short journeys (Atkinson, 1992). Recommendations for space allowance based on body size also do not allow for the effects of crowding on the ability of animals to avoid aggressive interactions. Formulas for the calculation of minimum space are typically based upon the weights of the animals, as this is often the only measure of size available.

There is some concern that increasing the space allowance during transport may increase the likelihood that animals will lose balance or fall over when the truck brakes or turns. However, research with beef cattle has found the opposite: the number of times beef cattle lost balance or fell, and the subsequent degree of bruising was higher at high stocking densities compared to lower stocking densities. Reducing stocking density further had much less of an effect (Tarrant et al., 1992). The major problem with the high stocking density appeared to be the fact that the steers were less able to adopt their preferred orientation (i.e. perpendicular relative to the direction of travel), suggesting that recommended stocking densities for transport of cattle need to be rethought. As well as floor area per animals, space allowance should consider the volume available for animals in order to ensure adequate ventilation (Scientific Committee on Animal Health and Animal Welfare, 2002), but this aspect of space allowance has not been studied in cattle.

It would seem logical that longer journeys should pose more of a threat to animal welfare than short journeys, and much of the concern with transport hinges on the increasing durations of journeys that animals undergo as a result of changes in rearing and slaughtering practices, especially the reduction in the number of slaughter plants (Knowles, 1999; Swanson and Morrow-Tesch, 2001; Scientific Committee on Animal Health and Animal Welfare, 2002; Petherick, 2005) and increased international trade in live cattle. In general, the duration of the journey is more important than the distance travelled (Knowles, 1999) and longer journeys have more impact on animals, as might be expected. Research has found a greater loss of body weight for longer journeys in beef steers (Warriss et al., 1995; Gallo et al., 2003; Ribble et al., 1995). Some of these effects can be overcome by electrolyte therapy (Schaefer et al., 1997). Other studies show little effect of journey duration. One study showed that the risk of developing pneumonia at the feedlot was independent of the distance transported (Ribble et al., 1995). Physiological responses to transport have been found to be independent of journey duration for beef steers (Warriss et al., 1995) or for 2- to 4-week-old calves (Knowles et al., 1997). An effect of journey duration may be difficult to isolate, since the effects will depend greatly upon how the transport is done (i.e. whether or not food is supplied, whether or not the animals can lay down, the speed and care of driving, etc.). Furthermore, there have been relatively few studies that have examined the effects of journeys of more than 24h.

After long journeys, animals may be placed in lairage either before being slaughtered or before a subsequent journey. The purpose of the lairage is to allow the animals to recover, and food and water are provided. However, the advantages of allowing lairage or rest stops will vary depending on the facilities provided. For example, mixing bulls with steers and heifers at lairage can increase the amount of mounting behaviour that occurs, as well as the subsequent bruising (Jarvis et al., 1995). In general, studies on lairage for beef cattle prior to slaughter show few positive effects for animal welfare (Tadich et al., 2005). Indeed, lairage can be counterproductive; Gallo et al. (2003) found that cattle transported for up to 16h showed greater losses of body weight when kept in lairage and that this weight loss was greater following 24h of lairage than for 3, 6, or 12h. Longer periods in lairage were also associated with poorer meat quality. Fighting of unfamiliar beef cattle during lairage can also be responsible for considerable bruising (Fordyce et al., 1985; Blackshaw et al., 1987). These results throw doubt on the value of lairage for improving animal welfare and Gallo et al. (2003) suggest that animals should be slaughtered as soon as possible after arrival at the slaughterhouse.

It seems likely that the effect of transport on the welfare will vary with the type of truck used, the number of stops and turns, the care of the driving, etc. (Scientific Committee on Animal Health and Animal Welfare, 2002). Most falls and losses of balance by beef cattle occur during breaking, gear changes, and turning (Tarrant et al., 1992). Unfortunately, there is little published research on these topics for cattle. Some information on truck design and ventilation is provided by Tarrant and Grandin (2000).

6.3 *Markets*

Transported cattle will often pass through markets or auctions before continuing on to their final destination. These invariably involve loading and unloading, exposure to novel surroundings and often unfamiliar animals, and increased chance of rough handling with clear detrimental effects on welfare. In beef cattle, the incidence of bruising is higher for cattle that pass through markets rather than for those travelling directly to slaughter (Jarvis et al., 1995) and cattle that pass through markets show greater signs of fatigue, dehydration, and food deprivation (Jarvis et al., 1996). Case studies indicate that passing through a market increases the chance of respiratory diseases in veal calves (Palechek et al., 1987). Alternatives to markets, such as video auctions (Knowles, 1999) or direct marketing would seem to present many advantages for animal welfare.

6.4 *Animals Unfit to Travel*

Much of the concern has focused on animals not fit for transport, including those that are ill or injured, and pregnant, lactating or very young animals (Scientific Committee on Animal Health and Animal Welfare, 2002). This is a topic best dealt with through action rather than through research: this would include adequate inspection of animals before, during and after transport (Scientific Committee on Animal Health and Animal Welfare, 2002), and development of better audit systems and greater accountability. There appears to have been relatively little research directed at this topic, except in the case of young calves.

Many countries impose clear limits on the age at which young calves can be transported, but recently there has been an increase in the transport of 1-day-old dairy calves in some countries, such as the USA (Grandin, 2002; Moore et al., 2002). The transport of very young calves raises obvious concerns as the young animals are deprived of food and water for lengthy periods at an age when the animal is particularly vulnerable. Young calves also spend much more of the day lying down than adult cattle, and transport can result in them standing for unusually long periods (Knowles et al., 1997, 1999). Surveys in the UK have shown that the mortality rates of young calves following transport can be high, and are higher with younger calves (Knowles, 1995). Early studies in Europe found that mortality rates for young calves varied between 1% and 30% (Knowles, 1995). In the USA 10–20% of 1-day-old calves die during or soon after transport (Moore et al., 2002), and smaller calves have a greater chance of dying (Moore et al., 2002). We see these rates of mortality as unacceptable and strongly recommend new action by the cattle industry to address this issue.

7 Slaughter

A chapter on short-term or acute challenges to animal welfare may seem a curious place to discuss slaughter, since its consequences are permanent. Moreover, any effect of slaughter on animal welfare is indeed brief; death removes the problem. However, there is a high degree of concern about how much animals might suffer during slaughter, and most jurisdictions have laws and regulations that serve to promote “humane slaughter”. This is the case even in the USA and Canada that have tended to avoid dealing with animal welfare through legislative mechanisms.

One of the most common questions people have about slaughter is whether the animals know they are going to die. There is no evidence that they do. Furthermore, the actual act of killing may have far less impact on animal suffering than the handling and manipulation that occurs before the killing (Grandin, 2001a, b). The general issues associated with the welfare of animals at the time of slaughter have been dealt with in a number of excellent reviews (Gregory, 1998; Grandin, 2000; Grandin, 2001a, b) and the interested reader is urged to consult these for a more complete picture than we provide. Concerns about animal welfare focus on the issues of how the animals are handled prior to slaughter, how they are restrained while being slaughtered, and whether or not they are conscious at the time of slaughter, with the majority of research focusing on the last of these.

To reduce the suffering during death it is widely accepted that animals should be unconscious, so animals are usually stunned before they are slaughtered. The stunning method used should cause as little suffering as possible, and should ensure that the animal remains unconscious until death. The captive bolt pistol is the most common method of stunning cattle. Application of the bolt either to the front of the head or just behind the horns leads to a rapid loss of consciousness, judged by EEG recordings (Lambooy and Spanyaard, 1981). Even with such extensive brain damage, some animals continue to show evoked responses in their EEG (Daly et al., 1987), although what degree of consciousness this indicates is not certain. Although the captive bolt appears an acceptable way of stunning cattle under ideal conditions, it can be misapplied in commercial settings (Grandin, 2000, 2001b). One concern with captive bolt stunning is that it may lead to contamination of the carcass with brain tissue, increasing the risk of transmission of BSE (Schmidt et al., 1999). Whole head electrical stunning appears to be an acceptable alternative, leading to a rapid loss of consciousness (Bager et al., 1992; Wotton et al., 2000). The main issue here concerns the placement of the electrodes, and the duration and intensity of the current used (Wotton et al., 2000).

One issue that has attracted considerable attention is ritual slaughter, where religious practices require that animals be alive at the time of slaughter. This tends to prohibit use of captive bolt stunning, since this may itself kill the animal. Electric stunning, from which the animal may recover, is acceptable to some religious authorities; however, others require that the animal be conscious. Animals become unconscious 20–126 sec after throat cutting (Daly et al., 1988), but stunning immediately after throat cutting may help reduce this time (Petty et al., 1994). Regardless, as Grandin (2001a, b) points out, the way the animal is handled and restrained prior to slaughter will likely have a greater impact on their welfare than whether or not they are stunned. For example, restraint devices in which the animal is inverted before the throat is cut prolongs the time (by several minutes) required for slaughter and the amount of struggling by the animals (Dunn, 1990).

Under the leadership of Grandin, research into animal welfare during slaughter has shifted away from examination of different techniques of stunning to a focus on auditing the performance of actual slaughter plants operating under commercial conditions (Grandin, 2000, 2006). Grandin has focused upon both the extent that cattle vocalize and the number of animals that are properly stunned before being hoisted onto the bleeding rail (2000, 2001a, b, 2006). Grandin's (2000) audit of US slaughter plants found that in only 81% of the plants were all of the animals properly stunned before being hoisted onto the bleeding rail. The failures in the other plants were due to a combination of poor equipment maintenance and inexperienced operators. However problems were more common with older bulls and cows, possibly because of their thicker skulls. In another audit of commercial beef slaughter plants in the USA, Canada, and Australia, Grandin (2000) scored the number of animals that vocalized during the handling and stunning procedures; at the best plants, less than 1% of the cattle vocalized while in the worst over 15% of the cattle vocalized. A high incidence of vocalization was associated with excessive use of electric prods. In some plants there was no use of electric prods seen and few cattle vocalized. In others, over 95% of the cattle were prodded and vocalization was common. The most common reason for the use of electric prods was a reluctance of the cattle to move into poorly designed races or stunning boxes. Improving the design of the equipment resulted in a significant decrease in the incidence of vocalizations. Interestingly, the line speed was not a factor influencing vocalizations. More recent audits have shown distinct improvements in most measures of cattle welfare (Grandin, 2006). These studies show how some simple measures can be used under commercial conditions to identify precise problems with the operations of the slaughter plants. The incidence of bruising on carcasses may also provide a way of assessing handling techniques and pre-slaughter facilities (Jarvis et al., 1995) although these measures are more time consuming to take.

8 Conclusions

The topics covered in this chapter have received considerable attention, both from those concerned with animal welfare and from researchers. In part this is because the effect on the animal's welfare is often obvious: few witnessing a calf bellowing during branding would doubt that it is suffering from pain. The effect of painful procedures on animal welfare has proved relatively easy to determine, perhaps because the effects themselves are acute and more easily measured. As this chapter discusses, we have relatively good knowledge of the degree of suffering caused by these procedures and the effectiveness of the pain mitigation strategies or alternative procedures that have been proposed. Animal welfare science has also made much progress in providing direction for improving the welfare of animals affected. For example, we reviewed research showing that tail docking of dairy cattle can likely be abandoned as a practice (it does not provide the anticipated benefits in udder health), that the pain due to dehorning can be reduced using a simple procedure that also makes the chore easier for producers to perform, and that weaning stress in beef calves can be much reduced by allowing calves continued contact with the cow even when the calf can no longer suckle. Although this represents important progress, these short-term procedures likely have a limited overall effect on animal welfare simply because they are short term. In the following chapter we turn to the effect of housing on animal welfare, where the effects are of much longer duration.

Chapter 6
Housing for Adult Cattle

1 Introduction

In this chapter, we look at the effect on animal welfare of the way that cattle are housed. Housing conditions affect animals throughout their life. Images of modern housing systems, such as veal calves in individual veal crates, have figured prominently in public concern about the welfare of farm animals. Many modern housing systems involve keeping animals indoors, in a restricted space, at high density, and often separated from other animals. In this chapter we concentrate on welfare issues that arise primarily from the physical and social (e.g. group versus individual housing, stocking density) aspects of the environment. Housing factors associated with feeding, such as feeder design, etc. are also discussed here. For convenience, we discuss housing for adult cattle (almost exclusively dairy cows) in this chapter, and housing for growing animals, including veal calves, dairy heifers, and beef cattle raised for meat production, in Chapter 7.

1.1 Diversity of Housing Systems

Cattle are housed in a wide range of ways, especially from a global perspective. The method of housing varies according to the reproductive state of the animal and geographical area, varying from country to country, and from region to region within countries (Figure 6.1). Even within a geographical region and for one type of animal, farmers still have a wide choice of housing techniques. The choice of housing is often made depending on the local climatic conditions, cost of building materials and construction, cost of labour, and availability of local materials. The way that the choice of housing affects animal welfare will also depend greatly on local factors.

Figure 6.1 From a worldwide perspective, adult dairy cattle are housed in a wide range of ways. These cover the small-scale, extensive systems found, for example, in mountainous areas of Europe, where the cattle are, effectively, free ranging for part of the year, to the fully confinement systems found, for example, in North America, where the animals have little or no access to the outside

1.2 Difficulty Comparing Housing Systems

An obvious question to ask is how the type of housing affects animal welfare. Do cows that have regular access to pasture fare better than those housed permanently indoors? Are cows living in groups better off than those in individual stalls? Do free stalls allow a better level of welfare for dairy cows than straw yards? In some cases, research has tried to compare housing systems per se. However, comparing housing systems that vary in many ways can be frustrating work. Systems often succeed or fail because of the details, such as the ways in which they are managed and the specific ways in which they are configured, which makes it hazardous to generalize from any one farm to the housing system in general. Housing systems differ in many respects, e.g. degree of social contact the cows have, indoor versus outdoor housing, spatial allowances, etc. and confounding of the various factors makes it very hard to determine which differences between housing systems are important. It is also difficult to determine if the differences between housing systems are due to intrinsic aspects of the housing system, or whether they result from a difference in factors that could be changed. For example, are differences between individual and group housing due to the presence of other animals (an intrinsic factor) or are they due to the different space allowances that tend to occur between group and individual housing? These problems are exacerbated by the lack of detail in the published research reports regarding the specific properties of the housing system, such as floor type and ventilation. All this makes it difficult to be confident that conclusions can be generalized to other farms using the same type of housing system. Moreover, housing systems can be expensive to install, so comparisons are generally unreplicated and one example of one system is compared with one example of another. Although such case studies can provide valuable insights, they are frequently presented as if they were replicated experiments.

Another research approach is to examine individual housing factors, such as floor type, that can vary within a housing system. Because confounding variables are generally avoided, both the science and the interpretation of the science are more straightforward. For this type of work, the aim is simply to determine how a given system can be improved, rather than evaluating the overall system.

2 Types of Housing for Lactating Dairy Cows

In many countries, lactating cows are kept at pasture, which may be permanent (typical in Australia and New Zealand), or seasonal (as in the cooler parts of North America and Europe). In the USA in particular, lactating dairy cows may also be kept in large outdoor enclosures with either a dirt or concrete surface (dry-lots), but with no access to pasture (Figure 6.2). Indoor housing, which may be seasonal or permanent (as in some zero-grazing systems), varies from loose housing in large bedded packs (Figure 6.3), to tie stalls (Figure 6.4), or free stalls (Figure 6.5). In tie stalls (also called stanchions), cows are restrained in a space that provides a lying and standing surface, as well as individual access to food and water. The cows are typically milked in the stall. In loose housing, cows have free access to an area for lying, which may consist either of stalls (also called cubicles) or a deep-bedded pack (commonly straw yards). The cows generally have free access to other areas

Figure 6.2 Many people imagine that cows spend most of their time at pasture. This is true for many countries, such as Australia and New Zealand. However, increasingly cows in North America and in some European countries have limited or even no access to pasture. Even when cows are allowed outside, they may be restricted to a dry-lot (right), which often has a mud or dirt floor. The reduction in cows' access to pasture has raised concern about their welfare, and in some European countries, legislation now requires that cows do have access to pasture. Research has shown that in general, cows with access to pasture are in better health than those that remain indoors. However, the difference between indoor housing and outdoor housing depends upon many factors, such as the quality of the ventilation, type of bedding used, etc. In addition, cows at pasture are not free of welfare problems. They can suffer from inclement weather, increased parasite load, and inadequate food, if stocking rates are too high

Figure 6.3 In one simple type of housing system, the cows are kept loose in a bedded area, which often has straw or wood shavings. Such systems allow the cows considerable freedom of movement and provide a comfortable surface for the cows to rest and walk on. However, much labour is needed to keep such systems clean and research has tended to find a higher incidence of mastitis in cows kept this way

for eating and drinking, and are usually milked in a central milking parlour. In general, tie stalls and free stalls are most common, but the relative use of different types of housing systems varies greatly from one region to another. In the USA, 31% of lactating cows are kept in free stalls and 53% in tie stalls (USDA, 2002). In northwest Germany, 57% of cows are in free stalls and 43% in tie stalls (Buenger et al., 2001). Automated milking systems, in which cows are milked by a robot, are becomingly increasingly common, especially in Europe. Although cows in automated milking systems are generally housed in free-stall systems, the management requirements of these systems can have quite different effects on the welfare of the animals (see Section 4 of this chapter).

An obvious first question concerns how well these different housing systems compare in terms of the effect on animal welfare. Tie stalls offer an animal a guaranteed place to lie down, ready access (with minimal competition) to feeding space and drinkers and the possibility of being fed a diet tailored to its individual needs. However, tie-stall housing limits how much the animal can move. If cows are also milked in the stall they may be tethered for months on end. In addition, there is a lack of opportunity for close physical contact between animals, combined with an inability to escape completely from aggressive neighbours. When tied in the stall, the animal cannot turn around and may not be able to groom all parts of its body.

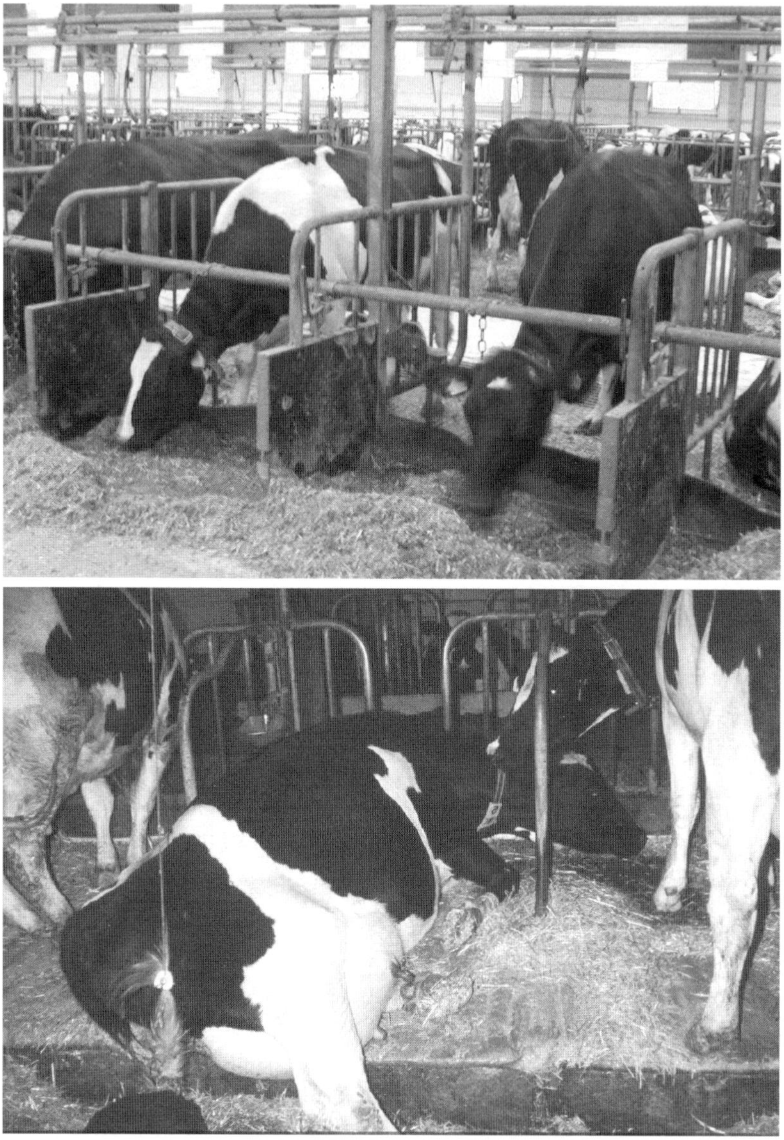

Figure 6.4 In tie-stall housing, each cow is kept tethered in its own stall, where it both rests and eats. This type of housing may be used only during the winter months, or all year-round. Cows are generally milked in their stalls. Cows in tie stalls do not have to compete over feed and can be fed an individual diet. However, the lack of opportunity for movement raises concern about animal welfare. In addition, badly designed stalls, for example, those that are too small, may interfere with resting behaviour

Figure 6.5 Free-stall or cubicle housing is probably the most widespread type of housing for lactating cows housed indoors in northern European and North American countries. The cows are provided with stalls in which to lie and separate feeding areas, between which they can move freely. Although free stalls might appear superior to tie stalls since they allow the cows to have greater freedom of movement, they appear to lead to an increased incidence of lameness. In contrast, free-stall housing is associated with a lower incidence of mastitis than straw-pack system. However, free-stall housing comes in a variety of styles and the effect of free stalls upon the welfare of the cows will depend upon many details of the system, such as the size and number of resting stalls, the types of bedding used, the feeding system, etc.

Loose housing with a bedded area overcomes the problem of mobility and allows physical contact between animals, but the animals are still housed in a restricted space and may not be able to escape completely from an aggressive dominant cow. Individual feeding is rarely possible so animals may need to compete for limited feed space, and they are required to eat a diet tailored to the average cow rather than to the individual. Cleaning of such systems can be difficult and the animals risk being dirty. Providing individual stalls in a free-stall system makes cleaning easier and the animals are generally cleaner. The stalls also provide a means for the animals to escape from aggression but there may only be a limited number of places to lie down creating another source of competition.

A number of studies have tried to directly compare the welfare of lactating cows in these different systems with a special focus on the frequency of health problems. A fairly consistent finding is that lameness and hoof problems are most prevalent in free-stall housing, although a great range in prevalence can be found within this system (Espejo and Endres, 2007). A higher occurrence of hoof problems has been found in free stalls than in straw yards in studies from the UK (Livesey et al., 1998; Whitaker et al., 2000; Webster, 2002; Haskell et al., 2006) and the Netherlands (Somers et al., 2003) and a higher occurrence of hoof problems in free stalls compared to tie stalls has been found in Sweden (Bergsten and Herlin, 1996) and the USA (Wells et al., 1999). The disadvantages of free stalls for hoof health appear to be most apparent around the time of calving: housing cows in bedded packs for the first 8 weeks of lactation greatly reduces the incidence of hoof lesions (Webster, 2002).

In contrast, udder health is generally worse in bedded pack systems than in free stalls (Faye et al., 1997; Peeler et al., 2000; Whitaker et al., 2000; Fregonesi and Leaver, 2001), although allowing the bedded pack to compost may reduce this effect (Barberg et al., 2007). There are fewer comparisons of the prevalence of mastitis in tie stalls: a survey of Norwegian dairy farms found a higher incidence of mastitis in tie stalls than in free stalls (Valde et al., 1997) and Swedish farmers who switched from tie stalls to free stalls experienced a reduction in the rate of clinical mastitis (Hultgren, 2002). Because of the differing effects on hoof health and udder health it is difficult to say which type of system is better for animal health.

Simply finding differences between housing systems in the incidence of health problems does not really tell us very much. First, we do not know what the main causal factors are contributing to health problems. For example, free stalls and tie stalls differ in how much exercise cows get, and the degree of social contact. Barns with free stalls often differ in design from barns with tie stalls, and the size of the stalls usually differs. Cows in free stalls usually stand on concrete floors, which exert more physical stress on the hooves than does straw or pasture, but the cows may also be spending more time standing on wet flooring, and competition over stalls may reduce lying time. Without more detailed studies, we do not know which of these factors is most important. Furthermore, how well cows do within any one type of system will depend greatly upon the details of that particular system, such as what size stalls were used, what type of flooring was present, and how the system was managed. For example, Fregonesi and Leaver (2002) suggested that problems of udder health in bedded packs were due primarily to problems of hygiene, which could be better controlled by improving bedding management. Some studies have examined behavioural differences when cows are kept in free stalls, tie stalls, or on bedded packs, but the problems mentioned above do not allow us to draw firm conclusions about the relative level of welfare using these measures. Since the behaviour of the cows will depend very much on the details of the systems, it is difficult to make generalizations about the behaviour in different systems. This is apparent in the contradictory results that are obtained. For example, Krohn and Munksgaard (1993) and Munksgaard and Simonsen (1995) reported that cows in tie stalls with some bedding or rubber mats spent longer lying down than cows either in a pen with concrete slatted floors or in bedded packs with access to pasture. In contrast, Haley et al. (2000) compared cows kept in large well-bedded packs versus narrow tie stalls with concrete surfaces, and found that cows spent approximately four more hours per day lying down when kept on the large well-bedded pens. The difference between the studies is likely due to the presence or absence of bedding in the tie stall: thus the cows' behaviour was more affected by the type of lying surface than by the type of housing system per se.

For these reasons we suggest that direct comparisons between different types of housing systems are generally not very informative. This is particularly true for behavioural measures, which we suggest are more likely related to the specific design features of the housing system, but is also true for health problems. Clearer assessments of the effects of housing on animal welfare come when we consider specific design features of the environment, such as flooring type, stall, and feeder design.

3 Access to Pasture

Most people imagine that cattle spend their days grazing on pasture. While this is an accurate depiction of cattle in many parts of the world, in Europe and North America particularly, dairy cows are increasingly housed indoors or on outdoor dry-lots with little or no access to pasture (zero-grazing). A survey of dairy farms in the USA found that less than 25% of lactating dairy cows had access to pasture (USDA, 2002). Zero-grazing management is also increasingly practised in some developing countries (e.g. Gitau et al., 1996), likely because higher milk production (though not necessarily more profitable production) can be achieved with controlled, grain feeding (Washburn et al., 2002). To some, this lack of access to pasture is seen as a threat to the well-being of dairy cows. Some countries have passed legislation making it mandatory to provide dairy cows with some access to pasture. However, what evidence is there that the welfare of cattle suffers if they cannot access pasture?

We must say at the outset that this is not an easy question to answer. As we discuss later, a number of studies report a higher incidence of welfare problems for lactating cows in zero-grazing systems, but what can we conclude from this? First, farms vary greatly in how much access to grazing the cows have. In some systems, cows are kept at pasture all of the time. In others they have access at certain times, for example, during the summer or during the day. In true zero-grazing systems, cows have no access to pasture. Thus, we need to know whether the effects on animal welfare are due to the complete absence of access to pasture or the occasional use of indoor housing. Furthermore, zero-grazing systems differ from pasture-based systems in many respects. For example, air quality may be different, as will the types of surface on which the cows walk and lie down, the stocking rates used, and even light to dark cycles. Cows on pasture tend to produce less milk and eat a different diet than cows without access to pasture (White et al., 2002). It is difficult to determine which of these factors may be responsible for any difference in welfare found. Furthermore, there is always the possibility that some modification of indoor housing systems (e.g. using a different type of ventilation or flooring) would raise the welfare of cows in zero-grazing systems to a similar or higher level than found at pasture. For these reasons, any results concerning the effect of grazing on the welfare of cattle must be interpreted with care and we must not assume that providing cows with access to pasture will automatically improve their welfare, or that a high level of animal welfare cannot be achieved in zero-grazing systems. Table 6.1 lists some of the potential threats to animal welfare from indoor and outdoor housing systems.

A number of epidemiological or experimental studies in different countries have found that lactating cows without access to pasture suffer from a higher incidence of a variety of maladies including mastitis (Waage et al., 1998; Barkema et al., 1999a, b; Washburn et al., 2002; White et al., 2002), metritis (Bruun et al., 2002), *Salmonella enterica* infections (Veling et al., 2002), dystocia, ketosis, and retained placenta (Bendixen et al., 1986). Thus, zero-grazing can be considered as a risk factor for all of these maladies. The effect can be quite large: in one experimental comparison, Washburn et al. (2002) reported that cows without access to pasture had 1.8 times as many clinical cases of mastitis and were eight times more likely to be culled for mastitis as cows at pasture. The studies do not allow us to identify the actual cause of the difference but providing cows with access to an exercise area or dry-lot does not seem sufficient (Bruun et al., 2002; Washburn et al., 2002).

Table 6.1 Some of the main potential threats to animal welfare from outdoor and indoor housing systems for lactating dairy cows. Although some access to pasture has generally been found to improve the welfare of cows, outdoor housing has some specific risks. Since the welfare of animals in any type of housing system will depend upon the details of the system and upon management, it is difficult to make generalizations about which type of housing system is better for animal welfare

Outdoor housing	Indoor housing
Inclement weather	Insufficient space
Parasites	Uncomfortable flooring
Poor walking tracks	Poor drainage
Long distances to walk	Uncomfortable stalls or lying areas
Poor-quality pasture	Poor ventilation
Lack of human supervision	Heat stress
Predation	Increased disease transmission

The most commonly reported welfare problem associated with restricted grazing is lameness (see Chapter 2). A large epidemiological survey of 4,516 dairy farms in the USA, found that a lack of access to pasture in winter was a significant risk factor for a high incidence of digital dermatitis, and that providing access to a dry-lot was not sufficient to overcome this risk. Nearly four times as many farms on which cows had no access to pasture had a high (>5%) incidence of dermatitis than farms on which cows were kept only on pasture (Wells et al., 1999). A smaller study in Chile (Rodriguez-Lainz et al., 1999) found supporting results: cattle housed permanently at pasture had a lower risk of digital dermatitis than cows housed in buildings for some of the year. This suggests that occasional or seasonal exposure to confinement housing may be as much of a risk factor as the complete absence of grazing (see Phillips, 1990). In countries as large as the USA and Chile, the type of housing used varies greatly depending on the region, and climatic differences could conceivably account for some of the effects of housing. However, similar effects of a lack of grazing have been noted in smaller countries, which do not have such climatic variation. An epidemiological survey of 86 dairy farms in the Netherlands (Somers et al., 2003) reported that all types of hoof disorders were more prevalent in cows in zero-grazing systems than among cows with some access to pasture. Again, the difference was substantial: the prevalence of severe cases of sole haemorrhage was twice as high with zero-grazing compared to other housing systems. Even where cows had some seasonal access to pasture, hoof disorders (but not digital dermatitis) were more prevalent during the period of indoor housing compared to the end of the period of summer access to pasture. In Kenya, where some dairy farmers use zero-grazing, the prevalence of lameness is lower where cows have some access to pasture (Gitau et al., 1996). Recent work indicates that even a short period of access to pasture can reduce lameness. When a matched sample of cows were either kept in a free-stall barn or moved out onto pasture, cows on pasture showed a dramatic improvement in gait over just 4 weeks of exposure (Figure 6.6; Hernandez-Mendo et al., 2007). However, some studies report an increased risk of digital dermatitis when cows do have access to pasture (Holzhauer et al., 2006), emphasising the difficulties of comparing housing systems per se.

Although, these studies indicate that lameness and hoof problems are less common for cattle with some access to pasture, they do not allow us to isolate the cause of this difference. Cows in indoor housing are more likely to be standing in manure and concrete, and eating more grain than cows at pasture, all of which increase the likelihood of lameness (see Chapter 2). One important goal for future research is to determine if indoor housing systems can be improved so as to obtain a similar level of lameness.

However, we should not assume from these findings that cattle at pasture are free of welfare problems. Cattle at pasture can be exposed to inclement weather, increased parasite load, inadequate energy intake, and high competition for feed if stocking rates are too high. Even at pasture, cattle can compete for food (e.g. Phillips and Rind, 2001) so that the degree of social competition can be one factor that influences the relative advantage of indoor versus outdoor housing. In general there is little information available on the welfare problems associated with outdoor

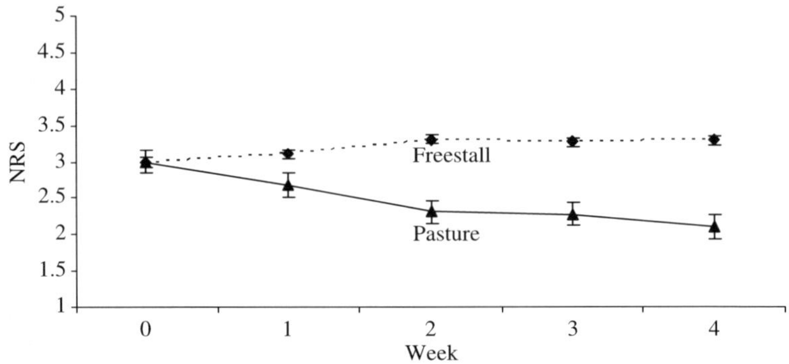

Figure 6.6 A matched sample of dairy cows, previously all housed in a free-stall barn, were randomly allocated in groups (n = 9 per treatment) to either continued housing in free stalls or continued housing on pasture. Changes in gait were assessed using numerical rating score (NRS ± SE) for 4 weeks after treatment was imposed, and showed a significant reduction in lameness (i.e. lower NRS) by the end of this period (from Hernandez-Mendo et al., 2007.)

rearing of cattle, and key issues are well summarized in two reviews on this topic (Hemsworth et al., 1995; Petherick, 2005). The effects of heat stress can be especially important for cows kept outside, especially when shade is not available. Much less is known about the effects of cold, although one study (Tucker et al., 2007) compared behaviour and cortisol responses in cows kept indoors or outdoors under wet and windy conditions in New Zealand during the winter. Cows kept outdoors spent less time lying down (likely because of the wet lying surface) and experienced higher cortisol levels. Moreover, the negative effects of the wet, windy conditions were most evident for cows that were low in body condition (i.e. relatively thin), indicating that poor food availability, typical of winter pasture, may aggravate the welfare effects of harsh climatic conditions. Clearly, environmental conditions such as extreme cold and hot weather, wind, and rain will play a significant role in the welfare of animals and new research is urgently required in this area.

4 Robotic Milking Systems

Over the past few years there has been growth in developed countries in the adoption of automated milking systems in which cows were milked using a robot (Figure 6.7). The adoption of this technology is driven by high labour costs and the desire for improved working conditions (fewer working hours), as well as the increased milk production that can result from more frequent milking. With the rising public distrust of farming practices that resemble "industrial" agriculture, public acceptance of robotic milking systems will depend partly on the impact of this technology on the welfare of the cattle. In robotic milking systems, milking is often

Figure 6.7 Automated milking systems are becoming increasingly popular in a number of countries, especially in northern Europe. In these systems, the milking machine is attached to the udder of the cow by a robot (left). The cows enter the automated milking system on their own in order to obtain a food reward. There is no evidence that such systems pose a threat to the welfare of the cows; in fact there may be a number of potential advantages, but this depends on how they are managed. Although these systems are sometimes called "voluntary", cows will often need to wait (right) to get in if the group size is too large. Perhaps the biggest concern, however, is the greater difficulty in detecting illness among the cows due to the reduced contact between people and cows that normally would occur during milking. This has led to an interest in automated health monitoring

described as "voluntary", perhaps providing an advantage for the animal. In reality, social factors present in a herd of animals (especially social dominance) can greatly influence when and how often cows can attend the robot milker. While many of the factors already discussed that affect animal welfare in ordinary free-stall housing will also be important in robotic milking systems, there are a number of welfare concerns that are specific for this type of technology.

One issue concerns the actual process of being milked. Is it more aversive for the cow to be milked by a machine rather than by a person? In one of the few studies of the physiological and behavioural responses of cattle to robotic milking, Hopster et al. (2002) compared primiparous cattle being milked in a robotic milking system with those being milked in a conventional milking parlour and found no evidence of a difference in the stress response. Cows milked by the robot tended to have lower blood concentrations of epinephrine and norepinephrine, suggesting lower sympathetic nervous system activation and lower heart rates. Although blood cortisol concentrations were slightly higher in cows milked by the robot, the values were within normal range and were well below those usually seen following acute stress. Cortisol concentrations rise normally during milking, and in this study cortisol concentrations were correlated with milking times, suggesting that the higher values in the robotic milking systems may have reflected the changed pattern of milking. There were no obvious behavioural signs of agitation during milking, and residual milk, often a sign of stress induced failure of milk ejection (see Chapter 3) did not differ between the two types of milking. Other studies have confirmed the lack of evidence that HPA axis activity is higher in robotic milking systems (Gygax et al., 2006). In general, there is little evidence of disturbed milk ejection or suppression of oxytocin release during milking in robotic milking systems.

One obvious consequence of robotic milking systems is that the opportunities for contact between people and cows that occur at milking are no longer there. This could affect animal welfare both by reducing the opportunities to check the health of the animals and by altering the relationship between the animals and the stockperson. There is little doubt that in robotic milking systems there is less opportunity for the stockperson to check the health of the animals. This is most apparent for diseases of the mammary gland, such as mastitis, but also for other maladies, such as lameness, since the twice daily moving of the animals to the milking parlour did, in theory at least, allow the stockperson to watch how the animals were walking. The reduced opportunity to check the health of cattle is probably one of the biggest concerns about robotic milking systems in terms of animal welfare, and has generated considerable interest in finding automated ways of monitoring animal health (e.g. de Mol and Woldt, 2001; Neveux et al., 2006; Pastel and Kujala, 2007). A second concern is the effect of the reduced contact between people and the cows on the human–animal relationship (see Chapter 9). One suggestion is that the cows may become more fearful of people (and hence more difficult to handle) since the twice-daily opportunities for contact between people and cows at milking are no longer there. Unfortunately, no studies have addressed this issue.

5 Stall Design

In this and the following sections we turn from a comparison of complete housing systems to examine the effects of specific features of the system that may influence animal welfare. Most of the research to date has focused upon improving free-stall systems and has focused upon the design of the stalls, and, to a lesser extent, the surfaces on which the cows walk and stand.

In free-stall and tie-stall systems, the level of animal welfare will depend upon the design of the stalls provided to the animals. The design features that make a good stall will vary somewhat between free stalls (where the stall is primarily a place to lie down) and tie stalls (where the stall is also the place where the cow must stand, eat, drink, and often be milked). Most research has focused upon the surface of the stall, whether bedding is used, and the dimensions of the stall. We deal with each of these in turn. First, however, we discuss some of the research techniques that have been used, and some of the pitfalls associated with each.

5.1 Measures of Cow Comfort

To assess the effects of barn design and management on cow comfort, researchers have generally used measures of both health and behaviour. Health measures have included measures of injuries, lameness, and udder health (which is particularly important for assessing the effect of bedding). Behavioural measures have included preference tests, in which animals can choose between two or more options, as well as observations of animals, when they are housed with only one option. Although the general issues regarding the use of indicators of animal welfare have already been discussed (Chapters 2 and 4), there are a number of issues that are specifically related to the assessment of stall design that we address later.

5.1.1 Health Measures

The health measures that are most relevant to stalls are those relating to the udder and lameness. The incidence of mastitis, particularly the environmental forms of infection, gives some information as to the degree of contact between cows and bacteria on the surface of the lying area. The incidence of contagious forms of mastitis is more likely affected by other factors such as sanitation of the parlour equipment and milking routine.

The incidence of lameness can provide some information about the adequacy of stall design, and several studies have shown a link between the design of the stalls and the incidence of hoof problems or lameness (Leonard et al., 1994; Faull et al., 1996; Haskell et al., 2006; Espejo and Endres, 2007). However, the relationship between lameness and the design of stalls is complex, and there are difficulties in using lameness or hoof health to assess stall design per se. In free stalls, the link between stall design and lameness occurs most probably because uncomfortable stalls will result in the cow spending more time standing, but the effect will depend also on the nature of the surface that cows use for standing (see Section 7 of this chapter). In tie stalls, this relationship is more complex, and likely results from a combination of the degree of wetness of the floor and changes in the time that the animals spend lying down. Since both udder health and hoof health can be strongly affected by factors other than the design of the stalls, their incidence does not allow us to pinpoint the precise nature of the problem.

Poorly designed stalls are likely to increase the risks of other injury to cows; for example, inadequate flooring can increase injuries to the knees and hocks (Figure 2.4), and stalls that are too small can increase the chance that cows will hit the bars of the stalls when rising or lying down. As described later, these types of injuries are more closely related to stall design than to other aspects of the housing and so are more useful than measures of mastitis or lameness for assessing the adequacy of the stall design. However, to make good use of injuries as a measure of cow comfort, we need to be aware of potential shortcomings with both how injuries are assessed and the interpretation of these data.

Injuries to the leg have been evaluated using qualitative methods of assessment (e.g. Weary and Taszkun, 2000) or quantitative measurements such as surface area of hair loss (Mowbray et al., 2003). The quantitative measurements have the advantage of being more repeatable, and more amenable to parametric statistical analyses, but taking these measures can be much more time consuming. The choice of the method of assessment should ultimately depend on how well it reflects the way that the injury actually affects the animal, either in terms of the pain experienced, or in predisposing the animal to other injuries, infections, or physical impairments such as abnormal gait (see Chapter 4). Unfortunately, for these leg injuries, and for many other types of injuries such as hoof lesions, little or no research is yet available to establish these links.

5.1.2 Preference Tests

An obvious first question to ask in creating better housing for cattle is what option do the cows themselves prefer? The advantages and disadvantages of preference tests have been discussed in Chapter 4, but there are some issues that are specifically related to cattle housing that we discuss here. Cows' preferences are of more importance in free-stall housing than in tie stalls since in the former the cows choose whether to lie or stand in a stall or elsewhere. One problem with free stalls is that cows sometimes choose to lie down in one of the alleyways instead of the stall, likely increasing the risk of mastitis (Kjoestad and Simensen, 2001). Hence, stalls that are designed to fit cows' choices are more likely to lead to appropriate use of the stalls by the cows. In tie stalls, cows have no choice about where they lie down, and so the value of choice tests comes primarily from whether or not they are able to make predictions about other effects of the stall on cow welfare.

Preference tests can be especially useful as a first phase in identifying features of housing systems that are important to the animals and can be a powerful source of insight into how cattle perceive aspects of their environment and how they rank the various options provided. However, designing and executing tests that avoid the pitfalls described in Chapter 4 requires thought and effort. The cost of building housing for lactating cows means that researchers normally use facilities that are already available. For example, Herlin (1997) compared cow preferences for three types of stall flooring by offering a group of 18 cows six stalls of each type. However, in an experiment like this there are two problems. Most obviously, the extent of preference may be underestimated as not all animals can chose the same option. Second, preferences of individual animals may not be independent (some cows may chose to associate with or avoid certain animals). Providing sufficient stalls to ensure that each cow in the group can have a free choice is an advantage, but this does not overcome the problem of social influences. One way to avoid these problems is to house cows individually, so that the animal's choices are independent of the choices of other cows and all cows have equal access to all options (e.g. Tucker et al., 2003; Figure 6.8).

5.1.3 Observations of Behaviour

An alternative means to assess stall design is to study the behaviour of the cows when they have access to only one type of stall. Since the primary purpose of free

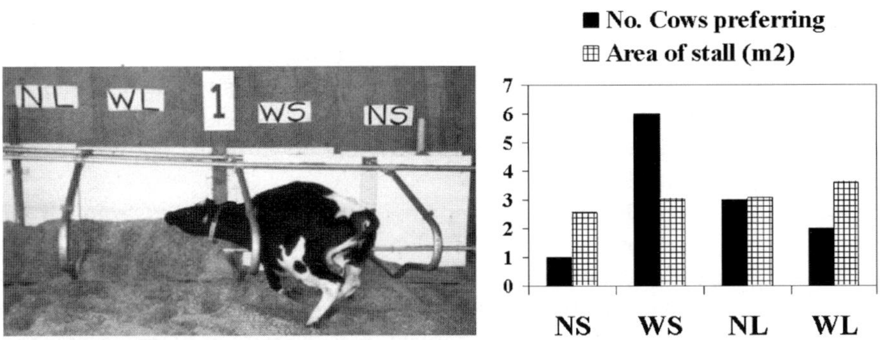

Figure 6.8 Tests of cows' preferences for certain stalls work best if each cow's choices are independent of those of other cows. Each cow must be able to choose whichever option it wishes. Tucker et al. (2003) tested cows' preferences for two stall lengths (S = short; W = wide,) and two stall widths (N = narrow; L = long). The design of the experiment ensured that cows' choices were independent. Cows were housed individually so that each cow was free to choose whichever stall it preferred and the choice of each cow would not be affected by the choice of other cows. The results showed that cows' preferences were affected more by the width than by the length or the total area of the stall

stalls is to provide a place for cows to lie down, observations have typically focused upon the resting behaviour of the cows.

Fine-grained studies consisting of detailed measures of behaviour, such as the specific movements associated with lying down or standing up can provide useful information about the adequacy of the stalls. Cattle get up and lie down in a quite stereotyped pattern of movements. Several reviews have described these basic behaviours (e.g. Lidfors, 1989; Anderson and Zurbrigg, 2003), and provide insights into how to measure the adequacy of stall design. Prior to lying down, cows appear to "inspect" the floor, which is apparent when the animal lowers its head and moves its head back and forth in a swinging motion. A reluctance to lie down may be apparent in an increase in the duration of this inspection phase (Müller et al., 1989). The actual process of lying down occurs in a series of three movements: cows begin by bending the front legs, then fall onto the front knees, and finally let their abdomen and hindquarters fall backwards (see Lidfors, 1989). A reluctance to lie down can be apparent in an incomplete sequence of behaviours, for example, the animal may place its weight on one knee, but instead of continuing the process, may then immediately rise again (Müller et al., 1989). Standing up begins with a lunge forward and upward with the head during which the animal rises onto its knees and breastbone. The animal then extends its head and neck upward while rising onto its hind legs, and then completes the movement by extending first one foreleg, and then the other. During this sequence the cow uses her front knees as a fulcrum, taking the weight off her hindquarters while she rises (Lidfors, 1989). The fact that during both getting up and lying down, the front knees bear a considerable portion of the cows' weight means that hard stall flooring is likely to increase the risk of injury to the front knees. Given the large weight of an adult cow, it is reasonable to suppose that the movements used when getting up and lying down have evolved as the most ergonomic methods of lifting the cow's weight so as to minimize the risk of injury to muscle and tendons. If this is the case, then aspects of stall design that inhibit or prevent a cow from getting up and lying down in the normal way are likely to increase the risk of injury.

One method of establishing appropriate stall dimensions is to examine the amount of space that cows take up when lying down or standing up. Ceballos et al. (2004) has used kinematic software and video recorded movements to accurately calculate displacements and velocities in three dimensions while cows lie down. These kinematic techniques are derived from the field of biomechanics that applies mechanical laws to the quantification of animal movement and forces. Ceballos and colleagues found that the movement envelope of cows was similar when lying down in an open pen versus a large free stall, but more work is needed to determine the effects of common stall and pen design on these lying and standing movements.

An alternative approach to assessing stall design is to observe animals when they are resting: such measures can include data on the actual amount of time the animal spends lying down. A reduced time spent lying because of inadequate stall design has been found to be associated with an increased incidence of hoof problems (Leonard et al., 1994; Faull et al., 1996). To determine which measures of usage are most appropriate in assessing cow comfort, it is important to develop an understanding of how a cow behaves when she is comfortable. Haley et al. (2000) used a simple comparison between a housing system considered "high comfort" (a large pen or box stall with mattresses) and one considered "low comfort" (a tie stall with concrete flooring). They measured many behaviours including lying, standing, and eating times, the number of times the cows stood up, and various leg and head positions during lying. Lying times were four hours longer and cows stood up and changed positions more often in the high-comfort housing. Cows also spent more time standing without eating in the low-comfort stalls. This study provides some insight into behavioural measures likely to change if a cow is uncomfortable, namely, time spent lying and standing, and the number of times she changes position between lying and standing.

5.1.4 Adequate Sampling for Behavioural Observations

Given that the frequency and duration of lying and standing are useful indicators of comfort, what is the best way of actually measuring these? The gold standard is to observe animals continuously for several 24-h periods, but this is labour intensive. In some cases less intensive observational methods can be used, but the accuracy of these methods will depend upon the nature of the behaviour. Characteristics of a behaviour that are likely to affect such accuracy include the amount time cattle spend performing the behaviour, the number of times they perform the behaviour, and the consistency of the behaviour over time, both within a 24-h period and across longer periods.

Intensive 24-h sampling using time-lapse video is difficult for many applied researchers studying commercial dairy farms, and some authors have suggested using much more convenient live sampling over limited time periods. For example, Cook et al. (2005) describe a number of indices that might be used during a one-time walk through of the barn: proportion of cows lying (number of cows lying in stalls/total number of cows in the group); "stall use index" (number of cows lying in stalls/total number of cows in the group not eating); and the "cow comfort index" (number of cows lying in stalls/total number of cows either lying or standing fully or partially in the stall). The question is whether these measures or related indices provide an accurate estimate of the behaviour of interest? The study by Cook et al. (2005) showed that none of these indices were able to accurately predict 24-h lying times as measured continuously from video, and were much affected by the time of day they were collected. These results indicate that the sampling frequency may be inadequate to reflect 24-h usage; namely, the time that the sample is taken may poorly reflect usage at other times. In addition, the location that the sample is taken may poorly reflect usage at other places in the barn.

Even when data are collected in the most complete and rigorous manner, correctly interpreting changes in the behaviour can be a challenge. For example, the time an animal spends engaged in one behaviour, such as lying down, will depend on the alternatives available to the cow and the other demands upon her time. For example, it is well known that high-producing animals have high metabolic requirements, and spend more time eating than low-producing cattle. This time spent eating seems to be at least partly at the expense of lying time, as higher-producing cattle also spend less time lying down (Fregonesi and Leaver, 2001). The effect of alternatives available to cows can also be important. For example, providing rubber flooring elsewhere in the pen increases the time that cows spend standing on this surface, and reduces the time they spend lying down in the free stall (Fregonesi et al., 2004; Tucker et al., 2006; Flower et al., 2007; Figure 6.9).

In both the case of the high-producing cow and the animal with access to rubber flooring, simplistically equating lower lying times with reduced comfort would not be advisable. Well-designed experiments and within-cow comparisons of treatment differences can help avoid such problems, but interpreting differences among farms is difficult.

The use of multifactorial indices can also be especially problematic. For example, the "cow comfort index" is the ratio of cows lying in stalls to cows lying and standing in stalls. Thus this measure may be affected by differences in barn and stall design that affect how much time cows spend lying down, and how much time that they spend standing either fully or partially in the stall. Cows require comfortable areas to lie down and to stand, but changes in stall design can affect these two behaviours differently. As reviewed later, changing stall surface and bedding depth can have strong effects on lying times, with little effect on standing, but changes in neck rail positioning affect standing with little effect on lying behaviour. Thus both aspects of design can affect the cow, but will have very different effects on indices based on the ratio of lying to standing. The take home message is that such measures must be treated with scepticism.

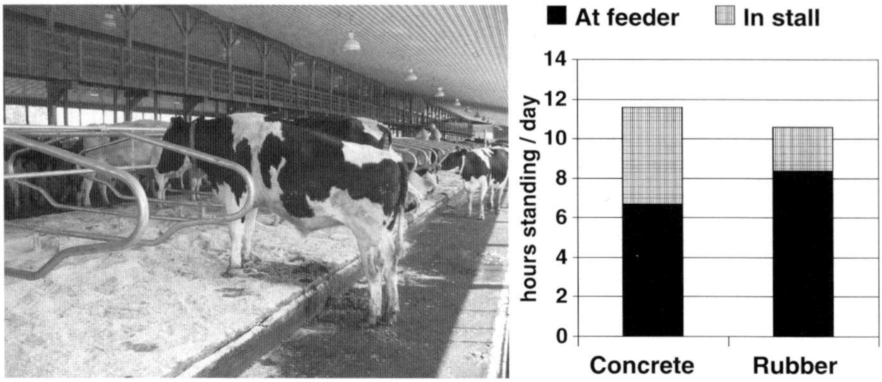

Figure 6.9 Although stalls are designed primarily to provide cows with a place to lie down, cows are often seen standing in them. Sometimes, this occurs with all four feet in the stall, but often cows are seen to stand with only the front feet in the stall. This position is associated with an increased risk of hoof injuries (Flower and Weary, 2006). Cows may stand in stalls because the stalls are too uncomfortable to lie in or because she has no where more comfortable to stand. The figure on the right shows the number of hours that the cows stood in front of the feed bunk or in the stalls when the floor in front of the feed bunk was concrete or covered with a soft rubber mat. With rubber mats the cows spend more time at the feed bunks and less time standing in the stalls (Figure based on data in Tucker et al., 2006.)

Lying, standing, and other behaviours can be assessed using three types of direct measure. The number of bouts over a period of interest (e.g. lying events/day), the average duration of these bouts (e.g. mean duration of lying bouts), and the combination of these measures (e.g. total lying time/day). Until now we have concentrated on this last measure, because sampling problems are reduced when estimating behaviours of long duration (like total lying time), compared to relatively infrequent events (like number of lying events). However, differences in total daily duration can be due to changes in either the number of bouts or the duration of these bouts, and these may be affected differently by design features. In some cases, differences in total lying time are closely associated with differences in the number of lying events but not the average duration of bouts, as in Tucker et al.'s (2003) comparison of mattresses with deeply bedded sawdust or sand surfaces. In other cases stall features can reduce the frequency of lying bouts but increase their average duration, as in Haley et al.'s (2001) study of cows in tie stalls with either concrete or mattress surfaces. Cows seemed less willing to lie down when kept on concrete, and stayed lying for longer periods, perhaps because of the discomfort associated with lying down and standing up on this surface compared with mattresses. Indeed, Rushen et al. (2007a) showed that cows with swollen front knees, an injury associated with the impact of lying down on hard surfaces, were most likely to reduce lying frequency when on concrete. Thus some features may affect the suitability of a surface for lying, but not for changing positions between lying and standing.

A final issue to consider is how much lying is too little, or how much standing is too much. Given that cows can change their lying behaviour depending upon other demands on their time budgets, and that they spend less time lying down when given a more comfortable location to stand, to what extent should we consider a reduction in lying time problematic? The effects of reduced lying time on dairy cattle have been assessed in several studies showing, for example, that cows are motivated to lie down for 12 to 13 hours each day (Jensen et al. 2005; Munksgaard et al. 2005). The most convincing evidence that a reduction in lying time is problematic comes from cases where the change in lying behaviour leads to physical injuries. For example, in some cases at least, reduced lying times necessarily result in increased time spent standing on concrete flooring, and prolonged standing on this surface is associated with an increased risk of sole lesions (e.g. Colam-Ainsworth et al., 1989).

5.2 Stall Surface and Bedding

Since one of the main purposes of supplying a cow with a stall is to provide her with a place to lie down, one obvious concern is whether the surface of the stall is sufficiently comfortable. This can depend upon the base of the floor, and the quality and quantity of bedding used. Traditionally, cows were kept on abundant quantities of straw bedding and this is still frequently used in straw yards and in some free stalls. The major advantage for cattle of straw bedding is that it provides a warm, soft surface on which to rest (Tuyttens, 2005). In some places, however, straw is not always easy or cheap to obtain and dairy farmers may use other forms of bedding, such as wood shavings, sawdust, or sand. The use of organic bedding tends to be associated with a higher incidence of clinical mastitis (Elbers et al., 1997; Barkema et al., 1999a, b; Wagner-Storch et al., 2003). The type of stall surface and bedding used are known to affect bacterial growth in the bedding and likely also udder health. Bacterial counts in bedding are usually lower when using inorganic bedding like sand versus organic bedding like straw or sawdust (Fairchild et al., 1982). Zdanowicz et al. (2004) showed that cows housed on sand bedding have fewer *Coliform* and *Klebsiella* bacteria on their teat ends compared to cows housed on sawdust bedding, but the reverse pattern was found for *Streptococci* bacteria. Given that the *Coliform* and *Klebsiella* are more difficult to treat than *Streptococci* infections, these authors recommended the use of sand bedding for udder health.

In some cases, farmers may dispense with bedding altogether and allow cows to lie directly on the floor of the stall. In some (unfortunate) cases, this will result in the cows lying on concrete, but the obvious problems with this option have led to the development of alternative, softer surfaces including rubber mats, crumbled rubber-filled mattresses, and waterbeds. An obvious question, therefore, is how important are bedding and the stall surface for dairy cows, and how do the different alternatives compare?

An epidemiological study found that the absence of bedding in both tie stalls and free stalls reduced the longevity of the cows, although the precise reasons for this were not clear (Buenger et al., 2001). A similar study in the UK found that reduced bedding in free stalls was associated with an increased incidence of lameness (Faull et al., 1996). For cows in tie stalls, sole lesions and hemorrhages are more prevalent among cows housed in stalls with concrete floors than cows in stalls with rubber mats (Bergsten, 1994; Bergsten and Frank, 1996). An insufficient amount of straw bedding has been identified as a risk factor for heel horn erosion (Philipot et al., 1994). For cows in free stalls, Vokey et al. (2001) found the highest incidence of lameness and of hoof lesions where stalls had bare concrete floors as opposed to sand bedding or softer rubber mats. The relationship with lameness reinforces the importance of good stall design for the welfare of cattle. However, the relationship is complex: poorly designed stalls may result in increased lameness since resting time is reduced and cows spend more time standing. Whether this will lead to lameness will depend upon what sorts of surfaces the cows are standing on.

As discussed earlier, to assess the surface of the stall, the two types of injuries that are most important are swelling of the front knees and the skin lesions that occur around the hock (tarsal joint). At some point during both getting up and lying down, a large proportion of the cow's weight is placed on the front knees. Consequently, it is not surprising that problems with stall design, especially floor softness, are apparent in injury or swelling to the front knees. Rushen et al. (2007a) followed cows housed in tie stalls either with a concrete floor or with a soft rubber mat and recorded the location and surface area of any cuts, abrasions, and patches of hair loss. The diameter of knees and hocks were also measured each week and any swelling or inflammation noted. When the total number of minor injuries (cuts, abrasions, and hairless patches) on each cow were summed over the 16 weeks of the study, there were no differences between the two types of flooring. However, the incidence of swollen knees and hocks was less than half on the soft rubber mats, a difference most evident on the knees of the front legs. Swollen knees and hocks usually come from the physical impact as the cow lies down and stands up, while abrasions and hair loss result from friction with the stall flooring. Therefore, the main advantage to softer stall flooring would seem to be in reducing the physical impact rather than reducing abrasiveness of the flooring.

The act of lying down or standing up may be more painful for cows kept on concrete due to swelling of the front knees. Rushen et al. (2007a) also reported that cows on softer mats showed an increased willingness both to lie down and stand up. In fact, cows kept on the softer flooring stood up and lay down almost twice as often as cows on concrete. When they did stand, they also stayed standing for longer before lying down again. Interpreting this behaviour as a sign of pain is supported by the fact that the cows that showed the greatest swelling of the front knees lay down for shorter periods of time compared to cows with less swelling of the knees. Although the advantages of softer flooring are thought to lie in providing some extra cushioning for the bony protrusions of the cow when she is lying, our results suggest that reduced physical impact on the front knees may be equally or more important.

A number of studies have now shown how stall features can contribute to the prevalence of hock lesions (see Figure 2.4), which is the other main type of injury related to stall design. A British study found a higher prevalence of lesions on farms using solid mats than on those farms using mattresses (Livesey et al., 2002). Although mattresses cause fewer injuries than solid mats, a series of experiments have shown that lesions are more prevalent on farms using mattresses than on those with deep-bedded stalls (Weary and Taszkun, 2000; Wechsler et al., 2000; Vokey et al., 2001) or bedded packs (Livesey et al., 2002). Mattresses remain popular among many dairy producers, and research is required to identify improved methods of managing stalls with mattresses, so as to reduce the risk of injuries. More fundamentally, we need a better understanding of how and when lesions are likely to develop in order to design housing systems that prevent lesions.

A study by Mowbray et al. (2003), showed how hock lesions develop over time, and how these can be reduced by changing stall design. Lesions were measured on the medial and lateral surfaces of the tarsal (hock) joint and on the dorsal, medial, and lateral surfaces of the tuber calcis (point of the hock). In one experiment, lactating cows were assigned to free stalls with either deep-bedded sand or geotextile mattresses on the day of calving. Stalls in the deep-bedded area had in excess of 20 cm of washed river sand over a dirt base. Stalls in the mattress area had a geotextile mattress covered with about 3 cm of kiln-dried sawdust bedding. Skin lesions on the hocks develop rapidly over the first 6 weeks of the study (Figure 6.10), but areas of both hair loss and skin breakage on the tarsal joint increased more rapidly for cows housed in stalls with geotextile mattresses than for cows using deep-bedded stalls.

Interestingly, the injuries on the tuber calcis showed the reverse pattern, with much larger areas of hair loss on cows using deep-bedded stalls than on those using mattresses. Similarly, Weary and Taszkun (2000) reported a higher prevalence of these lesions for cows using deep-bedded stalls than those using mattresses. In addition, they found that the difference in the total number of lesions was driven by the large number of lesions on the dorsal surface of the tuber calcis. The higher prevalence of these lesions on the dorsal surface is likely due to contact with the curb that can become exposed at the rear of the deep-bedded stalls; such contact is much less likely in stalls fitted with geotextile mattresses.

In a second experiment, cows were assigned to stalls with either deep sand bedding or geotextile mattresses recessed 5 cm below the curb, such that 3–5 cm of sand bedding could be maintained on the surface of the mattress. The mattresses recessed below the curb and covered with sand resulted in few lesions, and there was no difference in the mean area of lesions on the tarsal joint for cows using recessed mattresses and cows using stalls with deep sand bedding. However, lesions on the tuber calcis were again more of a problem for cows using the deep-bedded stalls. Weary and Taszkun (2000) argued that, although the surface of the mattress is not sufficiently abrasive to cause lesions, friction between the leg and mattress may

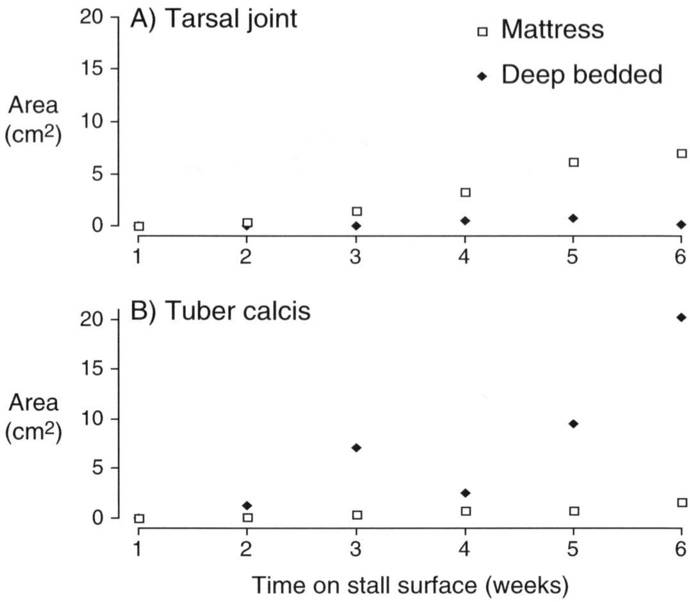

Figure 6.10 The mean area of hair loss on A) the tarsal joint (hock joint) and B) the tuber calcis (point of the hock). Data are shown separately for cows using stalls with geotextile mattresses, and those using stalls that were deep bedded with sand (Mowbray et al., 2003.)

cause heat to build up that reduces the strength of the skin. Mowbray et al. (2003) attempted to reduce the frictional injuries by adding limestone flour to bedding, but this was not effective in reducing injuries, perhaps because the flour did not reduce friction as anticipated. In addition to frictional heat, another possible cause of such skin lesions is pressure from body weight of the animal that reduces blood flow to skin over the area of contact with the lying surface (e.g. Bass and Phillips, 2007). Such pressure ulcers or "bed sores" would likely be affected by the addition of bedding, as more pliable materials support a greater proportion of the cow's surface area thus reducing pressure at any one point.

Bedding for dairy cattle has been the topic of several preference tests, all comparing different surfaces in different ways. Dairy cattle show clear preferences for softer stall surfaces (Tucker and Weary, 2004). For example, cows will preferentially lie down on deeply bedded surfaces with either sawdust or sand rather than mattresses with 2–3 cm of bedding (Tucker et al., 2003), and will select mattresses bedded with 7.5 kg of sawdust bedding over those with only 1 or 0 kg of sawdust (Tucker and Weary, 2004). Other work has shown that cows prefer heavily bedded stalls to lightly bedded mats, or solid surfaces like concrete, mats or wood (Jensen et al., 1988; Muller and Botha, 1997; Wagner-Storch et al., 2003). Only Manninen et al. (2002) found that cows avoided deep-bedded sand stalls. The cows in the study by Manninen et al. (2002) preferred straw-covered concrete and rubber mats, but this difference may have been due to the type of sand used or previous experience of the cows.

In studies where usage has been measured, cows spend more time lying down and lie down more often on softer, well-bedded surfaces. For example, Tucker and Weary (2004) found that lying time increased by 2.1 h /day and lying frequency by 2.4 events/day when cows had access to mattresses bedded with 7.5 kg of sawdust compared to mattresses without bedding. Previous studies have also reported increased lying times and/or lying frequencies depending upon the surfaces tested (e.g. Munksgaard and Simonsen, 1995; Haley et al., 2000).

5.3 *Stall Size and Configuration*

Cows seem to prefer lying down in more open areas, and physical barriers placed around the cow's lying area will likely make the area less likely to be used. This may be a suitable trade-off if the design features provide true benefits to the animal or the farmer, such as reduced defecation in the lying area. Structural elements of free and tie stalls, including the stall partitions and the neck rail or cow trainer positioned above the stall, are often assumed to provide such advantages but very little research is available to document this or show the effects of stall design on lying behaviour and other aspects of stall use.

Cattle stalls can vary in many aspects. For free stalls these features include the width between partitions, the shape of stall partitions, the length of the bed, the amount of lunge space, the height of the curb at the rear of the stall, the height and shape of the barrier ("brisket board") used to position the cow in the stall, and the height of the neck rail and its position relative to the stall curb. For each of these features there is a wide range of recommended specifications, largely based on the personal and divergent insights of dairy professionals.

It is clear that inadequate stall design can have negative effects on the animal welfare. Epidemiological studies in France (Philipot et al., 1994) and the UK (Faull et al., 1996; Haskell et al., 2006) have identified small stalls as being a risk factor for hoof problems and lameness. Frequent contacts between the cow's neck and the neck rail can be assessed by looking for wear marks on the underside of the neck rail. Similarly, Cook (2003) illustrates how poor positioning of mounting rails for stall dividers can be identified by the sound of cows hitting their chins on these structures when they lunge forward while standing up.

The little research to date on stall configuration suggests that smaller stalls reduce lying time. Tucker and colleagues performed a series of experiments on aspects of stall design, including stall width, length, and neck rail placement, and found that each of these features affects stall usage, although in different ways. Stall width clearly affected lying times: in two experiments stall width was increased (once from 112 to 132 cm, and once from 106 to 126 cm), and in both cases lying times increased by approximately 1 h/day (Tucker et al., 2004).

Cows are less likely to contact the stall partitions when kept in wider stalls (Blom et al., 1984) both while the cows are recumbent as well as when lying down or standing up. Cows in wider stalls also spend less time standing with just the front feet in the stall and more time standing fully in the stall (Tucker et al., 2004). The time standing half way in the stall can also be reduced (and the time spent standing fully in the stall increased), by increasing the available lunge space and raising the neck rail or placing it further from the rear curb (Tucker et al., 2005).

Increasing the time spent standing with all four hooves in the stall, and reducing the time spent standing partially in the stall, may also benefit cows by avoiding time on the relatively uncomfortable standing surface available in the alley (Stefanowska et al., 2001). However, larger stalls are more likely to be soiled by urine or faeces, and this can be a welfare concern both in terms of comfort and potentially through increased risk of udder infections (Tucker et al., 2004). However, increased soiling is due in part to the increased occupancy of these stalls, both lying and standing. Gaworski et al. (2003) reported a positive relationship between time spent standing and lying in the stall and the amount of faecal material in the free stall. In this sense, dirtiness of stalls should be seen as an indicator of good stall design!

Electric cow trainers in tie stalls and neck rails in free stalls are both intended to keep stalls clear by preventing defecation in the stall. Research has shown that cow trainers result in cleaner stalls, cleaner cows, and a lower incidence of heel horn erosion in the hind hooves (Bergsten and Pettersson, 1992; Zurbrigg et al., 2005). Lowering neck rails in free stalls increases time spent standing outside of the stall (Tucker et al., 2005), likely increasing cow's cleanliness but harming hoof health. Although efforts to improve stall cleanliness are generally motivated by the desire to improve udder health, there is little empirical evidence to support such a link. Indeed, the use of cow trainers is sometimes associated with a higher risk of mastitis, despite having cleaner stalls (Bakken, 1982; Oltenacu et al., 1998). Producers unwilling to invest more effort in stall maintenance will likely need to trade-off comfort for cleanliness. Changes to the lying area that make free stalls less suitable for standing need to be accompanied by other modifications to the barn that create comfortable standing surfaces outside of the stall.

5.4 Stall Location

Stall use can also be affected by where the stalls are in the barn (Wagner-Storch et al., 2003). Some experiments have found that cows rarely enter a given stall while seemingly identical stalls are occupied more than 80% of the available time. In one study, Gaworski et al. (2003) showed that stalls in the row closest to the feed alley were occupied 41% more frequently than were stalls in more distant rows. In addition, stalls located within the centre of each row were used 12% more often than those stalls located on the periphery of the row (i.e. either near a wall or fence). Natzke et al. (1982) also found that stalls on the periphery were used less than stalls in the interior of the row. These results suggest that certain stalls, particularly those farther from the feed bunk and on the periphery, are less desirable to dairy cattle perhaps because cows need to walk farther, or because they have to navigate past certain physical (e.g. narrow alleys) or social obstacles (e.g. dominant cows) on their way to the more distant stalls. Indeed, earlier work has indicated that the movements of subordinate animals are prevented by the location of dominant cows (Miller and Wood-Gush, 1991). Such factors may partly explain reduced user satisfaction and lower production in those barns with more rows of stalls (e.g. 6 vs. 4 row barns: Bewley et al., 2001).

Thus large differences in usage can occur even among identically configured stalls within the same barn. The fact that stalls within a pen vary in their popularity suggests that stall availability from the cows' perspective is not the same as from the producer's perspective – what looks to us as 1:1 cow-to-stall stocking density may seem considerably worse to the cows if they judge some stalls as unacceptable.

6 Surfaces for Standing and Walking

Dairy cows are heavy animals that spend much of their time standing and walking. The surfaces on which they perform these activities are thus an important component of the housing system. The hoof of the cow has evolved primarily for walking on surfaces like pasture. With the increased use of indoor housing, dairy cows are spending less time on pasture, a change associated with increased lameness (see Section 3 of this chapter). In tie stalls, cattle stand most often on the same surface on which they lie. In straw yards, the cattle stand and walk mainly on straw. In free stalls, the cows stand mainly on concrete, that is either solid or slatted. Concrete is durable, readily available, relatively easy to clean and inexpensive, but there is increasing evidence that concrete walking surfaces may be related to the increased incidence of lameness in dairy cows (Bergsten, 2001). Uncomfortable standing surfaces in front of feed bunks may reduce feeding time, especially for lame cows. Because of the obvious problems with concrete flooring, some dairy farmers are beginning to experiment with other materials, especially rubber mats placed in strategic areas such as in the milking parlour, commonly used walking areas, and in front of the feeders. In two experiments Tucker et al. (2006) examined the effects of softer surfaces in front on the feeding area, and found that cows spent more time standing in this area when softer surfaces were available. In one experiment in a facility where free-stall design may have been inadequate, cows also used the softer flooring to lie down, illustrating the complex ways in which the design of various components of cow housing can interact. Other experiments have also shown that providing cows with softer surfaces to stand upon will increase standing times in these areas (Fregonesi et al., 2004).

The importance of standing and walking surfaces for cattle welfare will vary with the type of housing system used. Cows that spend some time on pasture will be less affected by the type of flooring in the barn than cows that remain indoors all of the time. For cows in tie stalls, the same surface is used for standing and lying, but cows in free stalls have a choice as to where to lie or stand so there is likely to be a complex interaction between the type of flooring used in the stall, and the type of flooring used elsewhere.

6.1 Lameness

The most serious welfare problems associated with flooring surfaces are hoof-related diseases. Concrete flooring can increase the incidence of lameness by causing excessive and uneven wear of the hoof, by direct damage as a result of uneven surfaces or protrusions, by causing skin breaks that increase the risk of infectious diseases, and by increased impact forces that can increase the risk of damage to the corium. In general, management that results in the cows standing longer on concrete surfaces increases the risk of sole lesions (Colam-Ainsworth et al., 1989). A number of epidemiological surveys have shown convincingly that lameness is more prevalent where the cows walk mainly on concrete (Wells et al., 1995; Faull et al., 1996; Somers et al., 2003). Small-scale studies support the view that concrete flooring increases the risk of hoof lesions (e.g. Vokey et al., 2001; Vanegas et al., 2006), in part because of the greater force exerted on the hoof when the cows walk on hard concrete floors (Vander Tol et al. 2003; Franck and De Belie, 2006). In addition, digital dermatitis is more common on farms where cows walk on grooved concrete compared to farms using dirt, pasture, or smooth concrete (Wells et al., 1999), perhaps due to the abrasive properties of the flooring.

Aspects of the walking surfaces other than the material can influence lameness. Philipot et al. (1994) assessed risk factors for chronic and subacute laminitis as well as heel horn erosion and found that high steps (leading into the stall or into the milking parlour) and slopes were risk factors for sole lesions. Bell (2004) also found that steps and imperfections on the concrete flooring (such as cracks and holes) increased the risk of sole lesions.

Wet standing surfaces also increase lameness. Exposure to moisture softens the hoof leaving it prone to excessive wear or other damage (Borderas et al., 2004; Figure 6.11). Cows in free-stall housing often stand on wet surfaces. A survey in the USA found increased clinical lameness with wet flooring in stalls (Wells et al., 1995), and Philipot et al. (1994) found that wetness at the back of the stall increased the risk of heel horn erosions. Dermatitis is also more common when cows stand on wet surfaces, because of increased risk of transmission or softening of the hoof leading to more wear and abrasions increasing the risk of infection (Wells et al., 1999; Rodriguez-Lainz et al., 1999). Improving the drainage of urine and faeces at the back of the tie stall reduces the risk of several hoof problems, including dermatitis, heel horn erosion sole ulcer, and white line disease (Hultgren and Bergsten, 2001). Slatted floors, popular in European dairy barns, do allow much better drainage and are more likely to keep hooves dry. However, dairy cows kept in barns with

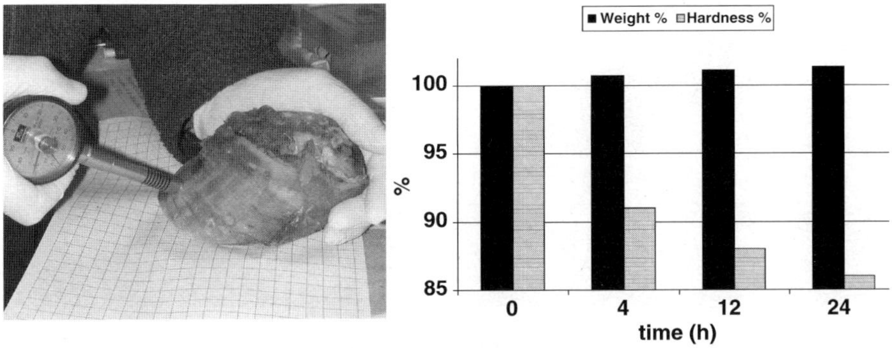

Figure 6.11 The claws of the hooves of the cows absorb water rapidly, gaining weight, and becoming softer in the process. This increases the risk of wear and erosion of the horn. Most of the changes occur during the first 4 h. The hardness of the claw can be measured with a durometer (left) The increase in the weight of the hoof indicates the amount of water absorbed. (Data taken from Borderas et al., 2004.)

slatted floors spend more time standing in the stall (Stefanowska et al., 2001), suggesting that they find the slatted floors uncomfortable.

6.2 *Walking Surfaces and the Movement of Animals*

Poor walking surfaces increase the chance of cattle slipping and falling, increasing the risk of injury, and reducing the willingness of cattle to walk quickly (Figure 6.12). Reluctance to walk increases the chance of human intervention, which can involve rough handling (see Chapter 9). The two most important properties of flooring that affect walking are the degree of traction (van der Tol et al. 2005) and softness. According to earlier research (Nilsson, 1992) walking surfaces must have a coefficient of friction between 0.4 and 0.5 to ensure that cows are able to walk easily without slipping; both concrete and hard rubber mats have coefficients of friction at the lower end of the range but softer rubber mats are superior (e.g. van der Tol et al., 2005; Rushen and de Passillé 2006). However, other studies (Phillips and Morris, 2000; Rushen and de Passillé, 2006) have examined cows walking on concrete floors that were either wet or dry or covered in slurry, and found that the coefficients of friction of the floor were not good predictors of the risk of slipping. The researchers questioned the extent to which standardized engineering measures of friction can appropriately characterize flooring for cattle. Direct observations of cattle walking on different surfaces appears to be a more promising approach.

Some studies have begun to examine how cow locomotion is affected by the walking surface. Flooring needs to have sufficient traction to avoid slips and falls (van der Tol et al. 2005), and accumulation of slurry can increase the risk of such accidents. Phillips and Morris (2000) found that when cows walked on concrete floors covered with slurry, they walked more slowly and with reduced stepping rates, but took

Figure 6.12 The surface on which cows walk or stand can have a large impact on their welfare, primarily by altering the likelihood of hoof injuries or lameness. Poor quality outdoor walking tracks (left) can increase the chance of hoof injury, especially if the cows are being hurried. Often, cows will prefer to walk on a softer surface (Telezhenko et al., 2007). The cow on the right is walking on a strip of rubber placed in the middle of the walkway. Such surfaces if sufficiently soft and providing good traction, increase the size of the cows' steps and hence walking speed

longer steps. Rushen and de Passillé (2006) found similar effects on the speed of walking and also noted that the incidence of slipping was considerably higher on slurry-covered concrete than on dry concrete. When allowed a choice, cows will generally avoid walking in a passageway where the floor is covered in slurry (Phillips and Morris, 2002).

In view of the obvious dangers of cattle slipping and falling, there have been a number of attempts to develop superior walking surfaces. Unfortunately, few of these have been systematically tested. Dairy cows prefer to walk on soft rubber floors than on concrete (Telezhenko et al., 2007). Phillips and Morris (2001) examined epoxy-covered floors that were of varying degrees of friction achieved by adding bauxite aggregates. Rather surprisingly, cows walked more slowly, but with longer strides on the higher-friction surfaces. In contrast, two other studies (Rushen and de Passillé, 2006; Flower et al., 2007) found that cows walked more quickly and with longer strides on floors when the degree of friction was increased by adding rubber mats that were both anti-slip and softer than concrete. It is possible that the epoxy grit used by Phillips and Morris (2001) made the floors uncomfortable. Preference tests showed that cows were able to sense the difference between the different floors but showed no preference for walking on the higher-friction material (Phillips and Morris, 2002).

Rushen and de Passillé (2006) noted that both the friction and softness of the flooring affected locomotion. Rushen and de Passillé (2006) filmed cows walking down specially constructed corridors with either concrete floors or soft, high-friction, rubber mats. Walking speed was higher on the rubber mats and cows slipped less often on this surface compared to concrete. When friction was increased but softness kept constant by adding a high-traction rubber mat to a concrete base, cows increased mean stride length and tended to increase walking speed, especially when crossing a gutter. When softness of the flooring was increased by adding crumbled rubber-filled geotextile mattresses under the high-traction mat, cows increased walking speed. Together, these results show that both traction and softness are important to the cows, and that methods of increasing friction that reduce the comfort of the flooring may have undesired effects.

Improvements in flooring must meet the needs of both the cow and the producer. Flooring surfaces must be economically affordable, durable and result in minimal foot and leg problems. Considerable research has now documented the problems with concrete flooring, and we agree with Guard's (2001) suggestion that standing and walking surfaces for dairy cows be constructed of something other than concrete. Both friction and softness are important for the cattle, and one should not be sought at the expense of the other. The problems of wet flooring and slurry accumulation are well documented: flooring needs to provide good drainage. Research is needed both to examine the advantages of currently available flooring surfaces and to identify optimum characteristics of the floors that can then inform the development of new materials.

7 Social Effects: Stocking Density, Social Dominance, Competition

How many animals should be housed within a given space? Although stocking density is usually described in terms of the area of space per animal, the issue of appropriate space is far more complex than can be adequately covered in this one measure. First, an increase in animal density will have quite different effects depending on whether it results from the addition of extra animals to a given area, or by a reduction in the area available for a given number of animals because the number of animals within a group has effects independent of space availability. In a large group of animals there are simply more individuals that need to be dealt with, leading to a greater variety in dominance relationships and more opportunity for aggressive encounters. Large groups can also place more demands on management, particularly if there is less time devoted to individual animals.

If we consider a fixed number of animals, then what competitive and non-competitive factors should be considered to determine the optimum stocking density? Non-competitive factors are those that do not depend upon interactions between animals, such as air quality (when a large number of animals are housed in an enclosed space) or the amount of manure deposited on the floor. These effects on animal welfare should be apparent in most animals. In contrast, competitive effects come about as a result of interactions between animals, particularly competitive interactions for limited resources such as to stalls, feeder space, or perhaps drinkers.

Competition is usually low in tie-stall systems so most research attention on stocking rate has focused on free-stall housing. The effect of competition can be apparent in the usual health and welfare measures, such as the incidence of various maladies. Although such measures may provide some information as to the severity of the problem, behavioural measures are often more useful in pinpointing the actual causes of the problem. The most relevant behaviours are those most closely related to the resources over which the animals are competing, such as resting time, feeding time, etc. or instances of actual aggressive behaviour, such as fights over feeding space or stalls.

One of the most contentious issues in free-stall housing concerns the number of cows to house together in a pen with a fixed number of stalls and feeder space. Usually, the density is expressed as the number of cows per stall (cow to stall ratio) or linear bunk space per cow. Since the purpose of the stall is to provide a clean, comfortable place for the cow to lie down, having one stall per cow would seem optimal. However, some reason that since dairy cows tend to lie down only about 12 h a day, it should be possible to have more cows than stalls and still have enough time for all the cows to lie down. Any resulting reduction in housing costs could provide obvious economic advantages to the producer. In a survey of modern free-stall dairy barns in Wisconsin, Bewley et al. (2001) report that only one-third of farmers keep their cows at a density of one cow per stall or less, while 15% of farmers overstock by more than 20%. These authors suggested that housing costs per cow are reduced by a third when farmers keep 30% more cows than there are stalls.

Do high stocking rates in free-stall barns reduce lying times? Research has shown clear reductions in lying time when the cow to stall ratio increases above 1.5. The average reductions in total lying time are around 1 h/day at a cow to stall ratio of 1.5 (Wierenga, 1990; Winckler et al., 2003) and up to 4 h/day at a cow to stall ratio of 2.0 (Friend et al., 1979). The effects are much larger for some cows, especially low-ranking cows. Even with a cow to stall ratio of 1.0, low-ranking cows spend less time lying in the stalls and more time standing in the passage ways (Galindo and Broom, 2000). Wierenga (1990) observed that lying times for low-ranking cows decreased by over 2.6 h/day at a cow to stall ratio of 1.5. Leonard et al. (1996) found that the average resting time when cows were housed at a cow to stall ratio of 2.0 was 7.5 h but that this varied between 2.7 and 11.9 h/day for individual cows. The low resting times were associated with an increase in the incidence of sole haemorrhages and lameness. The authors calculated that a reduction in lying time from 10 to 7 h/day was associated with a doubling of the incidence of sole haemorrhages, and a further doubling occurred when resting time was reduced to 5 h. Galindo and Broom (2000) also noted that lameness and hoof lesions were more among cows that spent more time standing. Thus, stocking rates above 1.5 cows per stall are likely to reduce resting time to the degree that is clearly associated with an increased risk of lameness.

At lower stocking rates, the effects are less evident. Compared to a cow to stall ratio of 1, no overall changes in average resting time have been reported for cow to stall ratios of 1.14 (Fregonesi and Leaver, 2002), 1.2 (Sugita et al., 1999), 1.25, and 1.33 (Wierenga, 1990). Fregonesi and Leaver (2002) reported no effect of a cow to stall ratio of 1.14 on locomotion score. However, one recent study showed clear effects on time spent lying and other behaviours when stocking density varied between 1 and 1.5 cows per stall (Fregonesi et al. 2007). When fewer stalls were available, time spent lying was reduced by 2 h per day, partly due to increased displacements of cows from stalls. When there are too few stalls, cows may also have to change the times that they rest: Wierenga (1990) noted that cows (especially subordinate cows) rested less often at night and more often during the evening at cow to stall ratios of 1.25 and 1.33. The effect of these changes on the welfare of the cows is difficult to judge.

Changes in stocking density can result in other changes besides resting behaviour. First, the incidence of aggressive interactions between cows can increase, even at a cow to stall ratio of 1.14 presumably because of competition over stalls (Fregonesi and Leaver, 2002). However, the largest change in behaviour is that cows spend less time standing in the stalls and more time standing in the walkways. At a cow to stall ratio of 1.25, cows spend an average of 70 min less each day standing in the stall (Wierenga, 1990), and increase time spent standing outside the stall. The effect of this change on the welfare of the cows will clearly depend on the quality of the standing surfaces elsewhere in the pen (Section 6 of this chapter).

Increases in stocking density have been found to have little effect on some other indicators of welfare. Fregonesi and Leaver (2002) reported no effect of a cow to stall ratio of 1.14 on locomotion score, body condition score, cleanliness, or rumination time. Feeding time or feed intake were not affected by stocking at cow to stall ratios of 1.14 (Fregonesi and Leaver (2002), 1.5 (Wierenga 1991), or 2.0 (Friend et al., 1979), even though these densities were associated with a proportional reduction in feeding space as well (see Chapter 9; see also DeVries et al., 2004). Perhaps because feed intake is maintained, studies have not reported an effect of stocking density on milk production, even up to cow to stall ratios of 2.0 (Friend et al., 1979; Fregonesi and Leaver, 2002). These studies tended to use a fairly low number of animals and observed them for short periods of time, reducing the chance of seeing effects on milk production.

Reduced feed bunk space availability has been shown to result in increased aggressive behaviour in cattle (Kondo et al., 1989). When feed bunk space is limited, increases in aggressive behaviour are thought to limit the ability of some cows to access feed at times when they want to. DeVries et al. (2004) tested if increasing space availability at the feed bunk improves access to feed and reduces social competition. Twenty-four lactating Holstein cows were each tested under two conditions: with 0.5 or 1.0 m of feed bunk space per cow. When animals had access to more bunk space there was at least 60% more space between animals (regardless of the number of cows at the feed bunk) and 57% fewer aggressive interactions while feeding. These changes in spacing and aggressive behaviour allowed cows to increase feeding activity especially during the 90 min after fresh feed was provided. During this period, cows with access to more feed bunk space increased time at the feeder by 24%, and this effect was strongest for subordinate cows.

In conclusion, research stocking rates in free-stall housing suggest that stocking at greater than one cow per stall can lead to increased competition between cows, a reduction in time spent lying and an increase in time spent standing. These effects will be most apparent in low ranking cows. At stocking densities less than 1.5, the research results are less clear. In general, most cows appear to adjust the time of the day that they rest to protect overall rest time. However, individual cows may have reduced rest times – any negative effects of competition will most likely be felt by the lowest-ranking cows. A number of factors are likely to influence the effect that stocking densities have on cows, and these need to be borne in mind when interpreting studies. The effects of overcrowding will depend upon the layout of the barn and the amount of time that cows have available to rest. For example, waiting times at milking vary considerably among farms, but can exceed 1 h per milking. Especially with three or more milkings per day, cows may have several hours less each day to eat and rest, likely increasing competition for feeding and lying spaces once cows are back in the pen.

8 Design of the Feeding Area

There are several aspects of the feeding environment that can influence the ability of cows to access feed, including the amount of available feed bunk space per animal and the physical design of the feeding area. One of the most obvious features of the feeding area is the physical barrier that separates the cow and the feed. The various barriers are all designed with the intention of allowing cows access to feed. Some designs may have other effects such as reducing the frequency of aggressive interactions at the feeder. For example, a feed line barrier that provides some separation between cows (e.g. headlocks; see Figure 6.13) might reduce competition and increase intake. Unfortunately, there is little scientific work addressing these topics.

Endres et al. (2005) compared the effects of two feed barrier systems (post-and-rail versus a headlock feed line barrier; see Figure 6.13) on the feeding and social behaviour. Although there was no difference in feed bunk attendance throughout the day, during periods of peak feeding activity (90 min after fresh feed delivery)

Figure 6.13 A post-and-rail (left) and a headlock (right) feeder for dairy cows

cows that had lower feeding times relative to group mates when using the post-and-rail barrier showed more similar feeding times to group mates when using the headlock barrier. There were also 21% fewer displacements at the feed bunk when cows accessed feed by the headlock barrier compared to the post-and-rail barrier. These results suggest that using a headlock barrier reduces aggression at the feed bunk and improves access to feed for socially subordinate cows during peak feeding periods.

Huzzey et al. (2006) tested the effects of stocking density at the feed bunk using both types of feed barrier. Cows were assigned to either headlock or post-and-rail barriers and tested at four stocking density treatments (0.81, 0.61, 0.41, and 0.21 m/cow, corresponding to 1.33, 1.00, 0.67, and 0.33 headlocks/cow). Daily feeding times were higher when using the post-and-rail barrier and feeding times decreased as stocking density increased regardless of barrier type. Cows were also displaced more often from the feeding area when the stocking density was increased, and this effect was greater for cows using the post-and-rail feed barrier (Figure 6.14).

Subordinate cows were more often displaced with the post-and-rail barrier design, particularly at high stocking densities. These results illustrate that some physical separation between adjacent cows can be used to reduce competition at

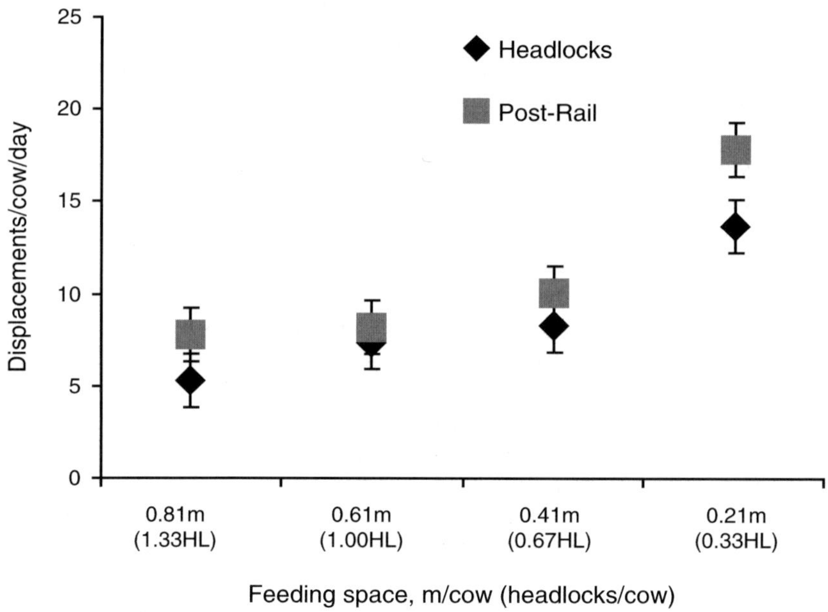

Figure 6.14 Mean daily displacements per cow at four different stocking density treatments when provided either a headlock or a post-and-rail feed barrier (From Huzzey et al., 2006.)

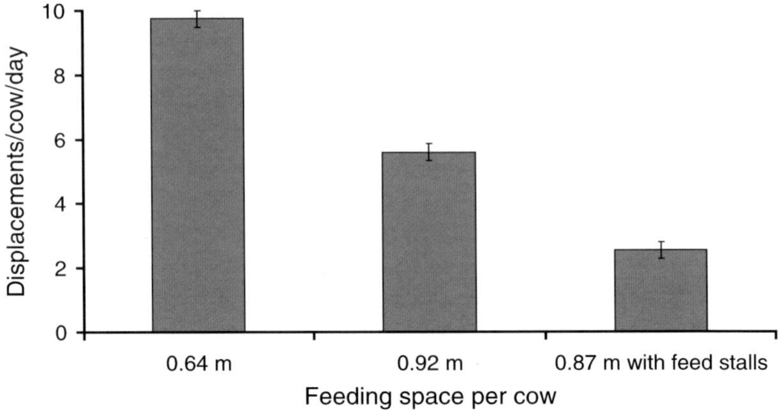

Figure 6.15 Daily number of displacements per cow at three different levels of feed bunk space (Adapted from DeVries and von Keyserlingk, 2006.)

the feed bunk. A less-aggressive environment at the feed bunk may also have long-term health and welfare benefits; cows that engage in higher number of aggressive interactions at the feed bunk may be at risk for hoof-health problems (Leonard et al., 1996).

In these two studies on feed barrier design (Endres et al., 2005; Huzzey et al., 2006) the use of a headlock reduced the incidence of displacements at the feed bunk. However, this barrier did not completely eliminate aggressive behaviour, indicating that the neck division does not provide full protection. DeVries and von Keyserlingk (2006) set out to determine if the addition of partitions (feed stalls) between the bodies of adjacent cows provides additional protection while feeding and allows for improved access to feed. Cows were tested with 0.64 m of feed bunk space/cow (representative of industry recommendations), 0.92 m of feed bunk space/cow, and feed stalls (0.87 m of feed bunk space/cow with feed stall partitions separating adjacent cows). When animals had access to more space, particularly with the feed stalls, there were far fewer displacements while feeding (Figure 6.15), and subordinate cows benefited the most from this reduction in displacements.

Reduced aggression at the feed bunk allowed cows to increase their daily feeding time and reduce the time they spent standing in the feeding area while not feeding. The feed stalls also caused a change in the displacement strategy at the feed bunk, forcing cows to initiate contact at the rear of the animal they were displacing rather than at the front or side. Despite this change in strategy, cows were less successful in displacing others. Based on these results, we conclude that the provision of more feed bunk space, particularly when combined with feed stalls, will improve access to feed and reduce competition at the feed bunk, particularly for subordinate cows.

9 Temperature Stress

Although temperature extremes are not as evident a source of welfare problems for cows as they are for other farm species, cattle are kept in parts of the world that regularly experience extreme cold or extreme heat. The lower critical temperature for lactating dairy cows has been estimated to be as low as −37°C, while the upper critical temperature may only be 25°C (Kadzere et al., 2002). This shows that heat stress is generally much more of a problem for the lactating cow than is cold stress (Figure 6.16).

Important dairy industries exist in several parts of the world with high temperatures, notably the USA, Brazil, and Israel, and considerable research has examined the effects of heat stress on dairy cows. Fortunately, this research has recently been reviewed (Kadzere et al., 2002; Collier et al., 2006) and Silanikove (2000) presents a thorough discussion of the implications of heat stress for the welfare of dairy cattle (see Chapter 7); a topic that has received far less attention than the effects of heat stress on milk yield or reproduction. Hyperthermia can easily occur in lactating cows, largely because of the considerable metabolic heat generated during lactation. Cattle breeds differ in their ability to tolerate heat stress, with *Bos indicus* being substantially more tolerant than *Bos taurus*, and Holstein cattle less tolerant than Jersey cattle (Kadzere et al., 2002).

Figure 6.16 Cattle are large animals with a relatively small surface area to volume ratio. When kept in cold conditions, they can also grow a thick coat. Together, these reduce heat loss. In addition, lactating cows can generate considerable metabolic heat. These factors lead cattle generally to be more susceptible to heat stress than cold stress

Heat stress occurs as a result of some combination of high ambient temperatures, direct or indirect exposure to the solar radiation, low air movement and high humidity. The housing in which cattle are kept can lead to heat stress by increasing the heat transfer from the environment to the animal and by interfering with the animals' thermoregulatory responses. Exposure to solar radiation is more of an issue for cattle housed outside, and low air movement more of an issue for cattle housed indoors. A number of experiments have shown the advantage of providing shade to dairy cattle housed outdoors (reviewed in Armstrong, 1994; Silanikove, 2000). For cattle housed indoors, appropriate ventilation, supplemented with water sprinklers, should aid the cattle to lose heat through increased evaporation (Armstrong, 1994). In practice, the benefits of sprinklers are not consistent; sometimes effects are evident (e.g. Keister et al., 2002) but other studies have produced mixed results (Chan et al., 1997; Thompson et al., 1999). Optimal positioning and use of sprinkler systems is an engineering problem that still awaits a solution. Speed of air movement is an important factor affecting heat loss in cattle (Turnpenny et al., 2000), but, while it is apparent that adequate ventilation inside buildings is essential, the best means of ventilating dairy barns is still an active topic for research. While ventilation and sprinklers should allow cattle to lose heat through evaporative cooling, cattle can also lose heat through direct conduction (Silanikove, 2000). In hot climates, cattle appear to show a preference for types of bedding that have good thermal conductivity and which promote conductive heat loss (Silanikove, 2000). The types of bedding that are the most appropriate for different climatic regions has not yet received much attention.

10 Conclusion

Cattle must live their lives in the housing systems that we provide for them, so the quality of this housing can have profound effects on their well-being. For the cattle industry, issues concerning housing are also strategic to address, as public concerns about animal welfare are often brought into focus with examples of unfortunate housing conditions. For example, the use of crates for veal calves, stalls for gestating sows, and battery cages for hens have all figured prominently in public criticisms of farming practices.

Housing has also been the focus of a growing body of research on cattle welfare, and a great deal has been learnt over recent years. Research comparing disparate systems is difficult to do well, and such comparisons are fraught with problems. However, careful comparisons within systems can be very useful in identifying specific weaknesses and how to address these. Research on housing is also made difficult by the fact that different housing methods are normally accompanied by changes in management that also have profound effects on the animals. Different housing systems also differ in the extent to which cattle interact socially, making competitive interactions much more of a welfare problem in, for example, loose housing compared to tie stalls.

For adult cattle, almost all research on housing has focused on dairy cows so this was also our focus for this chapter. Although the public may imagine that dairy cows in Europe and North America are commonly kept on pasture, this image is dated and continuous indoor housing is becoming the norm. Research has shown that some health problems, especially hoof health, are more prevalent with indoor housing, but research is still required to determine what aspects of pasture access provide benefits, and if these features could be provided indoors.

Three key aspects of indoor housing are areas for cows to lie down, feed, and stand. We describe some methodological pitfalls in research on cow comfort, but also show that powerful and suitable research approaches are available, particularly measures of physical injuries, preferences, and usage. The surface that cows lie down on is, in our view, the most important characteristic of the lying area. Cows prefer softer lying surfaces such as deep bedding. They are also less likely to sustain injuries and spend more time lying down in stalls with more bedding. Obviously, the lying surface needs to be maintained both to remove faecal matter and to keep an adequate lying surface – poorly maintained stalls are a health risk to cows and can much reduce stall occupancy. Any physical structures used in the lying area (such as stall partitions and neck rails in free stalls) need to be positioned to prevent injuries and allow for normal lying and standing behaviour. In general, physical structures within the stall reduce stall usage – producers need to realize that these are installed for their benefit and these tend to interfere with the cow's ability to use the stall. Interestingly, physical structures (like headlock feed barriers or partitions between feeding "stalls") seem to be beneficial at the feed area by reducing competition among cows.

Free-stall design is of concern not only because of consequences for cow's lying behaviour, but also because it can affect where cows stand and how much time they remain standing. Some aspects of stall design (such as narrow stalls and neck rails placed low or closer to the rear curb) can greatly increase the time cows spend standing fully or partially in alley, and the increased exposure to concrete and manure slurry can have negative effects on hoof health. New research on alternative flooring surfaces for cows is urgently required to address such effects. New flooring surfaces should be designed to provide improved traction, reducing the risk of slipping and likely aiding farm workers in moving cows and perhaps also in detecting heat. Softer flooring has also been shown to be important – cows prefer to stand on these surfaces compared to concrete and show improved gait when walking on softer materials.

The effects of milking systems on welfare have only been studied in relation to the development of automated milking systems. In general, these systems can provide welfare advantages to cows by allowing them more freedom to choose when and how often they will be milked. However, like any new system, much work still needs to be done to realize such welfare benefits without risk of other harms such as increased rates of udder infection.

Bringing animals indoors normally reduces space availability and increases the opportunity for animals to compete over resources such as food. Most research on this topic to date has focused on overstocking pens, as this is a popular management option especially in some regions. Research has documented negative consequences when cows are stocked at more than one cow to every available stall, especially for subordinate cows or lame cows. Bringing animals indoors can have the benefit of protecting them from the extremes of climate, but cows housed indoors can still face environmental challenges, perhaps most notably hyperthermia. A considerable body of research has now developed on means of reducing the risks of excess heat on lactating dairy cows, including the use of fans and sprinkler systems.

Chapter 7
Housing for Growing Animals

1 Introduction

In the previous chapter, we dealt with the effects of housing on the welfare of adult cattle. In this chapter, we deal with housing for growing cattle. We cover a diverse range of housing systems, ranging from the individual crate for veal calves, through cow–calf herds, to the outdoor feedlots of North American beef cattle. We cannot hope to cover all of the diverse welfare problems that exist within this range of housing systems. Instead, as in the previous chapters, we concentrate on those "hot" issues that have attracted most of the attention of researchers in animal welfare, such as housing for veal calves and indoor housing for beef production. As with the previous chapter, readers must remain aware that research comparing housing systems is notoriously difficult to interpret. The welfare of animals in any type of housing system will depend upon the details of the system, including the management and nutritional regimes. This makes it difficult to make general statements about the effects of housing systems per se on animal welfare.

2 Calving Area

The type of housing used affects the welfare of the young calf from the moment of birth (Figure 7.1). At birth, calves are particularly susceptible to temperature extremes, both cold and hot, and thus often require some form of shelter or shade. Riley et al. (2004) reported that Brahman and hybrid Brahman calves born on days where the temperature was less than 5.6°C were twice as likely to suffer from poor vigour and 1.6 times more likely to die than calves born on days when the ambient temperature was higher. In cold conditions calving indoors would seem preferable. In Quebec, perinatal calf mortality over winter can be very high (>10%) when calving outdoors, but indoor calving can reduce mortality by one third, as well as reduce the incidence of pneumonia (Ganaba et al., 1995). In warmer climates the season of calving has less effect on calf mortality (Sanderson and Dargatz, 2000), suggesting that shelter is less valuable. Calf mortality can

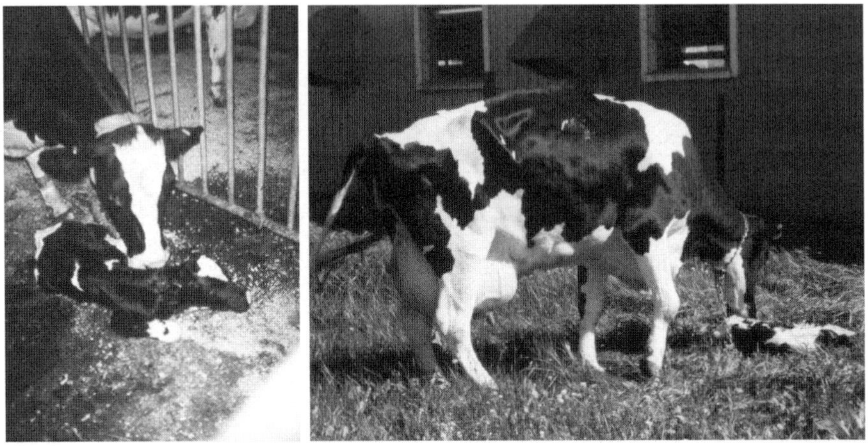

Figure 7.1 At birth, young calves are at their most vulnerable, being susceptible to cold and infection. To ensure their well-being, it is important to have a well-designed calving area that is clean and free from drafts. Allowing cattle to calve outdoors is generally associated with better hygiene and reduced periparturient morbidity and mortality. However, during inclement weather, the calves risk being exposed to cold. When cattle are housed indoors, most dairy farmers provide a separate "calving pen". Letting cows calve in the same areas as other cows increases the risk of disease transmission and calf mortality-

still be high in subtropical and tropical regions but it is not known to what extent the hot climatic conditions contribute to calf mortality. One study in Burkino Faso reported that 19% of live born calves die before the age of 12 months (Ganaba et al., 2002). Interestingly, these authors cited malnutrition as one cause of death of young calves (see Chapter 8) as cows from this region typically produce less than 1.6 L of milk per day and much of this is used for human consumption and thus not available for the calf.

Regardless of climate, calving in an enclosed space may still provide some advantages for the health of the calf by helping farm workers supervise calving and intervene if necessary. However, because of their undeveloped immune system, newborn calves are highly susceptible to infectious diseases, and the risk of infection is increased in enclosed areas. In the USA, the risk of various calf diseases (diarrhoea, respiratory problems, etc.) in beef herds is higher when calving takes place in a confined area, such as a pen, shed, or dry lot, compared to calving on pasture (Sanderson and Dargatz, 2000). In dairy herds, diarrhoea (Frank and Kaneene, 1993), respiratory problems (Svensson et al., 2003), and the risk of *Salmonella* infections (Losinger et al., 1995) are lower when calving in individual calving pens versus group settings. In maternity areas, removal of soiled bedding can also help reduce the incidence of diarrhoea (Frank and Kaneene, 1993).

In general there is little information available on the effects of outdoor housing on the welfare of calves born outside. Clearly, environmental conditions such as extreme cold and hot weather, wind, and rain will play a significant role in the welfare of the newborn and the growing young stock and we encourage future work in this area.

3 Housing Separately from Mother

One criticism of modern veal production and dairying concerns the raising and housing of calves separately from their mothers. In general, calves raised for beef production remain with their mothers until weaning at approximately 6 months of age. However, with veal and most dairy production systems, the calves are separated from the mothers soon after birth (Figure 7.2). Clearly, there are many factors that can affect the welfare of calves when they are separated from their mothers. Some of these factors are related to the nutrition of the calves and are dealt with in Chapter 8. However, some insights into the success of separate housing can be derived from comparisons of pre-weaning mortality rates.

Mortality rates tend to be similar in dairy and veal calves. Rates can be lower for beef cow–calf herds, although the difference is small with sizeable differences between surveys and countries (Table 7.1). There have been several large-scale epidemiological studies of pre-weaning calf mortality in beef cow–calf herds and dairy herds, but fewer epidemiological studies are available for veal calves and these estimates are more difficult to assess because of a greater use of culling and antibiotics on veal farms than typically occurs on dairy or beef farms. Bearing these limitations in mind, it would seem that separation from the mother at birth could contribute to an increased calf mortality. However, there is much variation between farms of a similar type. For example, Losinger and Heinrichs (1997) report that nearly half the US dairy farms surveyed had mortality rates between 0% and 6%, which is comparable to beef cow–calf herds and substantially less than the national average of 9.4%. In Denmark, calf mortality on dairy farms has been reported to vary from 0.7% to 37.9% (Agerholm et al., 1993). Mortality rates on North American veal farms are reported to vary from 0% to 30% (Sargeant et al., 1994; Stull and McDonough, 1994). A recent study investigating the genetic and environmental influences on calf vigour at birth and mortality reported that management of non-genetic factors would result in far greater improvements in calf vigour and mortality than genetic selection (Riley et al., 2004). For example, these authors report that Brahman or Brahman cross calves with difficult births had 2.59 times greater odds of poor birth vigour and 12.9 times greater odds of death before weaning than calves born with no dystocia. Thus, it would seem that improved housing and management can reduce mortality of dairy and veal calves to a level that we would expect were the calves to have remained with their mothers. The relatively small impact of such early separation of the calf from its mother may reflect the relatively precocial state of the calf at birth.

A number of studies have reported behavioural consequences of early separation of calves from their mother, which may reflect welfare problems. However, these effects are probably as much due to nutrition as housing, and so we consider these in more detail in Chapter 8.

Figure 7.2 In most developed countries, most calves raised for veal (above) or dairy production (below) are separated from their mothers soon after birth and housed in individual crates, pens, or hutches. This early separation and the use of individual housing (that provides no real social contact and only limited opportunities for movement) have been the focus of criticism from animal welfare groups. European Union legislation prohibits the use of individual housing for calves over 8 weeks of age. However, many farmers insist that individual housing for calves reduces the risk of disease

Table 7.1 Published estimates of rates of pre-weaning mortality of calves kept for beef, dairy, and veal production

Type of production	Country	Mortality rate (%)
Beef (cow–calf)	Canada	5.6[1]
	Switzerland	6.3[2]
	USA	3.7[3]
Dairy	USA	6.3[4]
	USA	9.4[5]
	UK	7.8[6]
	Denmark	5.4[7]
	Sweden	2.6[8]
White veal (milk fed)	Canada	8.8[9+]
	The Netherlands	5.4[10+]
	USA	5.8[11+]
Red veal (grain fed)	Canada	6.2[12+]

[1] Dutil et al., 1999; [2] Busato et al., 1997; [3] Sanderson and Dargatz, 2000; [4] Wells et al., 1996; [5] Losinger and Heinrichs, 1997; [6] Esslemont and Kossaibati, 1996; [7] Agerholm et al., 1993; [8] Olsson et al., 1993; [9] Sargeant et al., 1994; [10] Smits and de Wilt, 1991; [11] Stull and McDonough, 1994; [12] Sargeant et al., 1994; [+] Combined mortality and culling rates.

4 Housing for Unweaned Calves

4.1 Introduction

One of the most contentious issues in the raising of calves, and one that has attracted considerable criticism from the public at large as well as animal welfare groups, is the use of individual housing for unweaned calves. This is most common in the case of surplus male dairy calves raised for veal production, but is also common for dairy replacement heifers (Figure 7.2). Calves are housed in individual pens that prevent social contact, limit opportunities for movement and, in the case of veal production, may be too small to allow the calves to turn around. In some cases, calves are tethered. Largely in response to public pressure, the countries of the European Union (EU) effectively banned individual housing for calves over 8 weeks of age, stimulating interest in group housing systems (Figure 7.3). Individual housing continues to be widely used in North America (Stull and McDonough, 1994), although a number of veal producers in this region are now adopting group housing. Although the public focus has been upon calves kept for veal production, individual housing in a limited space is widely used in dairy production for unweaned replacement dairy calves. For example, the latest survey from the USA shows that 58% of dairy farms keep unweaned heifers in individual housing (USDA, 2002). In the case of calves reared for dairy production, however, there tends to be a wider range of options used by farmers compared to veal production. Unweaned dairy calves can be found at pasture with their mothers, in large or small groups either at pasture or indoors, in indoor individual pens, or in outdoor

Figure 7.3 The development of group housing systems for veal calves (above) and dairy heifers (below) has been driven by concern about animal welfare, the cost advantages associated with reduced labour needs and legislation. Comparisons of group housing systems with individual housing are complicated by the fact that group housing systems can vary greatly in the size of groups, the stocking density, the type of bedding, quality of ventilation, etc., all of which can affect the welfare of the animals. Epidemiological studies of dairy herds in the USA and Sweden have shown that calf mortality increases when group size is larger than 6–10 calves. However, in smaller groups, calf mortality is similar to that in individual housing

"hutches" (Figure 7.2), which may be wood or plastic, with the calves either locked inside the hutch, or allowed access to a small run.

Individual housing can have advantages for animal welfare, the largest being the reduced transmission of infectious diseases as a result of physical contact between calves. Furthermore, individually housed calves may be easier to observe and treat for signs of illness. Aggression between calves and competition over resources such as food is also limited with individual housing. However, there are also potential disadvantages with this practice. Most obviously, the calves are denied most forms of social contact and movement is restricted by the limited physical space that is usually provided. A substantial body of research has now assessed these advantages and disadvantages for calf welfare. Again, as in the case of housing for adult cows, the research is often not easy to interpret. Often comparisons are made between housing systems that differ in many respects, such as use of bedding, indoor versus outdoor housing, space allowance, etc. This makes it difficult to determine which variable is really most important. In many cases, important variables, such as the quality of the ventilation, are not described making it difficult to know how well the results can be generalized. Given these problems, it is perhaps not surprising that different studies sometimes report quite different conclusions.

4.2 Health Effects

Although individual housing is often recommended as a means of reducing disease transmission between unweaned calves, the research that has examined the relative health of calves housed individually or in groups has produced conflicting results.

Early epidemiological studies of veal calf housing systems did detect some health problems associated with group housing, but the relationship between group housing and morbidity was not straightforward. Webster et al. (1985a, b) examined 14 veal farms that bought in male dairy calves for veal production. Some of the farms kept the calves in individual wooden crates while others kept them in straw-bedded pens. The mortality rate (up to 16 weeks of age) was higher for the group-housed calves (3.8%) than the individually housed ones (1.7%). It should be noted, however, that the mortality rate for the group-housed calves was lower than the 4.2% average, and at the low end of the range of 0–30% mortality that is reported for veal farms in the USA that use individual housing (Stull and McDonough, 1994). Thus the difference between farms that use the same type of housing system can be larger than the average difference between different types of housing systems. Clearly some farmers can keep unweaned calves together in a group and attain a level of health as good as is found with individual housing.

Webster et al. (1985b) also noted a higher incidence of respiratory disease in calves that had been brought in from other farms and group housed rather than individually housed. This was especially evident during the first 2 weeks on the unit (31% for group housed vs. 0% for individually housed), but the relative advantage of individual pens persisted until 10 weeks of age. However, when other farms were examined that housed dairy replacement calves born on the farm itself, the incidence of respiratory diseases was similar for group and individually housed animals. This suggests either that the effect of group housing on morbidity was specific to the types of group management used in veal production, or that there was some interaction between the use of group housing and the bringing in of calves from other farms. The incidence of gastrointestinal (GI) disorders showed similar complexity. During weeks 0–2 the probability of having calves with GI disorders was higher for farms that group housed veal calves (71% of farms) than for farms that individually housed the calves (29% of farms). However, this difference had disappeared by 6–10 weeks. For farms with farm-born dairy calves, the same difference was apparent, but only for the farms that fed warm milk. Farms that group housed the calves but fed them cold, acidified milk had the same, low incidence of GI problems as farms that individually housed the calves. Furthermore, from 2–6 weeks the situation was the reverse; the incidence of GI tract disorders was higher on farms that used individual housing.

Together, these results suggest that health problems associated with group housing of veal calves may be specific to the particular management on the farm and may interact with the diet of the calves. More recent studies of veal calves reared in modern group housing tend to report very good health status and similar or improved growth rates compared to individual housing (Andrighetto et al., 1999; Xiccato et al., 2002).

Several large-scale epidemiological studies of health disorders in unweaned dairy calves fail to show a clear advantage of individual housing, although there is clear evidence that health problems can occur when group size is large. Group housing is not associated with an increased chance of calves being infected by *E. coli* O157 (Rugbjerg et al., 2003), *Salmonella* (Losinger et al., 1995), or the protozoan parasite, *Cryptosporidium parvum* (Mohammed et al., 1999), although there is some evidence that group housing may increase the chance of Johne's disease. A large-scale study of 1,685 dairy farms in the USA found that farms with unweaned calves in groups of seven or higher were more likely to experience high calf mortality (>6%) than farms that individually housed calves. However, farms that kept the calves in groups of six or less had similar mortality rates to the farms with individually housed calves (Losinger and Heinrichs, 1997). The detrimental effect of large groups was also shown in a study of 122 dairy farms in Sweden (Svensson et al., 2003). The farms were classified according to whether the unweaned calves were kept individually, in small groups (3–8 calves fed milk manually) or large groups (6–30 calves fed with an automatic milk dispenser). The incidence of diarrheoa did not differ markedly between the type of housing, although the diarrheoa was rated as more severe (i.e. loss of weight or suppression of appetite for 2 days or more) in calves housed in large groups than for calves housed individually or in small groups. The incidence of respiratory disorders was twice as high in calves in large groups compared to calves in small groups or in individual pens. In both cases there was no difference between calves in small groups and those in individual pens. Subsequent analysis of body weights (Lundborg et al., 2003) showed that calves in small groups recorded the highest gains, those in large groups experienced the smallest gains, and individually housed calves had intermediate results. More recent work has confirmed a higher incidence of respiratory disease in large groups (Svensson and Liberg, 2006).

These large-scale epidemiological studies throw doubt on the claim that individual housing of unweaned calves is advantageous for their health, although they do indicate that there can be problems when the calves are housed in large groups (more than 6–8 animals). However, there are a number of weaknesses with these studies. First, they show average differences associated with the average farm, but this does not mean that such differences will inevitably be found on any one farm. For example, Kung et al. (1997) found fewer health problems on one farm when calves were kept in relatively large groups (12–15 calves) compared to calves kept in individual pens. Another problem with epidemiological studies is that use of group housing may be confounded with other management variables. For example, the group-housed veal calves studied by Webster et al. (1985b) were fed from a teat whereas the individually housed calves were bucket fed. The large groups studied by Svensson et al. (2003) differed from the small groups in the method of feeding (automated milk feeders vs. manual feeding) and in the age range of the calves: the authors suggest that the main disadvantage of the large groups was due to variation in calf age. Thus health effects reported in these studies may not be due to the size of the group per se.

To overcome these potential confounding effects, smaller-scale studies have isolated the effects of group housing by controlling for feeding or management. Hänninen et al. (2003) and Chua et al. (2002) examined the health and growth of calves kept either in individual pens or in group pens (with either two or four calves), but which were fed and managed identically. Neither study found a difference in growth rates, and Hänninen et al. (2003) found that the incidence of diarrhoea was actually lower in the group-housed calves. Viral pathogens are often responsible for causing enteric disease and the type of pen will have little effect on transmission of these organisms through the air (Wathes et al., 1988). In addition, some physical contact between calves still occurs at the end or top of pens or through slatted partitions of individual pens, allowing for disease transmission. Proper management of housing systems (cleanliness, adequate ventilation, feeding), as well as calf immunity, are likely more important than housing system for disease. Thus, these controlled studies support the larger epidemiological studies in showing that unweaned calves can be kept in small groups without increased health problems, providing that housing, feeding, and management is appropriate.

4.3 Behavioural Effects

The most obvious behavioural effects of individual housing, which raise concern about the welfare of the animals, are the lack of opportunities for the calves to engage in social interactions, and the limited ability of calves to move. The latter will depend on the size of the space available, but generally calves kept in groups have a larger total area available, even when the space per animal is the same. The individual housing typically used in dairy production does not allow sufficient room for the animal to run or jump, while in commercial veal production, the crates often do not allow sufficient space for the animal to turn around. On the other hand, individual housing will reduce the incidence of aggressive behaviour and competition over resources such as food, and can prevent cross-sucking (Chapters 4 and 8), all potential positives for animal welfare. Because of the importance of adequate rest and sleep for young calves (Chapter 4), researchers have also examined the effects of individual housing on the resting behaviour of calves; most concern in this regard has been over the use of veal crates that may be too small to allow the animals to adopt the normal resting postures (Figure 7.4).

4.3.1 Social Behaviour

How much social behaviour do calves show, and how does the absence of social behaviour in individually housed calves affect their welfare? In one of the most comprehensive attempts to assess the effects of housing on calf behaviour, Webster et al. (1985a) visited 70 farms that housed unweaned veal and dairy calves in a number of different systems. The behaviour of 193 veal calves and 183 replacement dairy calves reared in groups and of 59 veal calves and 359 replacement dairy calves reared in individual veal crates or pens was observed. Because of the large number of calves involved, the amount of time that could be spent observing their behaviour was brief, limiting the reliability of the data. The animals were observed during 15-sec scan samples that were repeated at 5 min intervals, for a total of 4 h/day (but only during the day time) once at weeks 2, 6, 10, and 14. The group-fed calves spent between 1% and 2% of the time playing, fighting, or mounting. This was similar to that observed for "suckler" calves reared with their mothers, and did not change greatly with the age of the calves. This behaviour was prevented due to limited space in the veal calves kept in individual crates, but occurred with similar durations in the individually housed dairy calves. No details were given on the layout of the pens, but presumably pens used to house the dairy calves allowed some contact between animals. Both individual and group-housed calves spent about 1% of the time grooming other calves, but suckler calves spent less time engaged in this behaviour perhaps because of the absence of grooming by the mothers. Interestingly, vocalizations occurred more frequently among the individually housed calves than among group-housed or suckler calves, perhaps as a result of the social isolation (Chapter 4).

A comprehensive observational study was carried out by Chua et al. (2002) who carried out detailed observations (observations over 24 h, once per week for 7 weeks) on calves that were housed either individually or in pairs. Again, the pair-housed calves spent about 2% of the time engaged in social contacts.

What do these results tell us? First, even very young calves engage in social behaviour, but this takes up a fairly small percent of their time (less than 1 h/day). Second, housing calves individually does not prevent social behaviour from occurring; some contact can still occur depending upon the layout of the pens. The fact that vocalizations are more common in individually housed animals, and that calves will work to gain access to social contact (Holm et al., 2002; Chapter 4), suggests that the social contact is important to them.

What are the consequences for the calves of the reduced opportunity for social contact when housed individually? An obvious possibility is that individually housed calves may not develop the social skills necessary to cope with group living

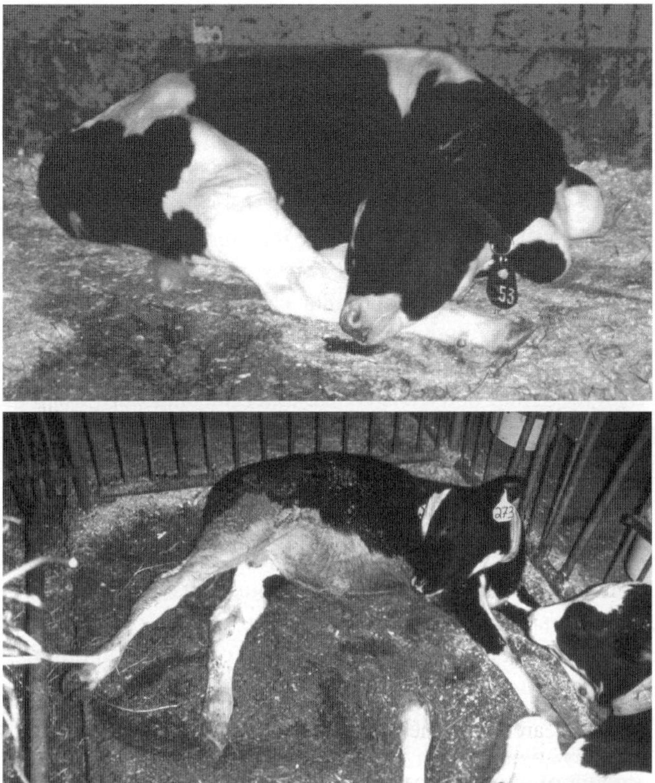

Figure 7.4 Calves adopt a number of different postures when resting. Postures in which the head is resting on the ground or on the calf's back or leg (top) are associated with rapid eye motion sleep. Postures in which the calf is lying on its side with its legs fully extended (middle) help thermoregulation. This is important for milk-fed calves that can generate much metabolic heat. Use of pens that are too small, such as on some veal farms (bottom), may prevent the calves from adopting these resting positions

later on, being either more or insufficiently aggressive or more fearful of other calves. Earlier research showed that unweaned calves that had been reared individually show more exploration of unfamiliar calves when they are given the opportunity to make social contact (Dellmeier et al., 1985). More recent research has replicated this finding and has attempted to unravel the motivational changes that underlie this effect. Jensen et al. (1997) kept calves either in single pens or group pens for 3 months. The calves were then subjected to an "open-field" test with an unfamiliar calf. The open-field test and the problems of its interpretation are discussed in more detail in Chapter 4. Individually housed calves had higher heart rates and showed longer latencies to approach the unfamiliar calf suggesting that these calves were more fearful. That this fearfulness was specifically of the unfamiliar calf, rather than of the enclosure itself, was shown by the fact that the individually housed calves took longer than the group-housed calves to enter the enclosure when the unfamiliar calf was present, but not when no calf was present. Although these results do indicate that individual housing makes calves fearful of other calves, the effects appear to be short lived: when the calves were retested after a further 3 months of similar housing the effects of early housing were no longer present.

A subsequent study (Jensen et al., 1999) found some longer-term effects. Calves that had been either individually housed or group housed for the first 3 months of life and then kept in similar tie-stall housing after weaning were tested at 26 weeks of age by introducing them into a small area that contained an unfamiliar heifer of the same age. The heifers that had been housed individually as unweaned calves sniffed and mounted the unfamiliar heifer less and engaged in less mock fighting (interpreted as play) than heifers that had been group housed. There was no difference in the amount of agonistic behaviour (butting or withdrawing). Other research has indicated that individual housing of dairy heifers reduced their ability to compete within groups later on (Broom and Leaver, 1978). Veissier et al. (1994) examined calves housed individually or in a group of eight animals until 14 weeks of age. At 14 weeks, all animals were placed in groups with unfamiliar animals. During the 2 h after mixing calves that had been individually housed showed more aggression (and less positive social behaviour such as playing or grooming) than the group-housed calves. However, when the mixing was repeated 5 weeks later, no differences were found. Thus individual rearing may reduce the calf's ability to cope with strange animals during initial encounters but this is likely overcome by a few weeks of group living.

In general, use of individual housing for unweaned calves does influence how calves react to other calves, but the effects appear temporary. We have little evidence that individual housing has long-term effects on social behaviour that significantly affect the calves' welfare.

4.3.2 Locomotion

Young, growing animals need exercise and there is considerable evidence from humans and laboratory animals that insufficient exercise can affect growth and health. Probably the greatest impact of individual housing for young calves on their welfare is in reduced opportunities for exercise. The amount of locomotion that calves show will obviously depend upon the amount of space available to them, but grouped calves usually have more total space available even though the amount of space per calf is the same. The presence of social companions can have a facilitating effect on locomotion: young calves play by running together from one end of the pen to the other. It is also likely that aggressive encounters among calves increases the amount of locomotion they show.

Although there are a number of reasons for expecting that calves kept in groups will, in general, have more opportunities to move around than individually housed calves, not all studies have found such a difference. For example, Webster et al. (1985a) noted that dairy and veal calves spent between 3% and 7% of their time in locomotion, regardless of whether they were housed in groups or in individual pens. That veal calves in crates showed as much locomotion as grouped animals in a pen may seem paradoxical. However, this study scored movements regardless of the type of locomotion: even veal calves in creates can pace backwards and forwards a few steps but this may not be equivalent to the running by grouped calves. Interestingly, suckler calves in a paddock spent about 10% of the day moving, suggesting that the space allowance even in group pens may be restricting the calves' locomotory behaviour. Since observations were only taken during daylight hours it is difficult to estimate how many hours per day the calves are active.

Other smaller-scale studies that have used more detailed observation have shown that individually housed animals tend to move less than grouped animals. Chua et al. (2002) compared individually housed calves and calves housed in pairs. The paired calves moved twice as much as the individually housed calves (1.43% vs. 0.64% of the day) despite the fact that the space allowance per animal was the same in each type (2.04 m/animal). Hänninen et al. (2003) compared individually housed calves (1.2 m/calf) with group-housed animals (4 calves with 8 m/calf) and found more movements among the group-housed animals (5.4% vs. 3.5%).

Unfortunately, the research that has examined locomotion in young calves does not provide enough information to enable us to draw firm conclusions about how the type of housing impacts on the calves' ability to move around or to assess the likely effects upon the animals' welfare. First, all studies to date have tended to lump together all forms of locomotion, so that taking a few steps backwards and forwards in a veal crate, is considered as the equivalent of running for several metres and jumping. It is unlikely that these different forms of locomotion are equally important for the calf. Furthermore, the amount of locomotion that occurs in pens will depend on the space provided, which may not be adequately described by the number of square metres per animal. For example, we suggest that calves are more likely to run and jump when housed in a long narrow space than they are in a square enclosure of the same dimensions. Calves may also have a greater incentive to move in some environments; for example, on pasture calves may move to access fresh grass, shade, or social companions.

What are the likely consequences for animal welfare of a reduced time spent moving? Individually housed calves will show more locomotion, especially running and jumping, when given the opportunity to do so (Dantzer et al., 1983; Dellmeier et al., 1985) suggesting that they are motivated to perform these behaviours. The most likely long-term effects on animal welfare would be poorer bone, muscle, and cardiovascular condition. We know little about the importance of exercise for the health of growing calves, but there is a large body of research showing the health benefits of exercise on other species. The effect that the housing environment of the calf has on later health of the cow needs more study.

4.3.3 Cross-Sucking and Aggression

One possible behavioural advantage of individual housing for unweaned calves is that this reduces the incidence of cross-sucking between animals. Cross-sucking can occur at a high frequency among group-housed, unweaned calves, although several studies have now shown that calves can be kept in groups with only a very low incidence of cross-sucking (discussed in Chapter 8). In any case, the incidence of cross-sucking appears to be more related to the way that the animals are fed (Chapter 8) and can be controlled by appropriate feeding techniques. In general, individual housing is neither necessary nor sufficient to prevent cross-sucking between calves.

Aggression also seems uncommon among unweaned calves (e.g. Webster et al., 1985a; Veissier et al., 2001) and its incidence would not seem sufficient to justify individual housing. However, group-housed calves may still displace each other from important resources, such as food (see von Keyserlingk et al., 2004) or favoured resting locations. The ways that grouping calves can affect feeding behaviour are addressed in Chapter 8.

4.3.4 Resting Time

In view of the importance to growing animals of adequate rest and sleep, individual and group housing have been compared a number of times in terms of how long the animals lay down and in what postures they do so. One concern that has been raised about the use of individual housing, such as veal crates, is that the space provided is too little to allow the calves to lie down in normal sleep postures, especially as they age (de Wilt, 1985; Figure 7.4). On the other hand, group housing of calves may result in calves' sleep being disturbed by other calves, and subordinate calves being displaced from preferred resting locations. Again, the amount of sleep and rest that calves show in any housing system will depend upon the details of the system, such as the actual space allowance, type of flooring, etc., and so caution is needed to interpret the results.

In their large-scale study, Webster et al. (1985a) found that suckler calves reared with their mothers at pasture lay down for about 50% of the time at 2 weeks of age, decreasing to about 37% of the time at week 14. It should be noted that observations were made during the day when calves tend to be most active, and so the figures underestimate the daily total time that calves lay down. However, the suckler calves provide a useful reference group to compare both group-housed and individually housed calves against. In general, group-housed dairy and veal calves lay down somewhat longer – about 60% of the time during week 2, but individually housed veal and dairy calves spent about the same amount of time lying down as the suckler calves. Thus, grouping calves need not interfere with their overall resting time, and while individual housing reduces total rest time somewhat, this is similar to that observed with suckler calves. The largest difference between the different housing systems, however, occurred with the time spent lying flat on the side (Figure 7.4). This occurred between 2% and 5% of the time for suckler calves and between 1% and 3% for group-housed calves. However, this posture was never seen in the veal calves kept in individual crates probably because it was prevented by the small size of the crates (less than 0.7-m wide). This issue is discussed in more detail in Section 3.4 of this chapter.

More recent work by von Keyserlingk et al. (2006) reported that calves show a distinctive diurnal pattern in lying behaviour (Figure 7.5), underlying the importance of 24-h observations. Chua et al. (2002) were able to follow calves 24 h/day, and both the individually housed calves and pair-housed calves were given the same space allowance (1.2 × 1.7 m), and this was adequate to allow the calves to adopt the normal resting postures. Flooring was also identical under the two conditions. In this study, individually housed calves were seen to lie down for 72% of the day, while pair-housed calves lay down for 70% of the day. We can conclude that resting time of calves is roughly similar whether group housed or individually housed, providing that the flooring is sufficiently comfortable and that the crate is wide enough so as to not physically prevent calves from adopting normal resting postures.

4.3.5 Conclusions

In conclusion, group housing for unweaned calves does not inevitably lead to increased health problems if the groups are small (7–10 animals) and well managed. Calves are motivated to seek the company of other calves and individual

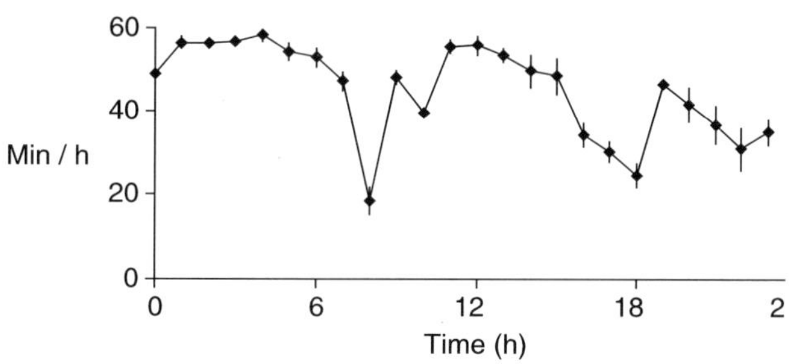

Figure 7.5 Diurnal pattern of lying time for milk-fed dairy calves (n = 6). These calves were 32 ± 4 days of age at the time of observation. Milk was available ad libitum from a teat, but fresh milk was provided twice daily at 0800 and 1800 h. (Adapted from von Keyserlingk et al., 2006.)

housing reduces the opportunities for social contact. However, the affect of this reduction on calf welfare is not well understood. Individual housing can influence how calves react to other calves, but the effects are temporary, and probably outweighed by the type of housing used after weaning. Because of the smaller total amount of space available to the calves, individual housing provides fewer opportunities for physical exercise, and this is likely to have a negative impact on the overall welfare of the calf. In general, calves rest adequately in both group and individual housing, provided that the calf has sufficient space to adopt the important resting postures. Behavioural problems often claimed to be associated with group housing, that is, increased aggression, competition, and cross-sucking can be controlled by appropriate management. However, group housing for unweaned calves will not always improve or reduce the welfare of veal calves; this will depend upon the details of the housing system, such as the group size, the spatial density, the use of bedding, as well as management factors, especially those aimed at protecting calf health.

4.4 Dimensions of Individual Housing for Calves

The consequences of individual housing for the welfare of animals will, of course, depend upon the amount of space provided. Although space availability can affect both veal calves and dairy replacement heifers, research to date has focused largely on the veal crate. Individual housing normally limits the calves' ability to walk or run, and it seems unlikely that under commercial conditions individual pens will ever be large enough to allow full expression of such behaviours. Hence, much of the research has focused simply on whether the individual pens are large enough to allow the animals to comfortably lie down.

Calves lie down in a variety of postures, either on their side or back, with the head supported by the neck or resting on the ground or body, and with the legs extended or not (Chapter 4; Figure 7.4). One of the most common complaints of the traditional veal crate is that it is too small to allow the calves, when older, to lie down with their legs outstretched. This is due not so much to the actual area of the crate as to its width, and this has been the focus of most research.

Veal calves housed in the traditional narrow crates appear to lie down as long as calves kept in group pens but are less likely to lie down with legs extended (de Wilt, 1985; Le Neindre, 1993; Stull and McDonough, 1994; Andrighetto et al., 1999) or flat on their sides (Webster et al., 1985a). According to Webster et al. (1985a), adopting this posture helps the calves lose heat, and since milk-fed calves generate considerable metabolic heat, their inability to adopt this posture may cause problems for thermoregulation. de Wilt (1985) reports that crated calves spend less time lying with their heads turned back over their bodies. This last posture may be important for calves to sleep properly and preventing calves from adopting these postures may result in some REM sleep deprivation (Hänninen, 2007).

Researchers have tried to determine what size of crate is necessary to allow the calves to adopt their normal resting postures. Detailed observations of the amount of space taken by calves when resting suggest that in order to lie with legs outstretched, calves weighing 70–210kg require 60–75-cm width (according to Van Putten, 1982), and calves weighing 170–300kg require crates 80–95-cm wide (according to Ketelaar-de-Lauwere and Smits, 1991). Webster et al. (1985a) also concluded that calves weighing more than 100kg should be kept in crates at least 85-cm wide. Tennessen and Whitney (1990) report that for 4-month-old calves (135kg), 60cm is the average width required to lie with the head turned back, although some calves require up to 70cm. However, Andrighetto et al. (1999) observed that calves in 60-cm-wide crates spent as long as grouped calves lying with their head laid back on their bodies.

Other work has indicated that even the larger of these sizes may be inadequate. Le Neindre (1993) noted that calves spent less time resting with all legs bent when kept in 65 versus 55cm crates at 13 weeks of age, but this difference disappeared at 17 weeks of age. At this age, time spent resting with all legs bent was only reduced when crates were 1.1m in width. Wilson et al. (1999) also noted that increasing the width of the crate from 56 to 76cm did not affect the amount of time spent in various lying postures, although calves in 56-cm-wide crates could not stretch one or more legs while lying down.

It seems that for veal calves to adopt their normal resting postures throughout the growth phase, crates need to be at least 1-m wide. However, a survey of veal farms in California found that nine out of ten calves were kept in crates that had a width that varied between 48 and 55cm (Stull and McDonough, 1994), and the Veal Quality Assurance Scheme of the American Veal Association now recommends a minimum width of about 66cm (Schnepper, 2001). Despite the improvements that have taken place in the size of individual pens for veal calves, it seems that these are still too small to allow calves to adopt their normal resting postures.

Small crates seem to have little effect on the growth of the calves. Terosky et al. (1997) found no growth differences for calves kept in crates that were 56-, 66-, or 76-cm wide. Either there is no effect of crate size, or all of the sizes studied were too small. Some studies have examined the effects of crate size on plasma levels of cortisol or immune parameters (e.g. Terosky et al., 1997; Wilson et al., 1999) but because of the difficulties interpreting these measures in terms of animal welfare (Chapter 3), we do not report these results here.

Much of the difficulty in assessing the effect of crate size upon the welfare of calves comes from the lack of information on why calves adopt the resting postures that they do. Although there are good reasons for assuming that some postures are important for thermoregulation or for the different phases of sleep (Figure 7.4), more research is needed on the functions of rest and sleep in calves, and the role of resting position in promoting adequate rest.

Finally, while most research has focused on the behavioural consequences of pen size, small pens can influence the welfare of calves in other ways. Recently, increased pens size has been found to be associated with a lower incidence of diarrhoea (Svensson and Liberg, 2006) and lower respiratory–disease–causing bacterial counts (Lago et al., 2006).

4.5 Bedding and Flooring

In the previous chapter we showed that the lying surface is particularly important in promoting cow comfort. Traditionally, farm animals were provided with some sort of organic material for bedding, usually straw, but the recent trend is to reduce the use of such bedding, primarily because of labour costs involved in cleaning and because of concerns about hygiene. For example, dirty bedding can increase the incidence of diarrheoa (Frank and Kaneene, 1993) and of *Cryptosporidium* infection (Mohammed et al., 1999). In some cases, animals are kept on bare concrete or wooden floors, usually slatted to allow the drainage of urine and feces. Concern about the effect of these surfaces on the welfare of calves has led to a number of studies.

For calves kept for white veal production, organic material tends not to be used because of the concern that bedding might be ingested and serve as a source of extra iron. Thus there has been considerable interest in finding alternative flooring for use in veal production. One alternative is slatted flooring, but veal calves kept on this surface suffer from knee injuries: Webster et al. (1985a) reported that 20% of veal calves kept on wooden slats had cut, swollen, or bruised knees, and far fewer injuries were observed on calves kept at pasture or on straw bedding. Milk-fed veal calves housed in elevated crates with wooden slatted floors were reported to have better growth rates and spend less time standing up than calves in crates with sloping rubber covered floors (Verga et al., 1985). However, in this study, the pens with the different floors were in different barns with different ventilation systems, etc., making this finding difficult to interpret. When given a choice, calves spend longer lying on hardwood slatted floors compared to rubber-coated synthetic plastic slats (Stefanowska et al., 2002). On both flooring surfaces calves often slipped when standing or walking (an average of one slip every 15 min). Hänninen et al. (2005) found no difference in growth rates or in the amount of time spent lying down between calves kept in individual pens with solid concrete floors or soft rubber mats. Nor did the type of flooring appear to affect measures of the HPA axis activity or secretion of growth hormone (Hänninen et al., 2006). The importance of the softness of flooring probably depends on the weight of the animals: the lighter weight of young calves may explain why concrete or wooden flooring do not have the same negative effects for calves as they do for adult cattle. However, a clean dry surface will still be important, and the thermal protection of bedding may be especially important for young calves kept in cool conditions.

4.6 Outdoor Versus Indoor Housing

As with adult animals, growing cattle are increasingly being housed indoors with little or no outdoor access. Indoor environments might be expected to have some disadvantages for calf health since many animals are sharing the same airspace, increasing the risk of airborne disease. On the other hand, outdoor-housed calves are exposed to greater variation in environmental conditions such as hot and cold temperatures, wind and rain. There is little research on the effects of exposure to wind and rain, but there are a few studies that have compared indoor and outdoor housing. Unfortunately, as discussed in the previous chapter, variation in the design of indoor housing systems and differences in management among systems makes it difficult to conclude much about their relative advantages. In particular, the advantages of indoor versus outdoor housing will vary greatly according to the type and quality of the ventilation system used and on the climate, making it difficult to draw conclusions.

In recent years, North American dairy producers have adopted outdoor hutches for dairy calves (Figure 7.2), and this is now the most common type of outdoor housing for dairy calves in the USA (USDA, 2002). A number of studies have compared outdoor hutches with indoor housing. Some studies have reported reduced disease and mortality and improved growth of dairy calves in outdoor hutches compared to indoor individual pens (McKnight, 1978; Fiems et al., 1998). However, one of these studies (McKnight, 1978) actually found lower growth in hutch-housed calves during winter months, and other work (Friend et al., 1985; Frank and Kaneene, 1993) has found no advantage to hutches in health and growth rates. One study (Kung et al., 1997) found higher morbidity (measured by days of medication) in dairy calves kept in hutches compared to group-housed calves indoors.

5 Housing of Weaned Dairy Cattle

After weaning, and (for heifers) before first calving, cattle are probably at their most resilient in terms of health status, suffering neither from the vulnerability of the unweaned calf, nor from the welfare problems associated with high milk production that afflict the adult dairy cow. Perhaps for this reason, the welfare of these animals has largely been overlooked. The main exception concerns beef cattle destined for slaughter, which is discussed in the next section.

The type of housing used for weaned cattle is highly variable, depending on the age of the animal and upon the type of production. Beef heifers kept for breeding are generally kept on pasture. For dairy heifers, the best available data come from the USA. Most farms (83%) provide weaned heifers with some access to outside areas, during some part of the year. Farms that provide outside access for weaned heifers are divided equally between those offering access to pasture and those offering access only to a dry lot (USDA, 2002). In contrast to unweaned heifers, weaned heifers that are kept indoors are kept mainly (88% of farms) in group pens, although a small number of farms keep them in tie stalls (7%) or individual pens (5%), or in free stalls (6%) (USDA, 2002).

The welfare of heifers housed in group pens will depend upon the details of the pen and housing. Few studies are available that have looked at these in any detail.

As in previous sections, space allocation is one obvious feature to consider. Aggression among heifers is not common, even when mixed with unfamiliar individuals (Veissier et al., 2001), but competition can occur. Grouping heifers by body weight may help limit these effects. Hindhede et al. (1999) found that weight gains of small heifers (130–250-kg body weight) were increased if animals were housed in small groups of similar weight compared to in larger groups mixed with larger animals. The opposite effect was found for the larger heifers (250–380 kg) suggesting that these were benefiting from the competition with the younger animals.

The type of bedding used can be important in affecting resting times. Heifers lie down more quickly and easily when housed on straw bedding compared to concrete floors (Müller et al., 1989). Any bedding that is provided obviously requires careful attention. Incomplete or irregular removal of bedding has been found to increase the risk of infection by *Cryptosporidium parvum*, a protozoa causing diarrheoa, anorexia, and general loss of condition (Mohammed et al., 1999).

Pregnant heifers are likely to experience fewer problems than lactating cows when kept in similar housing systems, given their smaller size and more modest energetic demands. For example, pregnant heifers housed in tie stalls or free stalls appear to have fewer hoof lesions and a lower incidence of lameness than lactating cows in similar housing systems (Bergsten and Frank, 1996). Some earlier experience with the adult housing system likely provides advantages later in life. Problems in the use of free stalls, for example, can be reduced by training heifers with stalls, and encouraging animals to enter these with food rewards and extra bedding (O'Connell et al., 1993).

Apart from these studies, very little research has examined the effects of housing on the welfare of weaned heifers. In general, dairy researchers have paid relatively little attention to heifers, even for factors important for production (Mourits et al., 2000), and we encourage more research on the welfare and housing of these animals.

6 Beef Feedlots

6.1 Introduction

Beef cattle are kept in a wide range of housing systems, varying greatly from one geographical region to another (Figure 7.6). It is not the purpose of this book to discuss each type of housing system, or to elucidate all possible welfare problems associated with each. Recent publications contain useful descriptive information about beef production systems in Europe, Australia, and North America (USDA, 2000; Scientific Committee on Animal Health and Animal Welfare, 2001; Petherick, 2005). Research to date on the effects of housing on the welfare of beef cattle has focused on only a few issues.

Figure 7.6 Beef cattle can be found in a variety of types of housing. Most common are the pasture-based systems, which are found in many parts of the world and are the most common for cow–calf herds (top). Indoor housing is common in European countries (middle), and the outdoor feedlots are more typical in the USA and Canada (bottom)

In a number of countries, cattle raised for beef production are housed with their mothers while nursing (usually called "cow-calf" herds), but after weaning they are typically transported to large feedlots, mixed with cattle from other farms, and fed a high grain diet until they reach slaughter weight. These feedlots can vary from open, unsheltered areas, usually with a dirt floor and concrete or wooden feed bunks, to smaller, semi-enclosed pens, often with a concrete floor that may be solid or slatted. Within the feedlots, each pen can house tens or even hundreds of animals. The best data on feedlots comes from the Feedlot'99 survey done in the USA (USDA, 2000). In the USA, feedlots typically contain over 1,000 head, with 40% of feedlots containing over 8,000 head. The majority of animals are growing animals (56% steers and 42% heifers) from beef breeds: only 5% of animals are dairy breeds and only 2% are adult cows or bulls (USDA, 2000).

Given the large concentration of animals, and the apparently "unnatural" type of housing, it is not surprising that a number of people are concerned about the effect of these types of rearing systems on the welfare of the animals. Again, however, we must emphasize that extensive management of beef cattle can create its own set of welfare problems (Petherick, 2005). Other welfare concerns deal with acute treatments, such as branding or dehorning, or the way that the animals are handled, as discussed in Chapters 5 and 9.

Health problems and disease represent probably the largest source of welfare problems. The overall mortality (from arrival until marketing) is estimated to be between 1% and 2% (USDA, 2000; Loneragan et al., 2001), which in relative terms may seem low, but these are young adult animals that should be at a stage of life where they are least susceptible to illness. Cattle in North American feedlots suffer from a variety of health problems, the most common being respiratory disease (57% of deaths), followed by digestive disorders, such as acidosis, liver abscesses and bloat (Loneragan et al., 2001). The incidence of respiratory disease in feedlots is high but varies greatly (5–44%) between years (Snowder et al., 2006). Nearly 2% of feedlot cattle suffer from lameness (USDA, 2000). Feedlot cattle suffer from a variety of bacterial and parasite infections, such as *E. coli, Giardia, Cryptosporidium* (e.g. Smith et al., 2001; Ralston et al., 2003), and use of antimicrobial treatment is quite high: in the USA in 1999, 19% of feedlot cattle received injectable antimicrobials for disease treatment or preventative measures, while nearly 70% received antimicrobials in the food or water supply (USDA, 2000).

Respiratory disease appears to result from the stress from transport to the feedlot, as well as mixing with unfamiliar animals and unfamiliarity with the feedlot or the type of food (Loerch and Fluharty, 1999). The extent that the physical aspects of the feedlot may affect the incidence of this disease does not appear to have been examined. The digestive disorders most likely result from the ways that the animals are fed, especially the use of high grain diets (Galyean and Rivera, 2003) that are discussed in Chapter 8. However, Galyean and Rivera (2003) suggest that the incidence of digestive disorders can be influenced by the social and physical environment. For example, the risk of acidosis is increased when there is a large variation in feed intake, and this variation could partly be affected by social dominance or aggression. Research is now needed to examine these effects and how they might be minimized by changes in housing and management.

The welfare problems of feedlot cattle are most obviously related to the physical and social environment, including temperature extremes, dust, dirty, or muddy pens, inadequate feed or water space, and high stocking density (e.g. Grandin, 2002). However, other than work on temperature stress (especially heat stress), very little research has been published on these issues.

6.2 Temperature Stress

Beef feedlots are often found in warmer parts of the world, such as the south-western US. Given that the upper critical temperature for growing beef cattle is estimated at around 25°C (compared to a lower critical temperature of −35°C; Hahn, 1999) heat stress is more often a problem than cold stress. Cattle have a variety of behavioural and physiological mechanisms to adapt to changes in temperature, but marked increases in temperature that occur more quickly than the animals' ability to adapt can cause serious problems. Heat waves occurring every 2–3 years in the southern and central US have been responsible for the deaths of thousands of feedlot cattle (Mader, 2003); as well as these obvious cases, heat stress no doubt causes considerable suffering to many millions of cattle. Considerable research has examined how heat stress on cattle can best be measured, and in finding means to reduce heat stress.

The heat load on cattle will reflect the relative balance of heat intake and heat loss. Heat intake is mainly from absorption of solar radiation and consequently cattle with black hair appear to suffer more from heat than cattle with lighter hair (Mader et al., 2002; Da Silva et al., 2003; Brown-Brandl et al., 2006). Metabolic heat for the digestion of food can exacerbate the heat load, and one major means that cattle have of adapting to heat stress is to reduce feed intake. Heat loss occurs mainly through evaporative cooling so high humidity can interfere with evaporative heat loss, and the danger of heat stress is highest with a combination of high temperature and humidity, quantified in the temperature–humidity index (Hahn, 1999). However, heat stress will also depend upon wind speed and the amount of solar radiation, and effective monitoring systems need to include all of these, rather than relying on air temperature alone (Eigenberg et al., 2000).

The animals' behavioural and physiological responses to heat can be measured and used to assess the effect of heat stress on the animals. The most direct measure is of core body temperature, which is usually assessed via rectal or tympanic temperature. The latter is preferred since it reflects hypothalamic temperature. This is a central regulator of the animal's responses to temperature changes and can be measured easily with data loggers attached to the animal (Hahn, 1999). Evaporative heat loss can occur through panting, and respiration rate is often used to assess animals' responses to heat either visually or through automatic monitoring (Eigenberg et al., 2000). Hahn (1999) estimates that respiration rate of beef cattle increases 4.3 breathes/min for every 1°C increase in temperature. Eigenberg et al. (2000) reported a linear correlation ($r = 0.73$) between respiration rate and temperature change within the range of 14–34°C. The degree of panting can also be assessed through whether the mouth is kept shut or wide open (Mader et al., 2002).

From a production standpoint, the most important response to heat stress is a reduction of feed intake; however, this may not necessarily indicate a welfare problem if it allows the animals to adapt to the higher ambient temperatures by reducing the metabolic heat. With new technology, the time that the cattle spend at feed bunks can be monitored automatically and can detect reductions in feed intake due to weather conditions (Schwartzkopf-Genswein et al., 2003). A number of other behavioural changes also occur during heat stress: in particular, cattle appear to bunch up when temperatures rise (Mader et al., 2002). The function of this grouping behaviour is not clear. Under some conditions standing close together may provide some shade, but the close proximity of animals is also likely to reduce airflow and may increase heat load (Mader et al., 2002). Decreasing inter animal distances may simply be a general response to threat by herd animals. While changes in any of these parameters can be used to assess animals' responses to temperature changes, the extent to which such changes can be taken as evidence of a welfare problem has not been resolved.

Certain aspects of the housing system can mitigate or aggravate heat stress, as reviewed by Mader and Davis (2004). Since heat stress results mainly from the direct exposure of the animal to solar radiation, the most logical way to reduce heat stress is to provide shade (Figure 7.7). A number of studies have shown positive effects of shade. For example, one study reported that when cattle were housed without access to shade, respiration rate increased 6.4 breathes/min for each 1°C increase in temperature, but when shade was available respiration rate increased by 1.6 breath/degree (Eigenberg et al., 2000). Pusillo et al. (1991) found improved feed intake, weight gains, and food conversion efficiency in cattle in Iowa feedlots that were provided with 4.18 m^2 of shaded area per animal compared to cattle housed in a similar pen without shade. However, the shaded area was in the form of additional space, and this extra space may have had some effects independently of the shade provided. In a more controlled study, Mitlohner et al. (2002) provided beef heifers in a feedlot in Texas with 2.12 m^2/animal of shade using a galvanized steel roof. Animals with access to the shade showed improved feed intake, higher daily gains, higher resting times, and lower respiration rates. These differences were likely due to temperature differences between treatments: in the shaded areas ground temperatures were far lower when no shade was available. The cattle used the shaded areas but feeding time was not reduced. The authors also noted a reduction in the amount of agonistic behaviour and "bulling", but this was only apparent at certain times of the day. Examination of the carcasses after slaughter revealed a reduction in dark-cutting meat, perhaps due to a general reduction in stress or the reduced aggression and bulling. An earlier study showed similar advantages of shade on weight gain and feed intake for cattle housed in a feedlot with concrete slatted floors (Mitlohner et al., 2001). Although shade provides obvious benefits there seems to be reluctance in the beef industry to provide shade in feedlots. In the USA, only 14% of surveyed feedlots provided enough shade for all cattle and 65% provided no shade whatsoever (USDA, 2000).

Figure 7.7 For beef cattle in outdoor feedlots heat stress can be a major threat to animal welfare, and has been responsible for the deaths of many thousands of animals. Heat stress can be alleviated by providing some simple forms of shelter that protect the animals from direct solar radiation. However, even simple shade is often not provided, despite the demonstrated economic benefits

There has been very little research on what types of shading structures are most effective. The physical structures used will affect not only solar radiation, but also air flow, humidity, and other factors. For example, windbreaks can provide some morning and afternoon shade for animals, but can also inhibit air circulation (Mader et al., 1999).

An alternative approach to preventing heat stress is to increase heat loss. Heat loss is mainly via evaporative cooling, and structures that interfere with airflow can result in reduced weight gains and feed intake in summer months (Mader et al., 1997). Under some conditions evaporative heat loss can be increased by spraying cattle with water. Experiments in controlled conditions have shown that wetting cattle by overhead sprinklers can reduce tympanic temperatures during heat stress, although the temperatures rise rapidly once the sprinklers are turned off. Studies under more commercial conditions have not shown the same success. Mitlohner et al. (2001) found no effects on weight gain, feed intake, behaviour, or tympanic temperature of spraying a fine mist over the cattle (at a rate of 0.3 L/min). The authors suggested that the lack of effect may have been due to the very small size of the water droplets produced by the mister. Davis et al. (2003) turned on water sprinklers when morning temperatures predicted high heat during the remainder of the day. The sprinklers operated for 20 min every 1.5 h from 10:00 to 17:50. Tympanic temperatures were reduced, sometimes by almost a degree, at certain times of the day (generally between 14:00 and 19:00). Whether this effect was large enough to be of significance for the animals is not clear. There is some work also indicating that inconsistent cooling regimes may be detrimental to welfare. Gaughan et al. (2004) reported that heifers subjected to sprinkling as a means of cooling during heat stress conditions showed increased feed intake, decreased respiration rates and rectal temperatures on the days where the cooling treatments were applied. However, the cooled heifers tended to have increased rectal temperatures and had higher respiration rates compared to non-cooled heifers on the days immediately following the treatment period (when no cooling was provided). The authors attributed these findings to inadequate adaptation to the hot conditions.

Perhaps because of the lack of suitable research to support the use of sprinklers, let alone how best these should be managed, they remain somewhat of a rarity in US feedlots (13% of feedlots provide sprinklers to all animals; USDA, 2002). Chapter 6 discusses some of the issues associated with sprinklers used on dairy farms.

Some heat loss can also occur through conduction when animals lie down on a cool surface. One problem in feedlots without shade is that ground temperatures can be high, discouraging cattle from lying down (Mitlohner et al., 2002). Davis et al. (2003) reported that simply spraying the ground in feedlots with water reduced tympanic temperatures, but did not affect feed intake. Again, too little is known about the most effective times to provide cooler lying surfaces. Since metabolic heat can contribute to heat stress, heat stress can be alleviated by changing feeding routines so that the peak metabolic load does not correspond to peak external temperatures (e.g. Mader et al., 2002). This idea is discussed in Chapter 8.

In conclusion, heat stress can be a major welfare challenge for feedlot cattle in certain climatic zones. Heat load can be reduced by providing shade, but there is little research on the best ways of providing shade, let alone other methods of reducing heat stress such as sprinklers for evaporative cooling or modifying the lying surface to increase conductive heat loss. Perhaps because of the very low critical temperature for beef cattle (Hahn, 1999) there has been less concern about the effect of cold stress, and certainly less research. However, unexpected cold weather, such as early snow storms, have been responsible for the deaths of tens of thousands of feedlot cattle in the USA (Mader, 2003). Over 72% of the feedlots in the USA provide some form of windbreak to protect the cattle against strong winds (USDA, 2002), although research to date has failed to show any clear advantages to providing windbreaks (Mader et al., 1997).

6.3 Research on Other Welfare Problems in Feedlots

Compared to the effects of heat stress, other welfare effects of the physical or social environment in feedlots have received much less attention. The stocking density and group size within a pen would be expected to affect welfare, but surprisingly we were not able to find any published articles in scientific journals that have examined this in a controlled fashion. Pusillo et al. (1991) report higher feed intake, weight gains, and food conversion efficiencies in feedlot cattle housed at a density of 16 m^2/animal, compared to those housed at a density of 2 m^2/animal in a more enclosed pen. However, these pens differed in flooring, feeder, and water space, and degree of exposure to the outside, making it difficult to know which factors were important.

One type of social behaviour that has received some attention is "bulling", when one or more animals repeatedly mount other animals. Bullers (the animals that are mounted) can suffer from bruising and appear more likely to become sick and die from respiratory disease (Taylor et al., 1997b), although why this happens is not clear. Bulling is increased if intact males are present within the pen, and often peaks when new cattle are mixed into the pens (Taylor et al., 1997a). However, the effects of stocking density, or the relation with social dominance do not appear to have been adequately explored. Aggressive behaviour between animals and effects of social competition for feed and water would seem important in feedlots and most likely depend upon the density of animals within the pen; more research needs to be directed at these potential problems.

For cattle in outdoor feedlots, flies appear to be a particular problem. Increased exposure to biting flies such as stable flies (*Stomoxys calcitrans*) and horn flies (*Haemtobia irritans*) can reduce feed intake, weight gains, and increase rectal temperature (Presley et al., 1996; Catangui et al., 1997; Campbell et al., 2001). Campbell et al. (2001) have calculated that the presence of only a few stable flies reduces weight gain, which becomes economically significant when there are more than five stable flies on each animal. Flies can be controlled by spraying cows with insecticide or use of ear tags impregnated with insecticide. For example, work from the University of Nebraska reported an 80% reduction in horn flies on cows' ears tagged with tags impregnated with insecticides and pastured outdoors (Campbell et al., 2006). Stable flies breed mainly around feed areas (Skoda et al., 1991) and cleaning of feedlot pens and feed bunks can reduce their numbers (Thomas et al., 1996). Some form of fly control in feedlots is essential to maintain good welfare.

Mud in feedlots is often cited as a major welfare problem (e.g. Grandin, 2002); as well as increasing the incidence of fly problems, muddy pens are associated with an increased prevalence of *E. coli* O157:H7 in fecal samples (Smith et al., 2001). From research on dairy cattle, it seems likely that muddy pens increase the incidence of lameness and reduce resting times, but neither of these effects has been examined in beef feedlots.

6.4 Conclusions

In general, compared to other forms of animal production, there has been little research on welfare issues in beef feedlots. The relatively high incidence of respiratory diseases and liver abscesses suggest that these welfare problems are substantial. Most research has focused on welfare issues that are of major economic importance, such as heat stress and respiratory diseases. We know little about the incidence of behavioural problems, such as those associated with social competition that may have less obvious economic effects.

7 Indoor Housing of Beef Cattle

7.1 Introduction

The type of housing for beef cattle that has attracted most attention in terms of research on animal welfare, is the use of indoor housing, especially group housing on slatted floor pens. This is common in northern European countries, but less common in North America and elsewhere. The majority of research has focused upon two particular issues: the spatial density at which animals are housed and the effects of the slatted floors.

7.2 Social and Spatial Density

The effects of space allowance on the welfare and performance of fattening beef cattle in indoor housing have been reviewed by Ingvartsen and Andersen (1993) and the EC Scientific Committee on Animal Health and Animal Welfare (2001). The most obvious behavioural effect of restricted space is reduced time spent lying down, and this reduction is found whenever space allowance is reduced below 4 m^2/animal (Scientific Committee on Animal Health and Animal Welfare, 2001). Occasional studies also report increases in aggressive behaviour, increases in oral stereotypies, disturbances of animals lying down, or an increase in abnormal movements when getting up and lying down. Many studies show a reduction in weight gain of fattening beef cattle (due to a combination of reduced feed intake and reduced feed conversion efficiency) when space allowance is decreased (Ingvartsen and Andersen, 1993; Scientific Committee on Animal Health and Animal Welfare, 2001). Ingvartsen and Andersen (1993) calculate that optimum growth rates of beef cattle weighing between 250 and 500 kg are achieved at space allowances about 4.7 m^2/animal. For cattle between 100 and 300 kg, Andersen et al. (1997) found a steady increase in feed conversion efficiency and growth when space allowance was increased from 1.8 to 3.1 m^2/animal. Although some studies report reduced growth rates when feeder space is reduced (Scientific Committee on Animal Health and Animal Welfare, 2001), reduced feeder space (to one space per five animals) has not been found to have consistent effects on growth rates, as long as animals are ad lib fed (Andersen et al., 1997). Reduced space allowances have also been reported to increase the incidence of tail tip necrosis, while mortality rates dropped from 2% to 0.5% as space allowance increased from 2.5 to 3.5 m^2/animal (Scientific Committee on Animal Health and Animal Welfare, 2001).

Other health problems have been less studied: Andersen et al. (1997) reported a high incidence of hoof and limb disorders and pneumonia but found no effect of space allowance (when varied between 1.8 and 3.1 m^2/animal). Expert veterinary opinion is that inadequate space allowance is an important risk factor for bovine respiratory disease (van der Fels-Klerx et al., 2000). Reduced space allowance leads to dirtier animals (Andersen et al., 1997). No consistent effects of space allowance on physiological or immune parameters have been found (Scientific Committee on Animal Health and Animal Welfare, 2001).

7.3 *Type of Flooring*

Given the importance of rest and sleep, the welfare of young animals in any type of housing system will likely be affected by the type of flooring on which they must stand and lie down. As discussed elsewhere in this book, farm animals are generally provided with some sort of organic material for bedding, but a recent trend is to reduce the use of such bedding, primarily because of labour costs involved in cleaning. In many cases, animals are kept on bare concrete floors, often slatted to allow the removal of urine and faeces. Concern about the effect of flooring on the welfare of the animals has led to a number of studies that have examined the effect of flooring type, with a particular emphasis upon slatted floors.

For young beef cattle, the EC Scientific Committee on Animal Health and Animal Welfare (2001) concluded that use of slatted floors, compared to the provision of straw bedding, resulted in reduced duration of lying down, reduced frequency of lying, more frequent interruptions of lying down, and an increase in abnormal movements when standing up or lying down (rising in front legs first or lying down first on hind quarters). This suggests that slatted floors make lying down and standing up uncomfortable possibly as a result of injuries to the legs, especially the carpal joints. Use of rubber mats reduces the incidence of abnormal standing up and lying down (Scientific Committee on Animal Health and Animal Welfare, 2001). Finishing beef steers allowed to choose between pairs of pens differing in lying surface ranked the floors in the following way: straw > sawdust > rubber mats > slats with rubber mats > concrete slats. The preference was most evident in the time spent lying down on the different surfaces, although a similar trend was seen in time spent standing (Lowe et al., 2001b).

Two of the most serious welfare problems of intensively housed fattening bulls are leg injuries and lameness and tail tip necrosis. Bulls kept on slatted floors were found to have higher mortality and culling rates (6%) compared to bulls kept on straw (2%) and a higher incidence of lameness (16% vs. 14%). Use of rubber mats reduces the incidence of injuries to the legs. Use of sloping solid concrete floors, as opposed to slats, slightly reduces the incidence of mortality and culling rates (to 4%) but appears to increase the incidence of lameness (Scientific Committee on Animal Health and Animal Welfare, 2001). Necrosis of the tip of the tail, probably resulting from tramping on the tail, is more common when concrete slatted floors are used compared to straw bedding (Scientific Committee on Animal Health and Animal Welfare, 2001). Use of slatted concrete floors, whether or not they are covered with rubber, results in dirtier animals than straw bedding (Lowe et al., 2001a), although feed intake and growth are usually not affected by flooring (Ingvartsen and Andersen, 1993; Hindhede et al., 1999; Lowe et al., 2001a). There appears to be no work to date on the effect of flooring type on locomotion in growing cattle and we encourage future work in this area.

In conclusion, use of fully slatted concrete floors leads to a reduction in the welfare of fattening bulls. Straw bedding appears to be the best surface, but other soft flooring surfaces such as rubber mats do provide some improvements. Increasingly, research on flooring and lying surfaces is being conducted for intensively housed dairy cows, and many of the findings from this work may also be applicable for these indoor-housed fattening bulls.

8 Conclusion

In this chapter on the housing of growing cattle we continue to argue that welfare in any type of housing will depend upon the details of the system, including the management and nutritional regimes. Research to date on the welfare effects of housing for growing cattle has focused on two areas of concern: confinement housing for the milk-fed dairy and veal calf, and indoor housing for growing beef cattle. For the unweaned dairy calf, much research has concentrated on the effects of group housing. This has shown that keeping calves in groups can provide calves with increased space and opportunities for social interactions. Group housing can result in increased risk of disease, but this risk can be minimized with small groups. For growing cattle reared indoors, disease and injuries are minimized when using low stocking densities and soft flooring materials (such as rubber) or bedding. In other respects, the effects of housing on the welfare of growing cattle have been under-researched. In particular, very little is known about how the well-being of weaned calves and feedlot cattle is affected by the various ways that they are housed. Other research priorities include work on the function of different resting postures for calves housed in stalls, on effective designs for shade cattle kept outdoors.

Chapter 8
Feeding and Nutrition

1 Introduction

Obtaining adequate food and water is essential for survival. Even when intakes are sufficient to keep the animal alive, low intakes can cause hunger or thirst, and chronically low intakes can threaten immune function. An inadequate balance of food types or nutrients can cause problems, as can supplying food in a way that does not meet the animal's behavioural requirements.

In many modern production systems, cattle are completely reliant on people to provide them with food and water. The science of animal nutrition has focused on providing diets with the correct amount of energy and the correct balance of essential nutrients to maximize growth or milk production. However, a considerable body of research into feeding motivation of animals (see Chapter 4) shows that the motivational controls on feeding are complex and that freedom from hunger requires more than proper nutrient intake. Unfortunately, very little is known about the controls of feeding motivation in cattle. We argue that research in this area has not sufficiently taken into account the impact of particular diets and feeding practices upon the welfare of the animals.

The focus of this chapter is not cattle nutrition per se. Moreover, we do not attempt to cover all aspects of how feeding and nutrition could affect animal welfare. Rather we discuss some ways in which feeding practices can have major effects on cattle welfare. Topics include provision of milk and colostrum to young calves, the process of weaning, feeding issues associated with veal calf production and problems associated with diets aimed to increase production of intensively managed dairy and beef cattle.

2 Unweaned Calves

The nutritional challenges facing milk-fed calves differ between dairy or veal production, in which calves are most commonly separated from their mothers at birth and fed milk by people, and calves raised for beef production, in which calves commonly obtain milk exclusively from their mothers. As mentioned in Chapter 2, the incidence of disease and mortality among milk-fed dairy and veal calves can be high. The nutrition of calves and the feeding management system used have an important influence on their health and welfare.

2.1 Colostrum

Colostrum feeding can have a large influence on calf health. Colostrum is the milk secreted by the cow during the first 24–36h after birth (Weaver et al., 2000). The importance of an adequate intake of colostrum has long been known but surveys continue to report that large numbers of dairy calves still receive either inadequate or minimally adequate levels (USDA, 2002; McGuirk and Collins, 2004). An inadequate intake of colostrum is not a problem only for dairy calves: Filteau et al. (2003) found evidence of inadequate colostrum intake in 19% of suckled beef calves.

An inadequate intake of colostrum can influence the health of calves in many ways. Most obviously, colostrum is the only source of energy and nutrients for the newborn calf and starvation of the newborn remains a serious welfare problem in extensively managed cattle (Mellor and Stafford, 2004). Furthermore, colostrum includes substances that increase the metabolism and digestive processes of the calf. Hormones and growth factors are present in colostrum, stimulating protein synthesis, cell division, and growth (Blum and Hammon, 2000).

Perhaps the greatest importance of colostrum is in providing the calf immune protection. Although the calf is born with a functional immune system and it is able to react to certain antigens, the immune system is considered as "naive" because it does not yet operate at an optimal level (Franklin, 2004). Colostrum contains antibodies, known as immunoglobulins (Ig) that are large glycoprotein molecules that constitute the main protection against diseases. The Ig contained in colostrum are absorbed into the calf's blood and this mechanism for acquiring immunity from colostrum is known as passive transfer. The Ig obtained in this way protect the calf until its own immune system becomes fully functional at around 3–6 weeks of age (Franklin, 2004).

The ability of the calf to defend itself against infectious diseases is directly related to the amount, quality, and timing of colostrum intake. The result of inadequate colostrum intake is a low concentration of circulating Ig in the blood of the calf, a condition known as "failure of passive transfer" (FPT). FPT can be defined as a calf's blood-serum concentration of IgG less than 10.0g/L (McGuirk and Collins, 2004). A number of studies have documented the close association between inadequate colostrum intake, FTP and increased mortality or morbidity of both beef and dairy calves (Rea et al., 1996; Filteau et al., 2003; Dewell et al., 2006). According to Wells et al. (1996), 31% of calf deaths during the first 3 weeks of life could have been prevented if colostrum feeding had been adequate (Wells et al., 1996). Even where death is avoided, there can be long-term effects of an inadequate colostrum intake: calves with FTP have lower body weights 6 months later (Dewell et al., 2006).

Colostrum-derived antibodies can remain active for many months (Munoz-Zanzi et al., 2002). Clearly, an inadequate intake of colostrum and too low levels of Ig in the blood represent a major risk factor for poor welfare of newborn calves.

Any environmental, nutritional, or management factors that affect colostrum intake increase the risk to the calf's health. Factors that influence the passive transfer of Ig from colostrum to the calf have been well documented (Weaver et al., 2000; McGuirk and Collins, 2004). The success of passive transfer is subject to two main limitations: (1) the small intestine of calves at birth is permeable to the Ig present in colostrum, but this permeability is gradually lost within the first 24 h of life (Weaver et al., 2000) and, (2) the amount of Ig in the colostrum varies with age, parity, health, and other factors, such as the nutrition of the pregnant cow (Quigley and Drewry, 1998). Generally, the recommendations are that the calf must have its first meal of colostrum before 12 h post-partum, the Ig content of the colostrum must be of high quality, and the calf must receive 4–6 L of colostrum in the first 12-h post-partum (Davis and Drackley, 1998). The timing of the first intake of colostrum is particularly important: the transfer of Ig across the gut epithelium of the calf is optimal in the first 4 h and decreases 12 h after birth (Weaver et al., 2000). Even a 30 min delay has been found to reduce the concentration of Ig in the calf (Rajala and Castrén, 1995).

Most beef cattle obtain their colostrum through suckling their mother and many dairy farmers also leave the calf with the cow for a period of time to allow it to suckle colostrum (Figure 8.1).

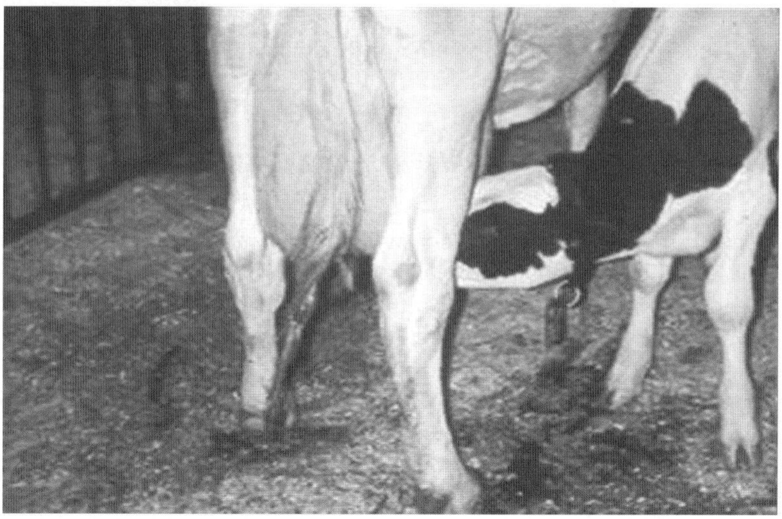

Figure 8.1 A calf suckling from its mother. This appears the most natural way for calves to feed and calves reared by their mothers are often in better health than those reared artificially, although this is not invariably the case. However, problems can occur. Allowing calves to suckle is not a reliable way of ensuring adequate colostrum intake. Cows vary in the quality of the colostrum they produce and calves may have difficulties suckling. Calves that are not also hand-fed colostrum have been shown to have a greater risk of failed passive transfer and a greater risk of diarrhoea

Unfortunately, this is not a reliable way of ensuring adequate colostrum intake. Franklin et al. (2003) compared calves that were allowed to suckle freely from their mothers with calves that were removed from their mothers and fed colostrum by bottle. The concentrations of serum proteins (a way of estimating Ig content) were lower in the nursed calves. This reflects a combination of factors, such as differences between cows in the Ig content of colostrum and differences between calves in their success at suckling. When dairy calves were left with the cow for 24 h, almost a half were found to suffer from FTP (Wesselink et al., 1999). Perhaps as a result of the low colostrum intake, calves that obtain their colostrum only by suckling suffer from a higher incidence of diarrhoea (Svensson et al., 2003). Some hand-feeding of colostrum therefore is essential to ensure the welfare of the calves. However, care must be taken to ensure that the colostrum fed this way has not become contaminated: some recent surveys report high levels of bacterial contamination of colostrum fed to dairy calves (e.g. McMartin et al., 2006).

Furthermore, there is some intriguing evidence that the presence of the mother may actually help the calf absorb Ig from the colostrum. Stott et al. (1979) found higher Ig absorption in suckling calves than bucket-fed calves, even when the amount and timing of colostrum ingestion was controlled for. The authors suggest that some factor in colostrum that speeds Ig absorption was lost when colostrum was stored. An alternative explanation, however, is that the presence of the dam alone may influence the calf's ability to absorb Ig from colostrum: Fallon et al. (1989) and Selman et al. (1970) found that bucket-fed calves absorbed more Ig from colostrum if the dam was present. Some producers tube feed colostrum in order to ensure adequate intakes in calves. No research has addressed if this tube feeding affects Ig absorption or has other affects on calf welfare.

2.2 How Much Milk Should Calves be Fed?

The welfare of milk-fed calves depends on how much milk they drink and how they obtain the milk. Extensively reared calves can die from insufficient milk intake (Mellor and Stafford, 2004). Although this rarely occurs among intensively managed calves, inadequate intakes can reduce immune function, and calves fed low volumes of milk often lose or fail to gain weight during the first weeks of life (Hammon et al., 2002; Jasper and Weary, 2002). Inadequate milk consumption is normally more of a problem for calves that are reared apart from their mothers than for calves that can suckle from the cow.

Under natural conditions, cows leave their calves in groups from about 2 weeks of age and usually continue to nurse calves for more than 6 months (Phillips, 1993). On most intensive dairy farms, however, calves are separated from their mothers within 24 h of birth and then fed milk by bucket or bottle until 4–10 weeks of age. A number of studies report lower incidence of mortality and morbidity among suckling calves than among calves reared separately from their dam (Rajala and Castrén, 1995; Webster et al. 1985b; Krohn, 2001; Chapter 2).

In some cases, the problems of hand-fed calves arise from the quality of the milk or milk replacer that they are fed. A recent report found lower morbidity and mortality among calves fed whole milk than calves fed milk replacer, although this difference may also have been due to the calves fed whole milk receiving greater quantities (Godden et al., 2005). Whole milk has a higher energy content and better balance of nutrients than some commercial milk replacers (Davis and Drackley, 1998). Furthermore, whole milk contains a variety of hormones and growth factors that may be important for calf health but which are not incorporated into milk replacer.

The quantity of milk fed may also be inadequate. Milk-fed dairy calves are usually fed a small amount of milk, which is substantially less than what they normally drink during a nursing or when milk is available ad lib (e.g. Jasper and Weary, 2002; Hammon et al., 2002; Hepola, 2003). For example, de Passillé and Rushen (2006a) showed that dairy calves that were allowed to nurse from their mothers drank 6–14L per day while in North America it is common for dairy calves to receive only 10–15% of their body weight in milk, which is roughly 4–6L of milk a day. The optimal amount of milk for a calf will obviously vary with a number of factors such as ambient temperature (e.g. Schrama et al., 1993). However, there is increasing evidence that the amount of milk generally fed to calves is not sufficient to satisfy their hunger and ensure optimal growth and development. A series of recent studies (Appleby et al., 2001; Diaz et al., 2001; Jasper and Weary, 2002; Khan et al., 2007) have now shown that the growth rates of calves can be greatly increased by feeding higher amounts of milk (Figure 8.2).

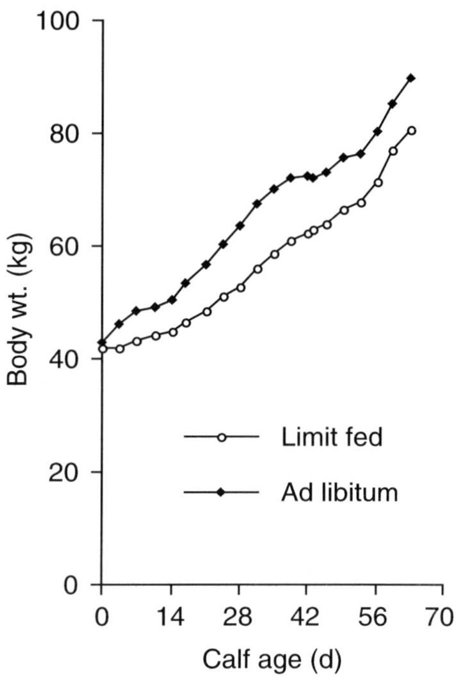

Figure 8.2 When dairy calves are provided *ad libitum* access to milk they typically drink much more than the 4–6 L per day these animals are conventionally limit fed. For example, when Jasper and Weary (2002) provided *ad libitum* milk calves averaged almost 10L per day, resulting in much higher weight gains for these calves, and this growth advantage was maintained even after weaning at 35 days of age (Figure adapted from Jasper and Weary, 2002.)

These improved weight gains are often associated with improved feed conversion efficiency (Van Amburgh et al., 1999; Diaz et al., 2001), although some studies have reported lower feed conversion efficiencies (Hammon et al., 2002; Hepola, 2003). Higher growth rates of *ad libitum* fed calves occur during the first weeks of life (Hammon et al., 2002), a period when conventionally fed calves show little weight gain and when health risks are high. When weaned off milk, calves fed higher quantities of milk can show a slight check in growth over the following week (Hepola, 2003), but normal growth soon recommences and they maintain their weight advantage over the conventionally fed calves (Jasper and Weary, 2002). Furthermore, there is evidence that the low amounts of milk do not adequately decrease feeding motivation, and probably leave the calves feeling hungry. Calves fed small amounts of milk a day make many more visits to the milk feeder, suggesting that they are still hungry (Hammon et al., 2002; Jensen and Holm, 2003; De Paula Vieira et al., 2007).

There is, however, some controversy regarding the effects of higher milk rations on diarrhoea. Quigley et al. (2006) claimed that calves fed additional amounts of milk replacer had a longer duration of diarrhoea episodes. However, in this study, calves that did not voluntarily drink all the milk allocated were force fed, and force feeding itself in known to increase the risk of succumbing to disease (Johnson, 1998). Furthermore, many other studies have shown that the incidence of diarrhoea is not affected by milk intake (Appleby et al., 2001; Jasper and Weary, 2002; Chua et al., 2002; Diaz et al., 2001; Hammon et al., 2002). In fact, Khan et al., (2007) recently showed that feeding higher quantities of milk actually reduced the incidence of diarrhoea.

In response to the welfare concern that calves fed restricted amounts are hungry, a growing body of research has focused on determining the effects of feeding more milk as well as exploring alternative feeding systems that allow calves to express more natural sucking behaviours.

2.3 How Should Milk be Fed?

On dairy farms in industrialized countries, calves are typically provided milk from a bucket and thus are unable to perform their natural sucking behaviour (See Chapter 4). Calves can also be fed milk through a teat, allowing the calves to suck. Teat-based milk feeding systems vary from simple arrangements where calves drink from teat bottles or buckets fitted with a teat, through feeding stations with multiple teats connected to a milk reservoir, to computer-controlled feeders (Figure 8.3).

Research has documented a number of potential advantages to allowing calves to suck for their milk. First, sucking behaviour itself appears to contribute to satiety (Rushen and de Passillé, 1995; Chapter 4) and influences the secretion of insulin and CCK, hormones that have been shown to be important for digestive function and satiety (de Passillé et al., 1993; Lupoli et al., 2001). Calves that suck for their milk have been shown to lie down sooner and sleep for longer than calves drinking from a bucket (Veissier et al., 2002). There is some evidence that heart rates are also

Figure 8.3 Feeding systems for dairy calves range from simple bucket feeders (above), which are cheap and easy to clean but do not allow the calves to suck for their milk, to complex computer-controlled milk feeders (below), which are expensive but reduce the labour necessary to look after calves and allow calves to drink milk in a more natural manner

lower among teat-fed calves (Veissier et al., 2002). Second, sucking milk through a teat has also been shown to reduce non-nutritive sucking (i.e. sucking on parts of the pen, etc.), partly because sucking milk from a teat increases overall feeding time (Appleby et al., 2001), especially if the teat has a small orifice to reduce flow rate (Haley et al., 1998). Group-housed calves normally show less cross-sucking when fed from a teat (see next section).

Computer-controlled milk and grain-feeding systems were developed in the early 1980s and are now widely available. There are many advantages of using automated feeders over conventional bucket systems, including increasing labour efficiency. Caring for group-housed calves on an automated milk feeding system requires less labour than when calves are housed individually (Kung et al., 1997; de Passillé et al., 2004), helping to offset the capital costs of the machines. Automated feeding systems facilitate the distribution of the total daily milk intake into small meals throughout the day, allowing a greater amount of milk to be fed without requiring the calf to drink a large amount at each meal. The pattern of drinking by the calves more closely resembles that seen during normal nursings (Senn et al., 2000). These systems can monitor the number and timing of visits, and the amount of milk consumed by each calf.

Some studies report a lower incidence of disease among calves fed with an automated milk feeding system (Kung et al., 1997). In contrast, other studies report higher mortality and morbidity, but this probably occurs because farmers tend to increase the size of the groups when these (relatively expensive) feeders are used (Svensson et al., 2003; Svensson and Liberg, 2006). Certainly the way that these feeders (or any group feeding systems) are managed can greatly influence their impact on calf welfare. Too many calves for the number of teats available increases social competition between calves for teats and can reduce milk intake (von Keyserlingk et al., 2004; Jensen, 2004). When calves are introduced into the group, milk consumption decreases temporarily (O'Driscoll et al., 2006), and there may be advantages to introducing calves into the group at an older age (Rasmussen et al., 2006). To improve the efficiency of milk-feeder systems, it is important to reduce the amount of time that each calf spends at the feeder in visits when it is not entitled to milk. Several studies have now shown that feeding larger amounts of milk substantially reduces the number of these "unrewarded" visits that calves make to the feeder (e.g. Hammon et al., 2002; Jensen and Holm, 2003; de Paula Vieira, 2007). Thus teat feeding can have advantages, but feeding systems need to be managed to avoid competition by keeping group size small, carefully managing the introduction of new calves, increasing the ratio of teats to calves, and feeding higher quantities of milk.

2.4 Cross-Sucking

One potential risk of grouping milk-fed calves is that the calves will suck each other, and this risk sometimes discourages dairy farmers from using group housing (Figure 8.4).

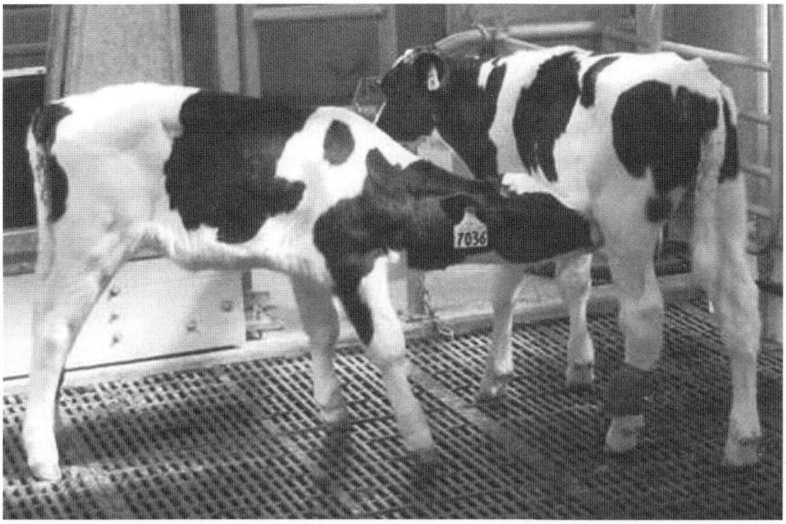

Figure 8.4 Milk-fed calves that are housed in groups occasionally suck each other; a behaviour known as cross-sucking. Although the effects of this on the calves' welfare are not clear, this behaviour discourages farmers from keeping calves together. However, cross-sucking can be controlled by appropriate management. The most important factors are to allow the calves adequate opportunities to suck the teat and to ensure they are fed enough milk so as to not feel hungry

Although some studies report problems of cross-sucking in group-housed calves (e.g. Keil and Langhans, 2001; Lidfors and Isberg, 2003; Margerison et al., 2003), others report only a low incidence of the behaviour (e.g. Mattiello et al., 2002). Cross-sucking can be controlled by appropriate feeding management. For example, automated milk feeders or other teat-feeding systems where the calves suck their milk through a teat can reduce cross-sucking (e.g. Loberg and Lidfors, 2001; Lidfors and Isberg, 2003). Even providing a dry teat after the meal (de Passillé and Rushen, 2006c) can substantially reduce cross-sucking. However, use of teat-feeding systems are not invariably associated with reduced cross-sucking (e.g. Veissier et al., 2002), suggesting that the details of feeding management are important. Automated milk feeding systems can be designed to allow the calves to suck the teat for a sufficient time after the meal to satisfy their sucking motivation. For example, cross-sucking is increased if teat-fed calves are not able to continue to suck the teat after milk ingestion (Jung and Lidfors, 2001) but a swing door (described in Jensen and Holm, 2003) can prevent calves from being displaced from the teat by other calves, and allows the calf to perform longer bouts of sucking on the teat. Slower rates of milk delivery can also reduce the incidence of cross-sucking (Loberg and Lidfors, 2001), likely for the same reason. Displacements of calves from the teat feeder can also be reduced by feeding the calves a larger quantity of milk (Jensen and Holm, 2003). Thus cross-sucking is largely preventable when calves are allowed adequate opportunities to suck on a teat for milk.

3 The Transition to Solid Feed

For feral or extensively managed cattle, the transition from milk to solid food is a gradual process that does not end until the calf is several months old, and this dietary transition is associated with a gradual reduction in interactions with the cow. In beef-suckler calves, weaning involves both separation from the mother and loss of milk, and typically occurs at 5–9 months. The separation from the mother can be abrupt, which can itself be a welfare problem; calves respond to this separation by vocalizing and increasing their movement (Veissier et al., 1989; Haley et al., 2005). Some recent research has examined how this emotional response to separation can be reduced. Letting the calf maintain visual contact with the mother is preferable to complete separation (Price et al., 2003). Two-step weaning, in which calves are prevented from suckling by a device attached to the muzzle, but are allowed to remain in contact with the mother greatly reduces the calves' responses when the mother is finally removed (Haley et al., 2005).

Weaning at older ages is presumed to be better for calf welfare since an older calf is less dependent on its mother as a source of food. Interestingly, studies on beef calves generally do not report any long-term effects on animal welfare of early separation from the mother (at around 3 months), at least as measured by weight gains (e.g. Myers et al., 1999; Arthington et al., 2005). However, there is a lack of research on early weaning that uses more subtle measures of welfare and little is known about the longer-term effects of weaning at different ages. Welfare problems are most evident when newly weaned beef calves are transported to a feed lot. These calves are forced to switch from familiar pasture-based diet with some milk intake to a high-grain-based diet in a feed bunk. In such situations, the problems arise from a combination of the diet change, and the difficulties that the calves have recognizing the new food source (Loerch and Fluharty, 2000; Walker et al., 2007). The calves are also confronted with additional stressors such as new social partners, new environment, and the transport to the feed lot.

The transition from milk to solid foods is also a welfare issue for dairy calves. This transition is often abrupt and usually occurs at a relatively early age of 1–3 months. In these circumstances, the withdrawal of milk is often associated with behavioural signs of distress, such as vocalizations and increased activity, as well as reductions in growth. The effects of weaning can be reduced if the calf is eating sufficient solids before milk is withdrawn. Calves typically consume little solid food before 3 weeks of age, likely due to their immature digestive system, and solid intakes continue to be low when calves are fed large amounts of milk (e.g. Jasper and Weary, 2002). Gradually reducing the milk ration over a number of days or weeks results in a rapid increase in solid intakes (Khan et al., 2007) and these gradual weaning methods are likely to be most useful for calves initially fed higher quantities of milk.

Dairy calves are typically weaned from milk onto a grain-based "calf starter". However, calves must later make another dietary transition to a forage-based diet, often again accompanied by changes in housing and social group. No published work that we are aware of has examined the welfare consequences of this later transition, and we suggest that understanding the full transition from milk to forage is an important topic for future research.

For calves raised for white veal production, welfare problems can arise from the lack of access to solid food. These calves are typically fed only milk for far longer than would occur under natural conditions. Welfare of the calves can be adversely affected both by anaemia (from the lack of iron in the diet) and from the lack of solid food itself. A recent report (EFSA Animal Health and Welfare Panel, 2006) reviews the welfare risks associated with diets used to raise veal calves.

4 Grazing

In many regions of the world, cattle are housed on pasture and must meet their nutritional requirements through the consumption of native herbages. Although, this provides a "natural" diet and freedom of movement, risks to animal welfare arise when either the availability or quality of the grass is low. Even brief periods of feed deprivation likely causes hunger in cows (Schutz et al., 2006). Longer-term effects of inadequate food intake include lost body weight and body condition (Stockdale, 2001). When the herbage quality is low in digestibility and protein quality, cattle are known to lose more than 10% of their weight (Ritter and Sorrenson, 1985). Poor body condition can also increase the risk of disease such as milk fever (Roche and Berry, 2006). The risks to welfare associated with poor grazing conditions should not be underestimated.

5 Beef Cattle in Feedlots

Some of the most serious risks to animal welfare within modern beef production systems arise from the way that cattle are managed in feedlots. The digestive system of cattle is evolved to efficiently digest forage but these diets do not allow cattle to grow at their maximum potential. To improve gains and profitability, heavy grain feeding has become the norm in feedlots (Figure 8.5). In most North American feedlots, 95% of the finishing diet is grain allowing for high weight gains (Castillo et al., 2004). However, the digestion of these diets challenges the health and welfare of the cattle.

The high starch content of feedlot rations results in rapid ruminal fermentation and produces high levels of volatile fatty acids (VFA) and a low ruminal pH. Rumen acidosis occurs when the pH of the rumen drops rapidly to between 5.0 and 5.5 (Boukila et al., 1995) and is thought to account for 25–30% of all feedlot deaths (Galyean and Rivera, 2003). Rumen acidosis also increases the risk of liver

Figure 8.5 In a number of countries, especially in North America, cattle raised for beef production are brought to market weight in large feedlots, where they are fed high-grain diets to maximize their growth rates. Overuse of these diets increases the risk of acidosis and can be one of the main risks to animal welfare within beef production systems

abscesses (Nagaraja and Chengappa, 1998), resulting in welfare problems even when death is avoided. Cattle with liver abscesses also have reduced intake, growth, feed efficiency, and carcass yield (Nagaraja and Chengappa, 1998). It is also common for feedlot cattle to suffer from subacute ruminal acidosis (SARA). In the hours following large meals of grain, cattle can sometimes be observed kicking at the belly, perhaps a sign of stomach pain (Radostits et al., 1994). Other behaviours displayed by feedlot cattle include panting, lethargy, a dull appearance, diarrhoea, pica (eating dirt), and reduced feed intake (Braun et al., 1992; Phy and Provenza, 1998; Stock and Britton, 2002). Feed intake fluctuates widely from day-to-day, likely due to animals reducing their feed intake to recover from the low rumen pH.

The risk of acidosis can be reduced through management techniques (Paton et al., 2006), including close monitoring of the feed bunk and using feeding practices that allow cattle to consume feed in smaller and more frequent meals (Schwartzkopf-Genswein et al., 2003). Gradually adapting cattle to high-grain diets over several weeks may help stabilize the microbial population of the rumen and avoid excessive accumulation of ruminal acids (Slyter, 1976; Radostits et al., 1994). An important question is whether cattle have the ability to choose an appropriate mix of forage and grain to maintain high growth rates while overcoming problems associated with acidosis (James and Kyriazakis, 2002). Some research has shown that cows will select feeds with high rumen-buffering capacity, which may help prevent SARA (Keunen et al., 2002; Beauchemin and Yang, 2005; Yang and Beauchemin, 2006).

6 Lactating Dairy Cattle

6.1 Acidosis and Feeding Practices

Indoor-housed lactating cows are generally fed a mix of concentrates and forage. These can be provided separately, which is predominant in Europe, or mixed together (in a "total mixed ration" – TMR), which is the predominant method in North America. Feeding concentrates increases the risk of welfare problems associated with acidosis, which is described in Section 5 of this chapter. SARA is considered by some to be one of the major threats to the welfare of lactating dairy cows (Shaver, 2002) and may affect 20% of lactating dairy cattle in early to mid lactation (Oetzel et al., 1999), although accurate diagnosis is difficult (Garrett et al., 1999; Duffield et al., 2004).

At present, the most reliable means of preventing SARA or acidosis is to use feeding methods that stimulate feeding behaviour or ensure a more even distribution of feed intake over the day (Cook et al., 2004). Providing TMR only once or twice a day may result in "slug feeding" when cows eat an excessive amount of feed during a short period of time. Cows are strongly attracted by the arrival of fresh food. For example, DeVries et al. (2005) found that when fresh feed was provided four times a day instead of twice, the time the cows spent feeding increased by 10 and 14 min. In addition, the daily feeding pattern changed, which resulted in cows having greater access to feed throughout the day (Figure 8.6).

This increased frequency of feed delivery also resulted in subordinate cows being displaced from the feeder less often, and thus having better access to feed, particularly

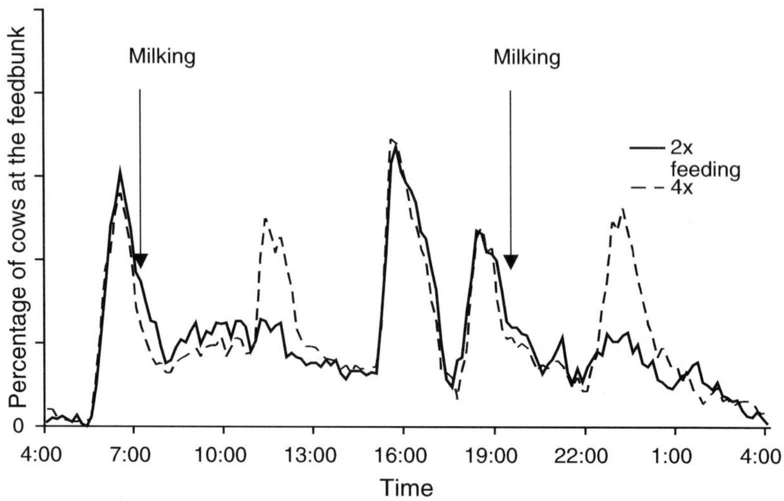

Figure 8.6 The number of cows at the feed bunk is highest following milking (at 07:00 and 19:00) and when fresh food is delivered either twice a day (05:30 and 15:15) or four times a day (05:30, 11:00, 15:15, and 22:30) (Redrawn from data in DeVries et al., 2005.)

fresh feed. Dairy cows have a tendency to sort through their feed (Figure 8.7) so cows that visit the feeder first consume richer feed than cows feeding later. DeVries et al. (2005) found that when the frequency of feed delivery was increased from once to twice a day, the amount of sorting by the cows was reduced. Thus feeding more often reduces variation in the time feeding and the quality of the diet cows ingest.

6.2 Health of the Transition Cow

The high milk production of the modern dairy cow places extra demands on her, leading to a high incidence of infectious and metabolic disease, particularly in the period around calving (transition). Gröhn et al. (2003) estimated that 30% of the cows developing a disease do so around calving. Metabolic diseases such as ketosis and clinical hypocalcaemia (milk fever) are common (see Chapter 2), which arise directly from inadequate or inappropriate nutrition (e.g. Rukkwamsuk et al., 1999). Susceptibility to certain infectious diseases also peaks at this time. The high prevalence of infectious disease during transition may also be due to inadequate nutrition, as these diseases often occur as a secondary illness to metabolic disease (Wentink et al., 1997; Reist et al., 2003). Inadequate nutrition may also contribute to the well-documented depression of the immune system that occurs around calving time (e.g. Mallard et al., 1998; Kehrli and Goff, 1989). The health status of the transition cow is a major issue for dairy production and the research on this topic has been reviewed many times, recently by Ingvartsen (2006).

Figure 8.7 Dairy cows fed a total mix ration (TMR) will often sort through the feed to eat the concentrates. If feed is not provided in a feed bunk, the cows will push the less attractive feed away. Farmers regularly push up this food. However, this sorting behaviour means that the composition of the TMR changes with time since the food was delivered. Cows that eat a long time after feed has been delivered will be eating a different diet than those that eat soon after delivery

The high demand for nutrients at the onset of lactation seems to be a prime factor leading to these diseases (Ingvartsen, 2006). Despite decades of nutritional and epidemiological work, the incidence of disease around the calving period in lactating dairy cattle remains high (see Chapter 2), underscoring the need for a more comprehensive understanding of the biological mechanisms associated with illness. Furthermore, we need to explore alternative feeding methods that minimize the incidence of illness. Ingvartsen (2006) discusses many of the possible means of avoiding or at least reducing the incidence of transition diseases. These include avoiding overfeeding cows prior to parturition and improving feed intake during the early post-partum period.

6.3 Ending Lactation

A typical lactation cycle of a cow in modern dairy production consists of approximately 300 days of lactation (classically 305 days) and a 40–60 days "dry" period. Although most cows would continue to produce milk beyond 305 days, having a dry period is important for the health of the cow (Oliver and Sordillo, 1988) and to ensure high milk production in subsequent lactations (Bachman and Schairer, 2003) and reproductive efficiency (Wilcox and Van Horn, 1999). Many different ways of bringing lactation to an end are used in practice (e.g. Natzke et al., 1975; Funk et al., 1982; Dingwell et al., 2002), all very different from the weaning process occurring under natural conditions where the calf gradually decreases both the frequency and size of milk feedings over a period of months (Krohn, 2001). Common dry off methods include reducing milking frequency, reducing the quality or quantity of food provided, and sometimes even reducing availability of water (Battaglia, 1998). Often these procedures are combined; for example, milking can be stopped abruptly at the same time as feed is restricted and access to water is prevented. Although abrupt cessation of milking is effective at stopping milk secretion (Browning et al., 1990; Brightling et al., 1998), it can result in higher rates of intramammary infections when compared to dry off through the use of intermittent milking (Bushe and Oliver, 1987; Natzke et al., 1975). The limited research to date on dry off has focused almost exclusively on udder health and there is almost no work addressing other welfare effects of dry off procedures. Preventing or limiting access to food or water raises obvious welfare concerns. Potentially less disruptive for the cows is continued *ad libitum* access to food, but with reduced energy content and digestibility; for example, by eliminating concentrates from the diet or offering only low-quality forage. One recent study (Valizaheh et al., 2007) switched cows from a 50% concentrate TMR to two different forage-based diets contrasting in quality. Cows switched to either diet reduced their feed intake and reduced milk production, but these effects were greatest for cows fed the low-quality forage. However, cows fed the low-quality diet also vocalized more often, suggesting that they were hungry. Thus, despite *ad libitum* availability, cows may not be able to digest sufficient quantities when fed poor-quality forages.

7 Water

Having access to a sufficient quantity of good-quality water is essential for the health and welfare of cattle. Water forms the largest component of an animal's body and is an essential nutrient required for all biological functions including temperature regulation, digestion, faecal development, and milk production. Cattle require large quantities of water: from 75 to 120 L per day for dairy cows and 25 to 70 L for growing beef cattle (National Research Council, 2000). Water consumption is closely tied to feed intake and high milk production is dependent upon the cows having access to large volumes of water.

Ambient temperature (Stockdale and King, 1983; Rouda et al., 1994) and relative humidity (Ali et al., 1994) can influence the requirements for consumption of water. When temperatures are high (i.e. greater than 30°C), water consumption increases as temperatures rise (Rouda et al., 1994; Molina and Tuero, 2000), but at lower temperatures cattle can maintain homeostasis without increasing water consumption. Water consumption is also affected by climatic factors: Stockdale and King (1983) found that daily water consumption was positively correlated with hours per day of sunshine. Day length may also influence water consumption as cattle rarely drink between sunset and sunrise (Sneva, 1970). Water intake can be affected by the method of water delivery. Trough design features such as the physical dimensions (Pinheiro Machado et al., 2004) and the rate of refill (Thomas, 1971; Andersson et al., 1984) will affect water consumption.

Water quality is an important factor affecting water consumption. If water contains compounds that diminish palatability, cattle will reduce their consumption (Embry et al., 1959; Weeth and Hunter, 1971; Patterson et al., 2002) or seek alternative sources (Digesti and Weeth, 1976). Through conditioned aversion, cattle may learn to limit their consumption of water containing compounds that result in gastrointestinal discomfort (Provenza, 1995; Ralphs and Provenza, 1999). Biological pollutants, including algae, manure and urine, lower the quality of drinking water, as do chemical pollutants such as minerals (Veenhuizen and Shurson, 1992; Carson, 2000). Under range conditions, cattle often drink from surface water. As surface water sources dry out, sulphate concentrations rise limiting intakes to the point that some cattle stop drinking, or experience signs of dehydration such as high faecal dry matter (Grout et al., 2006).

8 Conclusions

Problems with access to appropriate food and water pose some of the most severe threats to the welfare of cattle. We have reviewed a range of issues relating to both intensively and extensively managed cattle. Inadequate colostrum intake by calves remains an important cause of mortality in calves, and dairy calves are often fed restricted quantities of milk, limiting gains and likely causing hunger. Feeding increased amounts of milk greatly improves weight gains, especially during the first few weeks of life. Feeding this milk from a teat also facilitates natural sucking behaviour in calves, and reduces cross-sucking in grouped calves. Grouped calves will compete for access to a teat, but this competition can be reduced by providing calves more milk, and by increasing the ratio of teats to calves. Weaning calves from milk to solid-based diets is one of many dietary transitions faced by cattle, and these transitions can be difficult for cattle. Weaning in beef cattle can be eased by separating the social and nutritional components. For example, fence line and "two-step" weaning methods allow for continued contact between calf and cow but prevent nursing. Dairy calves are typically separated from the cow weeks before they are weaned from milk to solid food, but this nutritional weaning can still be an important source of distress, especially when calves are fed high rations of milk before weaning. New research is showing that gradually restricting milk increases solid food intake and reduces distress at weaning. Much research is still needed on how best to achieve this transition in diet, as well as the other dietary transitions that cattle face later in life, including changes in diet at calving and at dry off. Acidosis resulting from high-grain diets and the high metabolic requirements associated with the onset of lactation are issues for dairy and beef cattle. Although much research has been directed at defining nutrient requirements of cattle to optimize production, we are still largely ignorant of the factors that satisfy hunger in cattle or of the optimal methods of feeding that will ensure good health and welfare.

Chapter 9
Stockmanship and the Interactions between People and Cattle

1 Introduction

In the previous chapters, we discussed the way that the welfare of cattle can be affected by how they are housed and fed as well as by the short-term procedures that are done to them. However, one other important component of farming was omitted: the people who care for the animals. The decisions about how animals are housed, how they are fed, and how they are handled are made by people, and it is people who actually perform operations like tail docking, dehorning, etc. There are many ways in which the stockpeople or caretakers can affect the welfare of the animals in their care. The knowledge or technical competence of the stockperson can play a major role if it leads to improper choice of housing, poor feeding methods, or lack of appropriate treatment of illness, and the quality and diligence with which routine tasks are done can be also be important. In addition, a considerable body of research has now shown that the way that animals are handled by people can have a major effect on their welfare (Figure 9.1). The overall importance of stockmanship in ensuring good welfare of farm animals is emphasized in an excellent book by Hemsworth and Coleman (1998), and readers are encouraged to consult this. Until recently, it was true to say that the role of the stockperson or caretaker in affecting the welfare of farm animals had been neglected (Hemsworth, 2003), but this is beginning to change. In this chapter, we focus directly on the research that has examined the role of the stockperson in the welfare of cattle. The points of contact between animals and stockpeople vary substantially depending on the type of production, and so will the importance of stockmanship for animal welfare. Indoor-housed animals will, in general, be more dependent on human care than animals at pasture. The twice- or thrice-daily handling of dairy cows for milking results in far more contact with people than is normally found in extensive beef production systems. Consequently, it is not surprising that most research on stockmanship has been done with dairy cows and veal calves rather than with beef cattle.

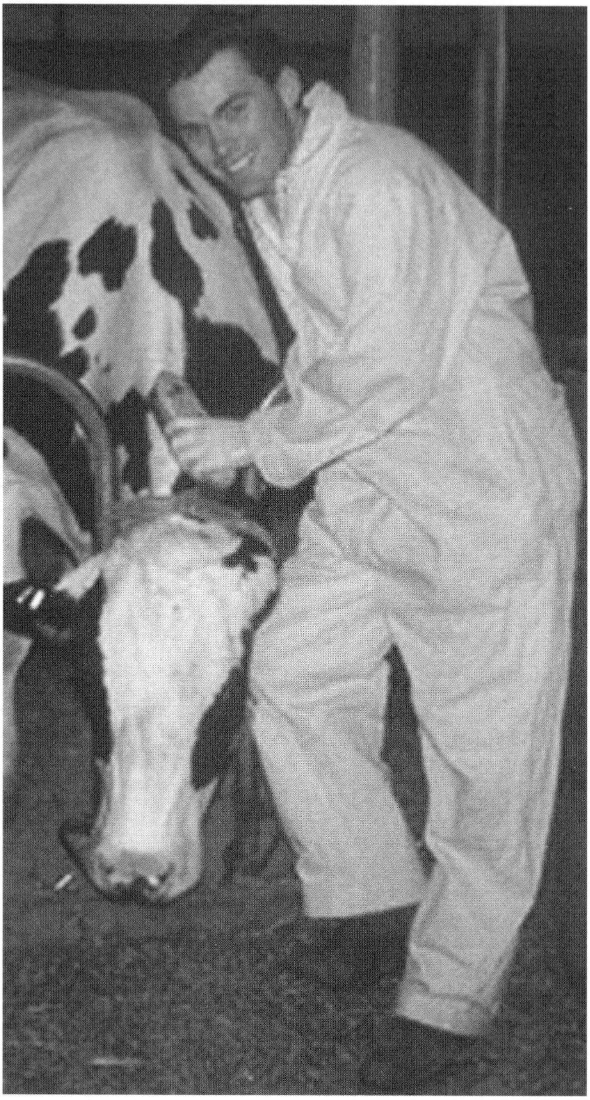

Figure 9.1 Differences between farms in the way that animals are handled can have a large effect on the animals' welfare. Dairy cattle that are roughly handled often become frightened of people, and this may result in substantial drop in milk production. The mere presence of an aversive handler during milking may be sufficient to cause the cows to "hold-back" milk, through a stress-induced suppression of oxytocin secretion (Rushen et al., 1999b). However, good stockmanship consists of much more than handling the animals properly. Diligent performance of routine animal care tasks, such as cleaning, is of major importance in ensuring good animal welfare (Lensink et al., 2001a)

2 Effect of Overall Stockmanship on Animal Welfare

The stockperson can most obviously affect the welfare of animals through the way that routine animal care tasks, such as feeding, cleaning, etc. are done. Despite growing recognition of its importance for good animal welfare, as well as good animal productivity, this component of stockmanship has been investigated scientifically and in detail in only a small number of studies. The quality of stockmanship, especially the care with which routine animal care tasks are performed, is responsible for some of the differences between farms and farmers in the level of animal welfare. For example, dairy farms in which calves are cared for by females tend to have lower calf mortality than farms where men are responsible for care of the calves (Losinger and Heinrichs, 1997).

In one of the most comprehensive studies of stockmanship, Lensink et al. (2001a) examined the role of stockmanship in affecting the health and productivity of veal calves on farms in France. Fifty veal farms were chosen that were operated by a single company, were located in the same region, and used similar management techniques, animal feed, etc. Veal calves were randomly distributed between the units. By removing obvious sources of variability in production, the researchers were able to more clearly isolate the effects of the quality of the stockmanship. Farmers were interviewed, and the researchers scored their attitudes to the animals, for example, whether or not they believed that calves were sensitive to human contact, and their attitudes to the work, for example, how important cleaning procedures were. The farms were also scored for cleanliness, and the performance of various management routines was noted. In addition, interactions between the farmers and the calves were observed, especially the extent that calves withdrew when the farmers approached, and the extent that the farmer engaged in positive interactions with the calves (i.e. petting or stroking calves, or letting them suck fingers, talking gently).

Sizeable correlations between these stockmanship variables and the production characteristics of the farm were noted. High-producing units (i.e. those with high daily weight gains, good food conversion efficiencies, and low mortality) had healthier calves, tended to be cleaner, had crates disinfected by an external company, had Sunday evening feedings of the calves, and were run by farmers whose own parents had managed a veal unit. This latter was thought to be important because it resulted in a greater experience in raising calves. The cleanliness of the barns accounted for 19% of the variance between units in daily weight gain and 22% of the variance between units in feed efficiency. The health of the calves was correlated with the attitudes of the farmers, for example, the more the farmer believed that calves were sensitive to human contact, and the more the farmer felt that cleaning was important, the better the health of the calves. The results show the importance of general stockmanship for the welfare and productivity of the calves, particularly in terms of the care taken in cleaning the facilities. In the researchers' own words: "Farmers who have a positive attitude towards animals and towards their work are more likely to obtain better production results, at least in part because of the better control of the calves' health. Also more positive attitudes towards animals lead to a more positive behaviour with them, which lowers animals' fear reactions to them. Hence by his/her behaviour with animals and his/her ability to control their health, the farmer plays also an important role in relation to the welfare of the calves." (p. 115).

As well as depending on specific attitudes and beliefs about the importance of routine care, the quality of stockmanship appears to be related to general personality attributes; Seabrook (1984) used a questionnaire to examine correlations between aspects of the stockperson's personality and the level of milk production on the farms where they were working. High- and low-producing stockpeople were found to differ in a number of personality attributes, with the high-producing stockpeople reported as being: "not easy going, considerate, not meek, patient, unsociable, not modest, independently minded, persevering, not talkative, confident, uncooperative, and suspicious of change"! However, it was not clear how these personality characteristics related to the care of the animals.

Despite the obvious importance of the quality of overall stockmanship to animal welfare, there have been few studies that have systematically examined the full range of stockmanship qualities in cattle production. However, one aspect of stockmanship has received considerable attention from researchers: the relationship between the way that animals are handled by people and the extent that the animals are frightened of people. The remainder of this chapter deals largely with this component of stockmanship.

3 Handling and Animals' Fearfulness of People

The extent that farm animals are frightened of the people that care for or handle them can have a marked effect on their welfare. Considerable research has shown that for most domestic animals, including cattle, the fear of people can be a major source of stress and a cause of lost production (reviewed in Hemsworth, 2003; de Passillé and Rushen, 2005; Waiblinger et al., 2006). There are marked differences between animals and between farms in the degree to which the animals are frightened of people. Although there are genetic differences between breeds and individual animals that account for some of the variance in the degree of fearfulness (see section 4.6), much of the animals' fear results from the way they are handled. Research is also suggesting that the personality and attitudes of the stockperson can influence the level of fear of animals, because such attitudes can influence how the stockperson handles the animals.

Hemsworth and Coleman (1998) present a simple, linear model of the relationship between the stockperson's beliefs and attitudes, their behaviour when handling animals and the impact this has on the animals (Figure 9.2). The stockperson's specific beliefs about animals have a direct influence on how the animals are handled. For example, a belief that cattle are not sensitive and difficult to move can lead the stockperson to use rough or aversive handling. Through a process of classical

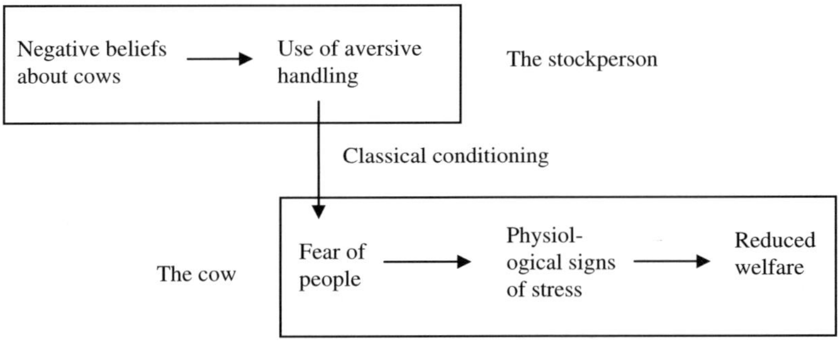

Figure 9.2 A simple model describing the causal link between a stockperson's beliefs about animals and the effect on their welfare and productivity (Adapted from Hemsworth and Coleman (1998)

conditioning, the animals learn to associate the aversive handling with the stockperson and become frightened of him or her. This, in turn, results in the physiological changes that typically occur when an animal is stressed, which have deleterious effects on the animal's welfare and productivity. Although such a simple model is unlikely to fully reflect the complexities of stockmanship, it is a useful way of conceptualizing the sequences of events leading from the stockperson's beliefs and attitudes, and it has received a strong degree of support from research.

This research has included both correlational studies, which describe the existing relationship between the animals' level of fearfulness and the type of handling used on farms, as well as experimental studies, in which animals are deliberately handled in particular ways and the resulting changes in fearfulness are noted. In both cases, it is essential that we have reliable and valid measures of how fearful cattle are of people, which is not always the case (de Passillé and Rushen, 2005; Waiblinger et al., 2006).

3.1 Dairy

Perhaps because dairy cows are handled more often than beef cattle, the bulk of the research on how handling methods affect the fearfulness of the animals has involved dairy cattle. The most frequent point of contact between dairy cows and people occurs at milking.

3.1.1 Correlational Studies

Differences between farms in the way dairy cattle are handled may explain some of the differences between farms in the level of productivity and the welfare of the animals. The milk yield of cows also differs according to who is milking them

(Hanna et al., 2006), which also may reflect the type of handling techniques used. Seabrook (1984) showed that the way that dairy cows were handled by the stock-person and the degree of fear the animals show towards the people can be a major factor underlying the differing productivity of different stockpeople. He observed the behaviour of cows that were being handled by "high-producing" stockpeople (i.e. stockpeople whose cows produced a large amount of milk) and compared this to cows handled by "low-producing" stockpeople. He found that the cows with the high-producing stock person were spoken to and touched more often, appeared less frightened and were more easily moved, and were more likely to approach the stockperson.

Recently, Breuer et al. (2000) found substantial correlations between the levels of milk production on dairy farms, the way that the animals were handled, and the degree of fear that the cows showed towards people. Thirty-one commercial dairy farms in Australia were visited and the stockpeople were observed while they moved and handled the animals during normal milkings. The extent that they used either gentle handling (consisting of stroking, patting, or resting the hand on the cows) or aversive handling (slapping, punching, tail-twisting, or hitting the animals with the hand or a stick) was observed. Also, the extent that the stockperson spoke or shouted at the cow, and also the average speed that the cows were moved from the pasture into the milking parlour were recorded.

The degree to which the milkers used aversive handling during milking varied widely between farms (Table 9.1). The percentage of all handling interactions in which the stockperson used highly aversive handling, for example, forceful slaps, hits, and tail twists, varied between 0% and 30%. The type of handling was found to be associated with the milk production of the farm. The annual milk yield was negatively correlated with the use of highly aversive handling techniques ($r = -0.39$).

Table 9.1 Results of a survey of Australian farms show a large difference between farms in many measures of how the stockperson handled the animals and in the cows' apparent fearfulness of the people. The table shows the time that the cow spent close to the test person during a test, the number of flinch/step/kick responses of the cow during milking, and the number of positive and negative tactile interactions and shouts used by the stockperson while moving the cow or milking and the speed with which cows were moved. The latter is important because impatient stockpeople may try to move cows too quickly. The average value for all farms, as well as the minimum and maximum value found on any one farm are shown (Data adapted from Table 1 in Breuer et al., 2000.)

Variable	Mean value	Minimum	Maximum
Time spent within 3 m of test person(s)	40	13	85
Number of flinch, step, and kick responses/cow/milking	0.21	0.02	0.54
Number of positive tactile interactions/cow/milking	0.11	0.02	0.40
Number of negative tactile interactions/cow/milking	0.32	0.09	0.84
Number of shouts/cow/milking	0.05	0	0.12
Speed in moving cows from pasture m/s	0.60	0.07	2.08

However, the use of positive handling techniques or moderately aversive handling (e.g. moderate slaps) was not correlated with milk yield.

The cows' fearfulness towards people was measured by observing the time that the cows spent close to a person in a specially constructed test enclosure. In addition, the behaviour of the cow was observed during milking, and each instance of flinching, kicking, or stepping (taken as signs of restlessness in the cow) was recorded. Average levels of fearfulness for each farm were calculated based on observations of 35–50 cows per farm. Again, large differences between farms were noted in the cows' degree of fearfulness: on the farm with the least fearful cows, the cows spent six times as long close to the person compared to the farm with the most fearful cows. Measures of cows' fearfulness were not significantly correlated with the behaviour of the stockperson. However, there were significant positive correlations between the frequency of flinches, kicks, and steps by the dairy cow and the use of harsh vocalizations by the stockperson and the speed with which the cows were moved into the pasture. Milk yield was significantly negatively correlated with the measure of fearfulness and with behavioural signs of restlessness by the cow during milking. A multiple regression analysis suggested that differences between farms in the degree of fearfulness and restlessness of the cows could account for 30% of the variance between farms in annual milk production, a sizeable figure!

A subsequent report from the same research group, this time using 66 dairy farms and taking measures of 30 cows per farm (Hemsworth et al., 2000), found similar results. Again there was a large difference between farms in the extent that the stockperson used aversive handling: the per cent of all handling interactions that involved highly aversive handling, varied between 0% and 38%. Again, the annual milk yield for the farm was negatively correlated with the frequency that aversive handling practices were used (Table 9.2). Rather surprisingly, milk yield was also negatively correlated with the use of apparently positive handling. This suggests that what we assume is positive for the cow may not be! This problem is discussed

Table 9.2 Correlations between the way the stockperson handled dairy cows, the fear the cows' showed towards the stockperson and the cows' tendency to flinch, step, and kick during milking (FSK), the annual milk yield and conception rate of the farm, and the concentrations of cortisol in the cows' milk. The asterix (*) indicates a significant correlation. The values are based on averages over 66 Australian diary farms and show the surprising magnitude of the effects on milk yield (Data are taken from Tables 5 and 6 in Hemsworth et al., 2000.)

	FSK	Milk yield	Milk cortisol	Conception rate
Stockpersons' handling of the cows:				
Number of positive tactile interactions	–0.24*	–0.25*	0.08	0.37*
Number of negative tactile interactions	–0.03	–0.36*	0.34*	–0.25
Total number of interactions with cow	–0.15	–0.41*	0.20	0.06
Cows' responses to people:				
Time within 3 m of person during test		–0.03	–0.38*	0.38*
Flight distance		–0.11	0.05	–0.09

in more detail in section 4.2. In fact, milk yield was most highly (negatively) correlated with the total number of tactile interactions between the cow and the stockperson and the regression model suggested this could account for 13% of the variance between farms in annual milk yield. The substantial amount of the variance in milk production shows the effect that stockmanship can have on the productivity of dairy cows. Negative correlations were also found between the stockperson's behaviour and conception rates: farms where the stockperson used a high frequency of negative interactions had fewer cows becoming pregnant after the first insemination.

The animals' fearfulness towards people was measured using the same technique as Breuer et al. (2000) except that the flight distance was also measured, defined as the distance that the cow first moved away from an approaching person. The animals' degree of fearfulness again varied widely between farms: average flight distance varied between 2 and 12 m. The degree of fearfulness shown towards people was correlated with the stockperson's handling of the cows: a greater use of negative handling techniques was associated with more fearful cows. More frequent positive handling was correlated with a lower flight distance and less flinching, stepping, and kicking by the cow during milking. In this study, the correlations between the fearfulness of the cattle and milk yield were smaller (Table 9.2) and not statistically significant. However, the researchers did find significant correlations with reproductive success: farms with more fearful cows had a significantly smaller proportion of cows conceiving at the first insemination. Differences between farms in the level of fearfulness accounted for 14% of the variance between farms in conception rates.

In support of the idea that aversive handling induces physiological changes in the animal associated with stress (Figure 9.2), Hemsworth et al. (2000) found significant positive correlations across farms between milk cortisol concentrations and the use of negative handling interactions and behavioural measures of cows fearfulness. This "chronic stress response" was considered responsible for the reduced production.

Many of these results were replicated by Waiblinger et al. (2002) who observed the behaviour of cows and stockpeople during milking on dairy farms in Austria. This study is valuable because it provides a replication under quite different circumstances. The dairy farms in Austria differed from the Australian farms in a number of respects: use of indoor housing rather than all year pasture, smaller herds (25–50 cows as opposed to 150–300) and different breeds (mainly Austrian Simmental rather than Holstein). Average measures for each farm were based on observations of at least 75% of the animals in the herd. Again, large differences were noted between farms in the way that the cows were handled by the stockpeople: the per cent of all interactions between the cows and the stockperson that involved negative interactions varied between 0% and 38%. The total number of tactile interactions between the stockperson and the cow varied between 0.5 and 13 per milking. Fearfulness of the cows was measured by approaching each cow in the barn and measuring the distance that the cow moved away. Again, there were large differences between farms, with the number of cows that did not move away when the experimenter approached varying from 2% to 48%. As in previous studies, a greater use of positive or gentle handling and a lower frequency of aversive handling were negatively correlated with the distance that the cow avoided the person: cows on farms where mainly gentle handling was used could be approached more closely. The overall milk production on the farm was negatively correlated ($r = -0.46$) with the use of aversive handling during milking.

Together, these studies clearly show that there is a large difference between dairy farms in the way that cattle are routinely handled, including the use of behaviour thought to be highly aversive to the cow. There are also large differences between farms in the extent that the cows appear to be frightened of people. Both the cows' fear of people and milk yield are affected by the way the cows are handled, with a high use of aversive handling being associated with fearful and lower yielding cows. Differences between farms in the fearfulness of the animals appear to account for a sizeable proportion of the differences between farms in milk production, indicating the significant economic costs associated with this particular welfare problem.

However, there are a number of limitations to these studies. Most important, they are correlational. All three report a large number of intercorrelations between variables which increases the risk of finding apparently significant correlations simply by chance. Multiple correlations make it difficult to determine what are the most important associations. Furthermore, it is difficult to infer the causal relationship. Both the increased fear and reduced milk yield may result from the aversive handling but other hypotheses are possible to imagine. Perhaps the cows are more fearful and low yielding because of genetic reasons, and their poor performance causes the farmer to become angry at them! Perhaps poor-quality stockpeople hit cows more often but are also careless at other stockmanship tasks, which have a more direct impact on production (e.g. Lensink et al., 2000; see Section 3.2). Use of multiple regressions can help isolate the most important relationships from a statistical point of view, but this still does not help us understand the causal relationships involved. For this, experimental studies are needed.

3.1.2 Experimental Studies

Direct evidence that the use of aversive handling causes the cows to become more fearful of people, and that this is responsible for the effects on milk yield, comes from small-scale experimental studies (Munksgaard et al., 1997; Rushen et al., 1999b). These experiments examined whether the same cows could be made fearful of one person but not of another. In these studies, dairy cows were handled repeatedly by two people, one of whom always handled the cows gently (talking softly, patting and stroking the cow, and occasionally giving feed rewards) while the other handled the cows aversively (hitting, shouting, and occasional use of a cattle prod). The cows' fear of each person was tested by each person standing in front of the cow's stall and measuring how closely the cow approached. The results showed that after repeated handling, the cows stood further from the aversive handler than from the gentle handler (Figure 9.3). Rushen et al. (1999b) also tested whether the degree of fearfulness elicited by the handling was sufficient to reduce milk yield,

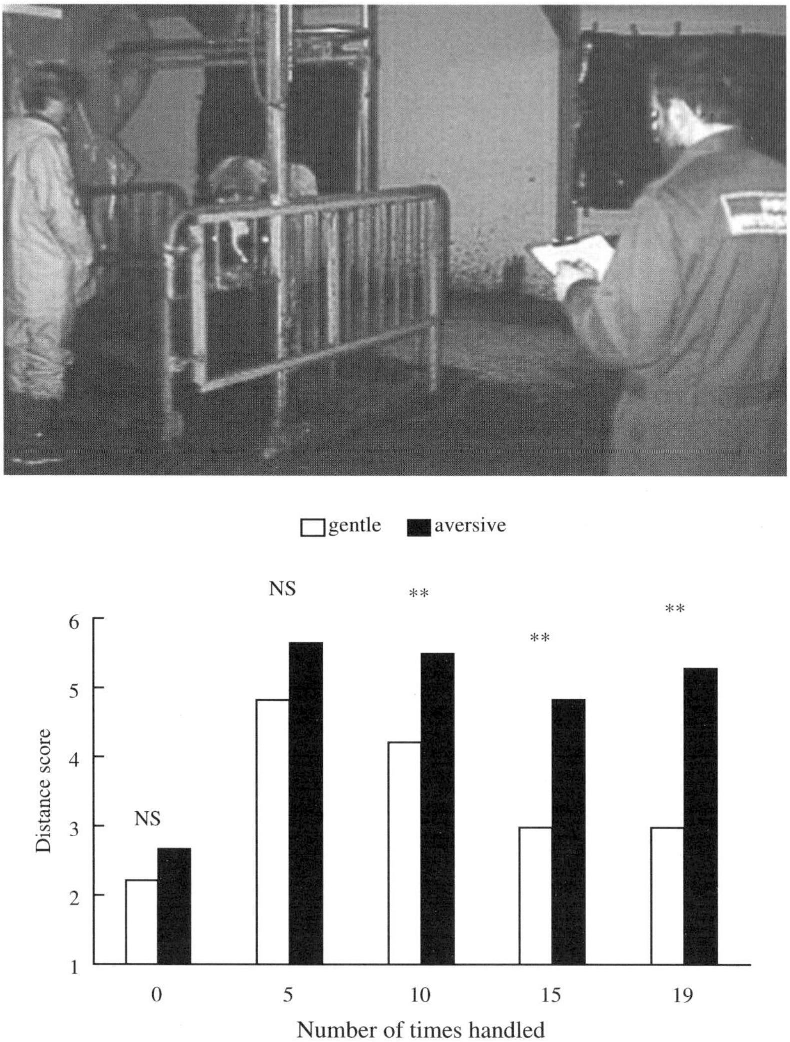

Figure 9.3 The cows' responses to the handlers were measured by having each handler stand in front of the cow's stall. The bar graph shows the average distance score of the cows when facing the gentle and the aversive handler (Adapted from Rushen et al., 1999b.)

by milking cows in the presence of the handlers. The gentle handler stood close to the cows for one milking and the aversive handler stood close to the cows for the other milking. Just the presence of the aversive handler during milking was sufficient to increase residual milk by 70% (which is a sign of a stress-induced block of milk ejection) and tended to reduce milk yield (Table 9.3). There were also some physiological signs of fear: the presence of the aversive handler increased heart rate during milking.

Table 9.3 The effects of the presence during a single milking of an aversive handler or a gentle handler. The asterix (*) indicates a significant difference (Data taken from Rushen et al., 1999b.)

	Aversive handler	Gentle handler
Milk yield (kg)	18.48	19.2
Residual milk (kg)	3.6*	2.1
Duration of udder preparation (min)	0.65	0.68
Kicks by cow/min during udder preparation	0*	0.93
Heart rate change during milking (b.p.m)	5.94*	3.42

However, the extent that the aversive handling affects milk yield may depend upon the type of handling used. A subsequent study (Munksgaard et al., 2001), which used a similar experimental protocol but less aversive handling, found the same effects upon the animals' withdrawal responses from the people but found no effects upon milk yield. These small-scale studies support the model presented in Figure 9.2 by showing that aversive handling will make dairy cows frightened of people, and that this fear may at times be sufficient to reduce milk yield.

Can these experimental results be generalized to explain the differences noted between commercial dairy farms? Evidence that they can come from a larger-scale intervention study by Hemsworth et al. (2002), who used a special training programme to improve the attitudes of stockpeople to dairy cows (described in more detail in section 4.4). Stockpeople at a total of 141 commercial farms were randomly allocated to the training programme or kept as controls. Following the training programme, the stockpeople were found to use a lower number of aversive handling treatments and a higher number of gentle or positive handling treatments. In addition, the flight distance of the cows was found to be lower on the farms on which the stockpeople had followed the training programmes. Furthermore, there was some evidence that improvements in the way dairy cows were handled improved milk yields and reduced cortisol concentrations in the milk (Hemsworth et al., 2002). Thus, the degree of fearfulness of cows appears to be a direct response to the type of handling practices used by the stockperson (Table 9.4).

3.2 Veal Calves

Compared to dairy cattle, fewer studies have examined the effects of handling on veal calf welfare. In their study of French veal farms, Lensink et al. (2001a) found that calves were less likely to withdraw when a person approached on veal farms where the farmer used positive behaviours towards the calves (e.g. petting, talking gently, letting calves suck their fingers, etc.) Furthermore, the frequency of positive contacts was negatively correlated with the mortality rate on the farm. However, the researchers found no clear evidence that the calves' fearfulness of people (as measured by their tendency to withdraw when people approached) was directly related to the health or growth rates of the calves. In contrast to the model presented in Figure 9.2,

Table 9.4 The results of an intervention designed to change the attitudes and beliefs of stockpeople about cows. The intervention resulted in a reduced use of aversive handling practices and a greater use of gentle handling. This resulted in a reduced flight distance of the cows as well as an increased milk yield (in one of the experiments) (Data adapted from Table 2 and Table 3 in Hemsworth et al., 2002.)

	Control	Intervention
Stockpersons' behaviour:		
– Use of mildly aversive handling techniques/cow/milking	0.43	0.24*
– Use of strongly aversive handling techniques/cow/milking	0.02	0.005*
– Use of gentle handling techniques/cow/milking	0.045	0.11*
Cows' responses		
– Flight distance m	4.49	4.16*
– Flinch, step, kick responses	0.1	0.13
– Milk cortisol nM/L	2.05	1.4
– Milk yield L/cow/month (experiment 2 n = 99 farms)	509	529

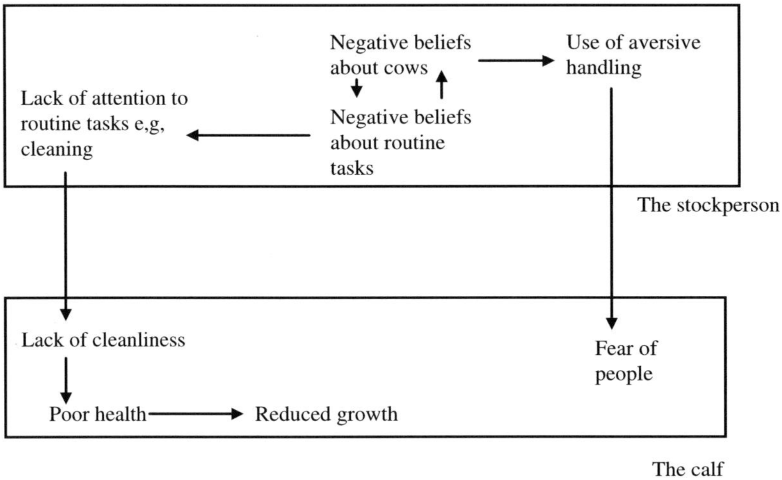

Figure 9.4 An alternative model of the relationship between stockperson's attitudes and beliefs and the impact on the calf. In this case, the largest effect upon the calf's growth occurs through a lack of attention to cleaning routines rather than through the animal becoming frightened of people. The calves' fear of people is a correlated response that could arise because stockpeople who have negative beliefs about the importance of routine care, may also have negative beliefs about the animals, and also use aversive handling techniques (Based on a model of Lensink, 2000.)

the authors suggested that the association between the type of handling and the performance of the calves may have occurred because of mutual correlations with other aspects of stockmanship, such as attention to cleaning routines etc (Figure 9.4). This research shows the need to be careful when interpreting correlational studies and the importance of combining these with experimental studies to better understand the causal relationships between the variables. A subsequent study on the same farms, showed that calves that came from farms where the farmers predominantly used positive behaviours towards them were easy to load on to trucks, showed lower heart rate when loaded, and had paler meat with lower pH when slaughtered, an economic advantage for veal farmers (Lensink et al., 2001b).

3.3 Beef Cattle

Problems of handling and the risk of injury to both animals and people seem to be of major concern to beef production. Beef cattle that are calmer when being handled have better meat quality, less bruising, and better growth rates (Blackshaw et al., 1987; Voisinet et al., 1997; Burrows and Dillon, 1997; King et al., 2006). However, research on beef cattle has tended to focus upon issues of temperament, and the genetic background of the animal (see Section 4.6.), and, while there are many practical recommendations about ways to handle beef cattle, detailed research studies on how handling techniques influence the animals' welfare are surprisingly few. The lack of research in this area may reflect the relatively infrequent contacts between people and beef cattle, certainly in extensive production systems. Clearly more research is needed on the effects of handling on the welfare of beef cattle.

4 How to Improve Stockmanship

Given the importance of stockmanship for animal welfare, it is necessary to find the most effective ways of improving stockmanship. Again, most research to date has focused upon the ways that cattle are physically handled by the stockperson, rather than other aspects of stockmanship.

4.1 Extra Contact With People

In some cases, especially for extensively managed beef cattle, animals may be fearful of people simply because they have not had sufficient contact with them. Increased handling or contact with people reduces animals' fear of people. For example, Waiblinger et al. (2004) showed that previous gentle handling helped calm cattle during subsequent veterinary examinations. However, cattle may be more prone to establish positive relationships with people when they are young, and several experiments have examined the effects of increased contact with people at an early age.

Increased gentle handling of younger cattle has been shown repeatedly to reduce their fearfulness towards people as adults (e.g. Boissy and Bouissou, 1988; Boivin et al., 1992a,b; Becker and Lobato, 1997; Figure 9.5). When the extra handling is

Figure 9.5 Increased handling of calves at an early age reduces their tendency to avoid people. However, there is no strong evidence for a sensitive period. Raising the calf with its mother appears to reduce the effectiveness of this early handling

done to young animals, the effects appear to be particularly long lasting. The reduction in fear has usually been noted several months later, and in some cases years later, without apparently needing to be reinforced by further handling (e.g. Boivin et al., 1992b). Some studies have examined the persistence of the effect by repeatedly testing animals as they age, and found little decrease with time, suggesting that the effect is permanent (e.g. Boivin et al., 1994; Boivin et al., 1992b).

An obvious question is whether cattle are more sensitive to human contact at some ages than at others. Although it is often suggested that a sensitive period exists for cattle, attempts to demonstrate such a sensitive period have had limited success. Boissy and Bouissou (1988) handled dairy heifers either at 0–3 months, 6–9 months, and 0–9 months of age. Reduced fearfulness and increased ease of handling was found for the group handled during 0–9 months of age, with less of an effect for the group handled during 6–9 months. The least effective period for handling was during the 0–3 months following birth. The authors concluded that extended pre-pubertal handling is more effective than short-term handling and that there appears not to be any critical period for this effect. In contrast, Krohn et al. (2001) found some evidence that 18 min per day of extra, gentle handling given to young dairy calves between 1 and 4 days of life was more successful than handling given between 6 and 14 days in reducing avoidance of people by calves up to 8 weeks of age. However, the effect varied according to the measure used: differences between the treatments were most apparent in the latency that the calves took to approach a person in the home pen. Flight distance from a person in an unfamiliar pen, however, was reduced equally by all three treatments. In summary, although there may be some age effects on the effectiveness of the extra handling, there is no strong evidence that calves have a marked sensitive period, unlike lambs (Markowitz et al., 1998)

The effect of early handling appears to vary somewhat according to how much contact calves have with conspecifics, particularly with their mothers and other calves (Figure 9.5). Close contact between people and bulls that are kept isolated from other bulls can substantially increase the aggression that the bulls show (Price and Wallach, 1990). However, whether this is because the bulls are "imprinting" on people, or whether they lack social restraints on their aggression to people as a result of their lack of social companions (suggested by Price and Wallach, 1990) is not clear. Cattle that are reared by their mothers are often more fearful of people than those reared by hand (Boivin et al., 1994), but this probably reflects a difference in the amount of contact with people. However, many studies have shown that extra handling of young animals is effective even if the animals are reared by their natural mothers (Boivin et al., 1992a), or if the animals are kept grouped with conspecifics (Boivin et al., 1994, 1992b). Creel and Albright (1988) hypothesized that calves weaned from their mothers and reared in isolation from other calves would "imprint" on people and be less fearful than calves reared in groups of other calves. However, they found no evidence that isolated calves approached people more readily, or were easier to handle. Although, there was some evidence of reduced flight distance of isolated calves, the authors concluded that social isolation from other calves does not make animals less fearful of people. More recently, Krohn et al. (2003) found that calves reared with their mothers had longer flight distances than calves reared individually, even though the amount of contact between the calves and people were similar. In summary, there is some evidence that socialization to conspecifics (either to a mother or to peers) interferes with socialization to people and may also reduce the effectiveness of extra handling in reducing animals' fear of people.

What type of contact is necessary to reduce fearfulness of people? Most handling procedures that have been used in experiments involve a mix of gentling and giving some kind of food reward. Whether or not the food rewards are necessary has not been fully resolved; that handling alone without additional feeding can function to reduce cattle's fearfulness of people has been shown a number of times (Boissy and Bouissou, 1988; Boivin et al., 1992a). However, Jago et al. (1999) found that handling young cattle, without providing a feed reward did not reduce the latency for the animals to react with people, whereas providing food either with or without extra handling were equally effective. In this study, the calves would often butt the person (a behaviour usually associated with feeding) suggesting that the calves were specifically associating the people with being fed. We do not yet know why, in some studies, food rewards are necessary while in others they are not. The age of the calves may be important, since in Jago et al.'s (1999) study the handling was imposed when the calves were very young (2–16 days old).

There may be several periods in the life of cattle when they are sensitive to increased contact with people. Following parturition, adult female cattle are sensitive to developing a social bond with their newly born calves, and may be more sensitive to human contact at this time also. Hemsworth et al. (1987) gave heifers extra handling at the time of their first calving, and found that the time taken to approach the experimenter was reduced. In a subsequent study, 5 min of extra gentle handling per week during the 3 months prior to first calving reducing the frequency of flinches, steps, and kicks during milking, and increase milk flow rate (a possible sign of reduced stress) during the first 20 weeks of lactation (Bertenshaw and Rowlinson, 2001).

4.2 *Identify Which Types of Handling are Aversive or Positive*

Since it is clear that the type of handling used by the stockperson has major effects upon the fearfulness of the animals, a necessary first step in improving the relationship between animals and the stockperson is to identify the particular behaviours that the animals find aversive or rewarding. Numerous recommendations have been made concerning which actions to use and avoid. A few cases are obvious, such as the effects of feeding and electric shock, but there has been little research on the effects of other treatments.

Recently, Pajor et al. (2000, 2003) used aversion-learning techniques (Chapter 4) to try to discover exactly which handling practices cattle do find aversive. In one study (Pajor et al., 2000) compared supposedly aversive treatments that are often used when moving cows, such as hitting with the hand, shouting, tail twisting (but not strongly enough to break the tail!), and use of an electric prod. Cattle were placed in a runway and at the end of this they were restrained and handled. The experimenters measured the speed that the animals moved down the runway as well as the effort required by the handler to move them. Based upon these measures, all treatments appeared to be aversive to some extent. However, hitting with the hand and twisting the tail did not differ significantly from the control treatment (of no handling), suggesting that the cattle perceived these treatments as relatively mild, although this conclusion likely depends upon the force used. Unsurprisingly, the use of the cattle prod was aversive but interestingly shouting appeared to be as aversive as the cattle prod. A subsequent study (Pajor et al., 2003), which allowed cattle to choose between pairs of treatments confirmed these results: cattle showed no preference for shouting over the electric prod, and no preference for no treatment over tail twisting (Figure 9.6). Other studies have also reported that cattle find shouting to be highly aversive (e.g. Waynert et al., 1999).

What types of handling do animals find positive or rewarding? That animals will approach people who feed them is not surprising, but can people also give animals "social rewards"? That gentle handling alone without additional feeding can function to reduce an animal's fearfulness of people has been shown for cattle (Boissy and Bouissou, 1988; Boivin et al., 1992a) suggesting that people can be

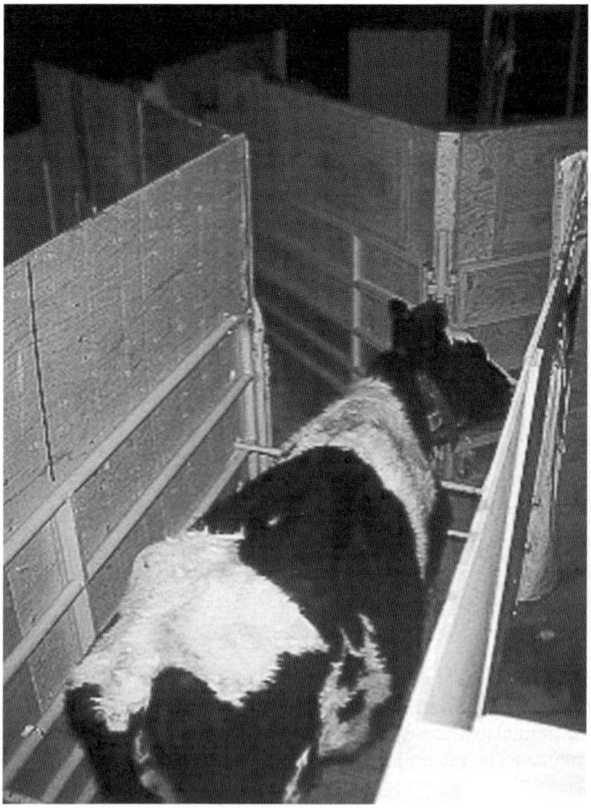

Figure 9.6 The figure shows a dairy heifer in a Y-maze about to make a choice between approaching two people, each of whom will handle the animal in a different way. The results of the cows' and heifers' choices between a number of pairs of handling treatments shows their preferences (Adapted from Pajor et al., 2003.)

a source of social rewards. However, attempts by Pajor et al. (2000, 2003) to find rewarding behaviours were less successful than attempts to identify aversive behaviours. Hand feeding was rewarding for adult cows but not for younger heifers, although the latter were attracted when food was presented in a bucket. Brushing, patting and stroking, and speaking in a gentle voice were not preferred by cattle relative to no treatment at all. In fact, some evidence was found that cows found brushing aversive, at least initially. Boivin et al. (1998) also found little evidence that cattle find physical contact with people to be rewarding. The fact that people and cows may not agree on what is rewarding (Figure 9.7) may explain some of the curious results reported earlier, such as finding that a high use of positive behaviours is associated with lower milk yields in dairy cows (Hemsworth et al., 2000).

Figure 9.7 People appear to enjoy close physical contact with animals and we assume that the animals must enjoy it as well. Gentle contact is certainly preferable than aversive handling and may help to reduce animals' fear of people. However, we have no clear evidence yet that cattle find this type of contact to be rewarding

4.3 Avoiding "Learned Aversions" to the Stockperson

At times, even the best stockperson will have to handle animals in an aversive way. Often this is done for the animal's own welfare, such as when giving injections or treating illness. Occasional use of such aversive procedures need not lead animals to become frightened of people (Lewis and Hurnik, 1998). However, one risk is that the animal, through a process of classical conditioning, will learn to associate the aversive handling with the particular individual, and develop a learned aversion and fear of that person (Hemsworth et al., 1996); that this can happen has been demonstrated many times (e.g. de Passillé et al., 1996; Munksgaard et al., 1997; Rushen et al., 1999c) and is often reported to be a problem for veterinarians. Learned fear of people can have serious effects on an animal's welfare, so means of preventing the development of this fear are needed. In some cases, ensuring sufficient positive or gentle contacts may conceal the occasional negative treatments (Hemsworth et al., 1996). One promising approach is to associate the treatment with a particular location, rather than a particular person (Rushen et al., 1998). Thus, learned fear in the animals may be reduced by performing aversive treatments in a special place rather than in the animal's home pen.

It may also be possible to mask the identity of the person in fairly simple ways. There is now clear evidence that cattle can tell different people apart (reviewed in

Rushen et al., 1999c; de Passillé and Rushen, 2005). What cues might the animals be using? Cattle have a reasonable degree of visual acuity and are capable of colour vision. Visual cues, especially those associated with clothing seem to be particularly important in recognition of people by cattle. Even very young calves can recognize people wearing different colour clothes (Rybarczyk et al., 2003). Munksgaard et al., (1997) and Rushen et al., (1999b) showed that cows that had learned to avoid a handler as a result of receiving aversive handling, no longer avoided that person when the colour of the clothing was changed. These studies show that under some circumstances a loss of recognition can occur following simple changes in the appearance of people, such as a change of clothes. It may be possible to take advantage of this association to reduce the occurrence of learned fears of particular individuals, for example, by wearing special coloured clothes when essential but aversive treatments are applied to animals. However, Boivin et al., (1998) showed that changes of clothing did not affect beef cattle's ability to distinguish familiar from unfamiliar people, perhaps because cattle can use other visual cues to recognize people. Taylor and Davis (1998) showed that cattle in fact could learn to distinguish people who wore the same

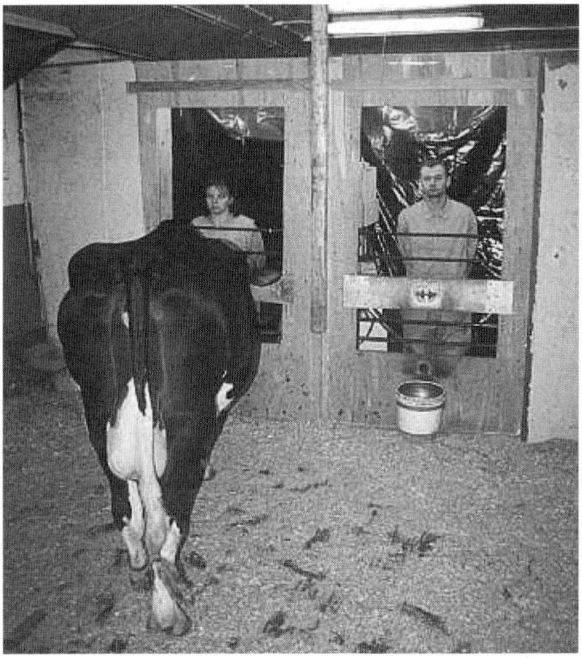

Figure 9.8 Discriminant operant conditioning can be used to test which cues cows use to recognize people. The cows needed to approach the correct person to obtain a food reward. Cows were correctly able to choose the person if they could tell their height or see their faces. However, if both the persons' faces were covered and there was no height difference, the cows were unable to recognize the people. This suggests that some cows may be able to recognize people by their faces, which may make it difficult to conceal the person's identity when doing aversive or painful husbandry procedures (From Rybarczyk et al., 2001.)

colour clothes and Rybarczyk et al. (2001) presented evidence that some cows can recognize people by their faces (Figure 9.8). Together, these results suggest that while clothing colour is used by cattle to distinguish people, it is by no means the only cue that they can use. Thus, while it may be possible to hide a person's identity under some circumstances, we should not underestimate the ability of cows to use quite subtle features to recognize people.

4.4 Altering Stockpersons' Attitudes

Clearly identifying which behaviours cattle find aversive and which they find positive or rewarding is an essential first step in making recommendations about improving ways of handling animals. However, the way stockpeople handle animals is likely to be a reflection of long-held beliefs about how animals need to be handled and attitudes towards animals in general; thus recommendations alone may not be sufficient to change the behaviour of the stockpeople (Hemsworth, 2003). The range of attitudes of stockpeople towards animals is discussed in Hemsworth and Coleman (1998), and we do not need to repeat this here. A considerable amount of research has now shown that the way that stockpeople handle animals is a reflection of specific beliefs of the stockperson, and that altering these beliefs may be an effective means of improving the way animals are handled (Hemsworth and Coleman, 1998; Hemsworth, 2003). Much of this research has involved pigs and as yet there are fewer studies that demonstrate these techniques for cattle.

Hemsworth et al. (2002) examined the effect of a "cognitive-behavioural intervention" on dairy farmers' attitudes towards cows. Over 100 commercial dairy farms were used, and the farmers were randomly allocated to a control group or to an intervention group. The intervention involved multimedia presentations emphasizing the research results that showed the negative effects of poor handling on the fearfulness and productivity of cattle, along with some clear examples of good and poor handling techniques. The intervention clearly improved the attitude of the farmers towards dairy cattle, specifically reducing the belief that considerable force was needed to move dairy cows. Visits to the farms showed that these changes in beliefs resulted in a reduced use of aversive handling techniques, reduced fearfulness, and tended to improve milk yield. This study clearly shows the potential for such interventions to improve at least one component of stockmanship, and to improve both the welfare and the productivity of the cattle.

4.5 Identifying Why Stockpeople Mishandle Animals

In addition to attitudes and general opinions on animals, situational factors can have a marked influence on the quality of stockmanship. Consequently, to improve the ways animals are handled, it helps to know what circumstances do lead to animals being handled roughly. From other professions, it is recognized that the diligence with which a job is done depends very much on the level of job satisfaction; low job satisfaction often results in careless work. Recognizing the importance of this for stockmanship, Seabrook and Wilkinson (2000) interviewed 238 dairy stockpeople in the UK to determine what factors affected their level of job satisfaction, focusing particularly upon the routine tasks stockpeople enjoyed and which they found most unpleasant. Importantly, the stockpeople clearly valued and enjoyed their interactions with their animals, and the nature of their interactions with animals was largely responsible for the difference between a "good day" and a "bad day". Milking was widely rated the most important routine task in dairying, and (fortunately!) was rated the most enjoyable. Maintaining herd health and welfare was considered of relatively low importance, especially by younger stockpeople. Stockpeople from higher-producing herds, however, rated herd health and welfare more important than those from lower-producing herds. Interestingly, the rated importance of herd health and welfare increased when stockpeople were re-interviewed after quality assurance schemes had been introduced. The authors felt that these may have led the stockpeople to alter their priorities. Despite its importance for maintaining good animal health and welfare, the task of foot trimming was disliked mainly because of inadequate equipment and holding facilities, and because it was generally considered a dangerous task. The authors concluded that the obvious dislike of foot trimming and cleaning routines shows the need to find improved means of cleaning cows and barns, for better designed parlours that facilitate cleaning, and for improved equipment and facilities for foot trimming. This type of research is clearly useful in showing what types of improvements are needed in order to improve job satisfaction and hence job quality (Figure 9.9).

Difficulties in moving cattle can be a major cause of frustration for stockpeople and this can be responsible for much of the rough handling that occurs. Finding better ways to move animals is likely to lead to significant reductions in the use of aversive handling. This could be achieved either through improved raceway design or by finding other means of improving animal movement; Ceballos and Weary (2002) found that providing small food rewards when dairy cows entered the milking parlour reduced the time the cows took to enter the parlour, and reduced the need for the stockperson to push the cows or use other aversive handling techniques.

4.6 Genetic Selection

Although the way that animals are handled by their caretakers has a major impact on the extent that they are fearful of people, differences among animals in their responses to people may also be influenced by genetics. For wild animals, we could expect natural selection to favour some degree of fear of people and other potential predators. As animals and humans became mutually dependent during the process of domestication, natural selection to maintain animals' fear of human beings would have relaxed. In fact, during the domestication process, animals most likely were selected for

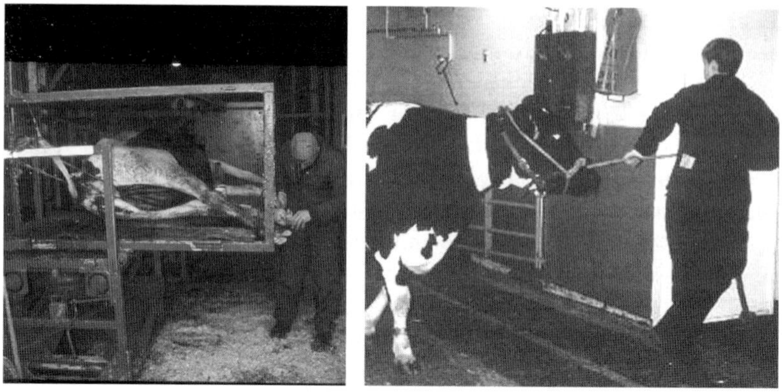

Figure 9.9 To improve stockmanship we need to pay attention to the level of job satisfaction of the people who care and handle the animals. To reduce rough handling of animals, we need to understand the situations in which people become rough with animals. If facilities are not well designed, handlers may have difficulty moving cattle and become frustrated with them. Other routine tasks, such as hoof trimming, are essential for good welfare but are rated as unpleasant by stockpeople. Designing better equipment for such tasks may result in them being carried out more often

docility and tameness by farmers seeking ease of handling (e.g. Price, 1998, 1999). Genetic contributions to such characteristics would most likely be expressed as differences among breeds and among individuals within a breed. However, because early experience, maternal effects, and rearing conditions of animals affect their level of fear of humans at later ages, experiments must be carefully designed to objectively estimate the relative importance of genetic and environmental influences on such characteristics (Price, 1998). Estimates of heritability of temperament in cattle vary widely, and may differ systematically between breeds (Gauly et al., 2001), but generally heritability is low to moderate (Burrow, 1997, 2001). Genetic differences in the behaviour of domestic animals, including their responses to people is of interest since differences in responses to handling appear to be linked with meat tenderness (King et al., 2006) and genetic selection for better "temperament" is an ongoing research area for beef cattle (Kadel et al., 2006).

Indeed, genetic selection for tameness probably continues long after animals have been domesticated, due to increased culling rates of animals that are difficult to handle. The increased mechanization and intensification of animal agriculture may have shifted the target of artificial selection towards efficiency of production rather than handling ease. For example, Ewbank (1993) claims that due to reduced handling in intensive dairy production, and the consequently reduced selection for tame animals, there has been an increased difficulty in handling dairy stock. It is also possible that some of the intensively selected productivity traits, like rapid growth and lean meat, are linked to certain negative handling traits, resulting in more nervous and aggressive animals that are difficult to handle (Grandin and Deesing, 1998). Turner and Lawrence (2007) provide some evidence that increased selection for maternal defensiveness in order to improve calf survival may indirectly lead to increased aggressiveness towards people.

4.7 Understanding the Nature of the "Human-Animal" Relationship: Social Companion or Predator?

At a more theoretical level, improvements in our ability to handle cattle will require that we have a better understanding of the fundamental nature of the relationship between people and cattle, especially how cattle "perceive" people. Often, the relationships between people and farm animals have been described in terms of predator–prey relations. However, the type of relationship that can develop between people and animals that are in constant contact can be much more subtle; and it has been claimed that such relationships are genuine social relationships, similar to what would be seen between conspecifics (Estep and Hetts, 1992). In a very interesting review, Tennessen and Hudson (1981) examined the different species which have been successfully domesticated, and found several similarities in their basic social structure. Domesticated species tended to have wild ancestors that lived in groups, rather than being solitary, with dominance hierarchies, rather than a territorial system. Studies of traditional herding societies (Lott and Hart, 1979) suggest that these traits may have aided the process of domestication by allowing people to enter into social relationships with the animals and so exploit their natural social behaviour.

It is often claimed that people can, and indeed should, establish dominance over their animals. One of the few detailed studies of how people appear to enter into genuine social relationships with animals, including dominance relationships, was Lott and Hart's (1979) study of the Fulani cattle herdsmen of sub-Saharan Nigeria. According to Lott and Hart, while "western" farmers tend to rely primarily on physical means to control, for example, fences or restraints, the Fulani control their animals primarily through their knowledge and exploitation of the cattle's natural social behaviour. Lott and Hart claim that the Fulani herdsmen control their herds by inserting themselves into the social system of the cattle, and they described some of the ways in which this is achieved. Interestingly, the Fulani appear quite aggressive in establishing dominance over their cattle, especially the bulls in the herd. They respond to any threats from the bull by attacking the bull or threatening it by yelling and waving a stick. The herdsmen continually reinforce their dominance over the other cattle by occasionally hitting them for no obvious reason, and by breaking up fights between other cattle. This routine use of physical aggression seems contrary to what is often argued: that good stockmen should not use physical force on their livestock. However, while in recent years, the emphasis has been put on the negative effects of animals' fear of people, we need to remember that it may also be important for the safety of the handlers to ensure some degree of dominance, especially over the larger farm animals, and, in some cases, a degree of physical force may be the most effective way of achieving this.

However, social dominance is not the only form of social relationship that the Fulani have with their cattle. The Fulani herdsmen also establish amicable relationships with their animals by spending considerable time walking among them, and scratching them on the head and neck, places where the cattle often groom themselves or each other. These affiliative behaviours are the same sort of social behaviours that the cattle use with each other. An affiliative relationship between people and animals might also be apparent in the ability of people to provide social support to animals, especially animals in social isolation. For example, human-reared lambs vocalize less in the presence of people than when alone and this effect is most apparent for the shepherd that had raised the lambs than for a stranger (Boivin et al., 1997). Recently, Rushen et al. (2001b) found that when cows under stress were brushed by a person they knew, there was some evidence of reduced heart rate as well as reduced defecation and vocalization, although

Figure 9.10 Cattle are occasionally seen to play with people suggesting that they may form genuine social relationships with them. This cow is butting the leg of the person; a behaviour that resembles the apparently playful headbutting that occurs between cattle. Such behaviours suggest that cattle may form genuine social bonds with people (Rushen et al., 2001a.)

there were no effects on plasma cortisol concentrations or the amount of milk retained. Thus in some cases the presence of people may provide some form of social comfort. A number of authors (e.g. Seabrook and Bartle, 1992) have suggested that handling would be improved if the handlers used "species-specific" behaviours, used by the animals themselves when establishing social bonds or social relationships (Figure 9.10). For example, the best places to touch animals could be determined by examining where animals groom each other. Unfortunately, except in the obvious case of feeding, we still know very little about what types of contact with people cattle find rewarding.

5 Conclusions

From the research reviewed in this chapter, and from extensive research done on other species of farm animals (reviewed in Hemsworth and Coleman, 1998) it is clear that the people who care for the animals can have a major, if not decisive role to play in affecting their welfare. Differences between stockpeople may be responsible for much of the differences between otherwise similar farms in the level of animal welfare and productivity. Research is beginning to provide concrete examples of how poor stockmanship can lead to poor animal welfare, and to show some of the ways that stockmanship can be improved. Notably these improvements, as well as improving animal welfare, often lead to substantial benefits to the farmers themselves, either through increased health and productivity of the animals, or through increased efficiency and safety of operations that involve handling the animals, for example, most evident at milking. Research to date has focused upon the more obvious aspects of stockmanship: how the animals are handled and how they become fearful of people. However, stockmanship involves much more, and the relationships that can develop between people and animals can be quite subtle. Research needs to consider a broader range of qualities associated with stockmanship.

Chapter 10
Conclusions

In the preceding pages we have provided an overview of research on the welfare of cattle, focusing on those aspects that have been well covered in the scientific literature. This research has dealt almost exclusively with the welfare issues associated with confinement systems that have been developed in Europe and North America. As animal production grows rapidly in Asia, South America and other regions of the world, there is a need to increase research that deals with the animal welfare issues within the production systems typically found in these regions. In some cases, these animal welfare issues may be the same as those in the European and North American confinement systems since these types of systems are sometimes imported wholesale into new regions. However, housing systems that function well in temperate climates may require special modifications to work in warmer areas. Furthermore, the breeds of cattle that are highly productive and profitable in temperate climate housing and management systems may experience different welfare challenges when brought onto farms in tropical and sub tropical regions. These include an increased risk of heat stress and exposure to new pathogens to which the animals may have limited resistance. Breeds that are highly productive in temperate regions may not cope as well with fluctuations in food and water availability that more hardy locally adapted breeds are able to withstand. In other cases, the traditional production methods may be maintained but expanded, or new methods of production developed. In both cases, animal welfare issues are likely to arise. It is our hope that as animal production grows in these regions, so does local research on animal welfare that will be able to address these concerns.

In Chapter 1, we reviewed various definitions of animal welfare and briefly discussed the social context providing some insights into current interest in farm animal welfare. The different players in the animal welfare debate often have different views of what animal welfare is, and this is partly responsible for disagreements, for example, between behavioural scientists and veterinarians, or between farmers and the public at large. We must recognize that animal welfare is multifaceted and that research must focus on all of the issues involved. Scientists, in particular, must resist the temptation to redefine animal welfare so as to make it more amenable to scientific investigation but less relevant to the concerns being expressed by other stakeholders. Much of the recent research into animal welfare has been driven by public concern and the (primarily European) legislation that has resulted from this. Consequently, many of the issues that are raised in this book are concerns that are already present in the public eye such as individual housing for calves and tie stalls for lactating cows. Other issues may be seen as threats to animal welfare by the public but the research suggests that the effect on animal welfare may not be as obvious as it appears. The early separation of the dairy calf from its mother is an example of this. Lastly, there are some serious threats to animal welfare that have not generated much attention among the public. Problems of acidosis due to feeding of high grain diets would be an example. Research into these latter topics has focused primarily on the economic losses associated with disease and has tended to pay less attention to the impact on animal welfare itself.

The first part of the book discusses some of the potential indicators that we might use to measure animal welfare, either in a research setting or in the field. Chapter 2 focuses on animal health and disease, and emphasizes the importance of including measures of disease and injuries in any assessment of cattle welfare. Despite the obvious importance of animal health to animal welfare, our ability to use health measures to assess animal welfare is limited by the difficulties in obtaining reliable and valid measures of the occurrence of various illnesses. Perhaps the most important message from this chapter is the urgent need for veterinarians and animal welfare researchers to collect better information about health problems on farms. An additional challenge comes from our limited understanding of how these different ailments really affect the animal both in the short term and after apparent recovery. In particular, relatively little research has attempted to examine how much suffering different health ailments cause to the animals. A second important area for new research in this area is better quantifying the longer term and economic impacts of disease. It is this type of data that can be most useful in increasing awareness of farmers to welfare problems on farms and can be used to explicitly address the economic return on investments that help improve animal welfare.

Much of the research into animal welfare involves taking measures of stress physiology. This often helps give a more scientific aura to a research field that some fear appears too subjective. Physiological measures are important in part because of their usefulness in identifying health issues at an early or "prepathological" stage before a disease has actually developed. Despite the importance of this aim, we lack knowledge of which physiological measures are most predictive of disease. Furthermore, as we argue in Chapter 3, there are serious difficulties in finding appropriate physiological measures of stress and separating out the effects of a genuine challenge to animal welfare from normal physiological fluctuations. Physiological measures are often thought to be a useful window into the emotional states of the animals and have been used in the assessment of pain and distress. We see some value in using these measures to identify the acute emotional responses to short-term procedures such as branding and dehorning (see Chapter 5); however, for assessing longer-term stressors such as those involved with housing systems (see Chapters 6 and 7) we see greater challenges involved with this approach. A decade ago many animal welfare researchers were hopeful that these physiological measures would provide "hard" scientific evidence to address welfare concerns. However, the findings reviewed in Chapter 3 illustrate that much care is needed in interpreting these measures. For these reasons we see limited practical value in this approach of assessing animal welfare.

While rooms could be filled with the books and articles that cover the health and nutrition of cattle, the scientific study of cattle behaviour is still in its infancy. Many of the most controversial issues in animal welfare have focused on the behaviour of animals. The ability to express natural behaviours is considered by many as an inherent part of animal welfare, above and beyond the relationship between these behaviours and the animal's health or emotional state. Although we are critical of the naive suggestion that "natural systems" are always better for animal welfare, we review evidence in Chapter 4 of how an improved understanding of natural behaviour can lead to important insights in the design and management of production systems for cattle. For example, in Chapter 8 we illustrate how understanding the way a calf interacts with the cow in natural systems can be used to improve the design of mechanical feeding systems for calves to allow for the more natural expression of feeding behaviour and much higher rates of growth. It is important to emphasise that mechanical feeding systems, varying from simple buckets to automated nipple feeders, look nothing like the "cow" but seem to meet the calf's needs in a system that is still amenable to intensive production. An improved knowledge of behaviour is important in that it may be useful as a window into the emotional states such as pain and distress. As with the physiological measures discussed above, some of the key challenges are to demonstrate that the behavioural measures are valid indicators of the underlying welfare state of the animals.

Such examples also illustrate a broader issue of how improved knowledge and training in animal behaviour may lead to further insights in the design of production systems and in the interpretation of scientific results in other disciplines, including nutrition and physiology. An ongoing challenge for future researchers is identifying which natural behaviours are truly important to the animals and need to be incorporated into the production and management systems that we design. While few would argue that sucking behaviour is important for calves, other behaviours like foraging or maternal behaviour in adult cattle still need to be studied. A particular area of interest is play behaviour in young animals that is often prevented in individual housing (see Chapter 7).

While the first four chapters of the book set the stage in terms of general issues of defining and assessing animal welfare, the second part (Chapters 5–10) addresses specific risks to animal welfare within beef and dairy production systems and ways in which these may be alleviated.

In many ways, the most obvious types of welfare concern are those that involve acute or short-term painful or frightening procedures, as addressed in Chapter 5. Although housing and other factors may end up having more important and longer-term consequences, it is these short-term procedures that are most likely to attract public attention and they are arguably those that are the most easiest to address. Over a decade of research, dozens of scientific publications have illustrated that dehorning is painful for calves and have demonstrated that a variety of interventions can be used to reduce this pain. Most obviously the use of polled sires can alleviate the need for the procedure. However, many producers have yet to adopt these solutions, suggesting that an ongoing challenge is increasing producer awareness and motivating changes in practice. Here the challenge for animal welfare researchers is to collect data that will be more useful to producers in making their decisions, for example, understanding the production consequences of these painful procedures and how these may affect the health of cattle. Moreover, researchers need to develop solutions that are relatively easy for producers to adopt such as those that reduce costs and labour. For example, breeding to a polled sire alleviates the pain associated with dehorning and reduces costs and labour for the producer.

We argue in Chapters 6 and 7 that the way animals are housed has the largest and longest lasting impact on the health and welfare of cattle. Although each of the topics addressed in this book have their own challenges in terms of research methodology, we show that some of the most serious flaws in experimental design come from comparisons among housing systems. Specifically, many authors have tried to draw general conclusions about housing features by comparing one instance of one design with a single instance of an alternative design. As many readers will understand, there can be much variation within any one housing system and the details in which these systems are managed will have as much affect on the animals as the system itself.

Cattle are social animals and in all areas of cattle production there is a large growing use of group housing systems. While group housing has advantages for animal welfare as it provides social contact and increased space for animals, it also comes with special challenges including increased risk of health problems and increased aggression. For example, when producers move from tie stalls to free stalls they frequently witness an increase in hoof pathologies. A challenge for researchers is to develop management and housing alternatives that overcome these problems. In this case, changes in the design of flooring systems may help reduce the risks of hoof disease in free-stall housing.

Work on housing is often followed with particular interest by producers and indeed much of the published research on cow comfort was motivated by specific requests from the cattle industry. This has meant that research findings in this area have found a ready audience and rapid uptake by many segments of the industry. More generally, this illustrates how close cooperation between researchers and producers can facilitate early adoption of improved practices. That said, there has been little study of factors leading to adoption of new technologies and practices on farm. For example, it is our impression that in North America ideas that can be marketed by an existing industry through the products and services that they offer are most likely to be adopted quickly. In a European context, research that best resonates with legislators may be most likely to be applied on farm.

The way that animals are fed can have important and long-term effects on their welfare. The effect of feeding and nutrition on animal welfare is often underappreciated by the general public and animal scientists and consequently has received little research attention in the published literature. The effect of diet composition on production traits has been well studied but how both the quality of the diet and the way it is delivered affect the welfare of cattle is much less understood. In Chapter 8 we review the effects of limit feeding and the lack of foraging opportunities. For example, diets fed to beef cattle that are high in soluble carbohydrates with little fibre are associated with an increased risk of acidosis and liver abscesses. We believe that this is the most serious welfare problem affecting feedlot cattle. Although, the role of specific nutrients on cattle health is reasonably well researched, this understanding is not always reflected in production practices. Of particular interest is colostrum feeding – although the health benefits for calves are well documented, many animals still are not provided adequate quantity or quality to achieve passive immunity. Despite the apparent ease with which this problem could be solved, we believe that failure of passive transfer due to inadequate colostrum feeding practices remains one of the most important welfare concerns for dairy calves.

As argued earlier, the details in how housing systems are managed greatly influences welfare. In Chapter 9 we discuss stockmanship and how interactions with people can affect cattle welfare. These interactions can be important both because of the immediate effects they can have on the animal, and because animals can learn to associate positive or negative interactions with specific people or contexts. Of particular interest to producers is evidence linking poor handling practices with lower production. Of special interest to social scientists is increased understanding of the attitudes of animal caretakers and how these attitudes affect treatment and care and how both attitudes and behaviour can be changed through training and education. We strongly advocate the increasing professionalization of stock people and encourage the cattle industry to provide and implement training procedures that will improve cattle well-being and job satisfaction and safety for the people that work with these animals.

There are several potential audiences for this book including scientists interested in research on the welfare of cattle and for this group we hope that we have identified some of the key weaknesses in previous research as well as some of the important questions that still need to be addressed. The second audience are students of agriculture and veterinary medicine that are becoming increasingly aware of animal welfare concerns and who need more information to address these concerns in their profession. The third audience is the regulatory community including producer organizations, retailers, and others that may be interested in assessing aspects of animal welfare on farms. This is a growing area of interest as well as an important challenge both practically and scientifically. There is a need to develop measures that can be scored reliably, are practical for the assessors, and provide useful information about the real state of welfare of the animals on that farm. Finally, producers, veterinarians, and extension personal that are visiting or working on farms will value some of the scientific results reviewed above, and will, we hope, be able to draw new insights into how better to manage, house, and handle the animals under their care.

References

Aeberhard, K., Bruckmaier, R. M., Kuepfer, U., and Blum, J. W. (2001). Milk yield and consumption, nutrition, body conformation traits, body condition scores, fertility and diseases in high-yielding dairy cows – Pt 1. *Journal of Veterinary Medicine, A, 48*, 97–110.

Agerholm, J. S., Basse, A., Krogh, H. V., Christensen, K., and Ronsholt, L. (1993). Abortion and calf mortality in Danish cattle herds. *Acta Veterinaria Scandinavica, 34*, 371–377.

Alban, L. (1995). Lameness in Danish dairy cows: Frequency and possible risk factors. *Preventive Veterinary Medicine, 22*, 213–225.

Alban, L., Agger, J. F., and Lawson, L. G. (1996). Lameness in tied Danish dairy cattle: The possible influence of housing systems, management, milk yield, and prior incidents of lameness. *Preventive Veterinary Medicine, 29*, 135–149.

Ali, S., Goonewardene, L. A., and Basarab, J. A. (1994). Estimating water consumption and factors affecting intake in grazing cattle. *Canadian Journal of Animal Science, 74*, 551–554.

Andersen, H. R., Jensen, L. R., Munksgaard, L., and Ingvartsen, K. L. (1997). Influence of floor space allowance and access sites to feed trough on the production of calves and young bulls and on the carcass and meat quality of young bulls. *Acta Agriculturae Scandinavica, Section A, Animal Science, 47*, 48–56.

Anderson, B. H., Watson, D. L., and Colditz, I. G. (1999). The effect of dexamethasone on some immunological parameters in cattle. *Veterinary Research Communications, 23*, 399.

Andersson, M., Schaar, J., and Wiktorsson, H. (1984). Effects of drinking water flow rates and social rank on performance and drinking behaviour of tied-up dairy cows. *Livestock Production Science, 11*, 599–610.

Anderson, N. G., & Zurbrigg, K. (2003). Auditing cow comfort – video behind barn doors. *Advances in Dairy Technology, 15*, 97–114.

Andrighetto, I., Gottardo, F., Andreoli, D., and Cozzi, G. (1999). Effect of type of housing on veal calf growth performance, behaviour and meat quality. *Livestock Production Science, 57*, 137–145.

Appleby, M. C., Weary, D. M., and Chua, B. (2001). Performance and feeding behaviour of calves on ad libitum milk from artificial teats. *Applied Animal Behaviour Science, 74*, 191–201.

Archer, J. (1973). Tests for emotionality in rats and mice: A review. *Animal Behaviour, 21*, 205–235.

Armstrong, D. V. (1994). Heat stress interaction with shade and cooling. *Journal of Dairy Science, 77*, 2044–2050.

Arthington, J. D., Eicher, S. D., Kunkle, W. E., and Martin, F. G. (2003). Effect of transportation and commingling on the acute-phase protein response, growth, and feed intake of newly weaned beef calves. *Journal of Animal Science, 81*, 1120–1125.

Arthington, J. D., Spears, J. W., and Miller, D. C. (2005). The effect of early weaning on feedlot performance and measures of stress in beef calves. *Journal of Animal Science, 83*, 933–939.

Atkinson, P. J. (1992). Investigation of the effects of transport and lairage on hydration state and resting behaviour of calves for export. *Veterinary Record, 130*, 413–416.

Aubert, A. (1999). Sickness and behaviour in animals: A motivational perspective. *Neuroscience and Biobehavioral Reviews, 23*, 1029–1036.

Bachman, K. C. and Schairer, M. L. (2003). Bovine studies on optimal length of dry periods. *Journal of Dairy Science, 86*, 3027–3037.

Bager, F., Braggins, T. J., Devine, C. E., and Graafhuis, A. E. (1992). Onset of insensibility at slaughter in calves: Effects of electroplectic seizure and exsanguination on spontaneous electrocortical activity and indices of cerebral metabolism. *Research in Veterinary Science, 52*, 162–173.

Bakken, G. (1982). The relationship between enviromental conditions and bovine udder disease in Norwegian dairy herds. *Acta Agriculturae Scandinavica, 32*, 23–31.

Balm, H. M. (1999). *Stress physiology in animals*. Sheffield, UK: Academic Press.

Barberg, A. E., Endres, M. I., Salfer, J. A., and Reneau, J. K. (2007). Performance and welfare of dairy cows in an alternative housing system in Minnesota. *Journal of Dairy Science, 90*, 1575–1583.

Bareille, N., Beaudeau, F., Billon, S., Robert, A., and Faverdin, P. (2003). Effects of health disorders on feed intake and milk production in dairy cows. *Livestock Production Science, 83*, 53–62.

Barham, A. R., Barham, B. L., Johnson, A. K., Allen, D. M., Blanton, J. R. J., and Miller, M. F. (2002). Effects of the transportation of beef cattle from the feedyard to the packing plant on prevalence levels of *Escherichia coli* O157 and *Salmonella spp. Journal of Food Protection, 65*, 280–283.

Barkema, H. W., Schukken, Y. H., Lam, T. J., Beiboer, M. L., Benedictus, G., and Brand, A. (1999a). Management practices associated with the incidence rate of clinical mastitis. *Journal of Dairy Science, 82*, 1643–1654.

Barkema, H. W., Van der Ploeg, J. D., Schukken, Y. H., Lam, T. J. G. M., Benedictus, G., and Brands, A. (1999b). Management style and its association with bulk milk somatic cell count and incidence rate of clinical mastitis. *Journal of Dairy Science, 82*, 1655–1663.

Barnett, J. L., Coleman, G. J., Hemsworth, P. H., Newman, E. A., Fewings-Hall, S., and Ziini, C. (1999). Tail docking and beliefs about the practice in the Victorian dairy industry. *Australian Veterinary Journal, 77*, 742–747.

Bartussek, H. (1999). A review of the animal needs index (ANI) for the assessment of animals' well-being in the housing systems for Austrian proprietary products and legislation. *Livestock Production Science, 61*, 179–192.

Bass, M. J. and Phillips, L. G. (2007). Pressure sores. *Current Problems in Surgery, 44*, 101–143.

Battaglia, R. A. (1998). *Handbook of livestock management*, 3rd edn. Upper Saddle River, NJ : Prentice-Hall.

Beauchemin, K. A. and Yang, W. Z. (2005). Effects of physically effective fiber on intake, chewing activity, and ruminal acidosis for dairy cows fed diets based on corn silage. *Journal of Dairy Science, 88*, 2117–2129.

Becker, B. G. and Lobato, J. F. P. (1997). Effect of gentle handling on the reactivity of zebu crossed calves to humans. *Applied Animal Behaviour Science, 53*, 219–224.

Bell, E. (2004). Description of claw horn lesions and associated risk factors in dairy cattle in the lower Fraser Valley, British Columbia. M.Sc. thesis, University of British Columbia.

Bellows, R. A. and Lammoglia, M. A. (2000). Effects of severity of dystocia on cold tolerance and serum concentrations of glucose and cortisol in neonatal beef calves. *Theriogenology, 53*, 803–813.

Bendixen, P. H., Vilson, B., Ekesbo, I., and Åstrand, D. B. (1986). Disease frequencies of tied zero-grazing dairy cows and of dairy cows on pasture during summer and tied during winter. *Preventive Veterinary Medicine, 4*, 291–306.

Benson, G. J. (2004). Pain in farm animals: Nature, recognition, and management. In: G. J. Benson and B. E. Rollins (eds), *The well-being of farm animals: Challenges and solutions* (pp. 61–84). Oxford: Blackwell.

Bergeron, R., Badnell-Waters, A. J., Lambton, S., and Mason, G. (2006). Stereotypic oral behaviour in captive ungulates: Foraging, diet and gastrointestinal function. In: G. Mason and J. Rushen (eds), *Stereotypic animal behaviour: Fundamentals and applications to welfare*, 2nd edn. (pp. 19–57). Wallingford, UK: CAB International.

Bergsten, C. (1994). Hemorrhages of the sole horn dairy-cows as a retrospective indicator of laminitis – an epidemiological study. *Acta Veterinaria Scandinavica, 35*, 55–66.

Bergsten, C. (2001). Effects of conformation and management system on hoof and leg diseases and lameness in dairy cows. *Veterinary Clinics of North America – Food Animal Practice, 17*, 1–23.

Bergsten, C. and Frank, B. (1996). Sole haemorrhages in tied heifers in early gestation as an indicator of laminitis: Effects of diet and flooring. *Acta Veterinary Scandinavia, 37*, 375–382.

Bergsten, C. and Herlin, A. H. (1996). Sole haemorrhages and heel horn erosion in dairy cows: The influence of housing system on their prevalence and severity. *Acta Veterinaria Scandanavia, 37*, 395–405.

Bergsten, C. and Pettersson, B. (1992). The cleanliness of cows tied in stalls and the health of their hooves as influenced by the use of electric trainers. *Preventive Veterinary Medicine, 13*, 229–238.

Bernues, A., Olaizola, A., and Corcoran, K. (2003). Labelling information demanded by European consumers and relationships with purchasing motives, quality and safety of meat. *Meat Science, 65*, 1095–1106.

Bertenshaw, C. and Rowlinson, P. (2001). The influence of positive human–animal interaction during rearing on the welfare and subsequent production of the dairy heifer. *Proceedings of the British Society of Animal Science, 17*.

Bewley, J., Palmer, R. W., and Jackson-Smith, D. B. (2001). An overview of experiences of Wisconsin dairy farmers who modernized their operations. *Journal of Dairy Science, 84*, 717–729.

Blackshaw, J. K., Blackshaw, A. W., and Kusano, T. (1987). Cattle behaviour in a sale yard and its potential to cause bruising. *Australian Journal of Experimental Agriculture, 27*, 753–757.

Blanchard, D. C. and Blanchard, R. J. (1988). Ethoexperimental approaches to the biology of emotion. *Annual Review of Psychology, 39*, 43–68.

Blecha, F. (2000). Immune responses to stress. In: G. P. Moberg and J. A. Mench (eds), *The Biology of animal stress: Basic principles and implications for animal welfare* (pp. 111–121). Wallingford, UK: CAB International.

Blecha, F., Boyles, S. L., and Riley, J. G. (1984). Shipping suppresses lymphocyte blastogenic responses in Angus and Brahman X Angus feeder calves. *Journal of Animal Science, 59*, 576–583.

Blom, J. Y., Konggaard, S. P., Larsson, J. G., Nielsen, K., Northevd, A., and Solfjeld, P. (1984). Electronic recording of pressure exerted by cows against structures in free-stall housing. *Applied Animal Behaviour Science, 13*, 41–46.

Blum, J. W. and Hammon, H. (2000). Colostrum effects on the gastrointestinal tract, and on nutritional, endocrine and metabolic parameters in neonatal calves. *Livestock Production Science, 66*, 151–159.

Boissy, A. and Bouissou, M. F. (1988). Effects of early handling on heifers' subsequent reactivity to humans and to unfamiliar situations. *Applied Animal Behaviour Science, 20*, 259–273.

Boissy, A. and Le Neindre, P. (1997). Behavioral, cardiac and cortisol responses to brief peer separation and reunion in cattle. *Physiology & Behavior, 61*, 693–699.

Boivin, X., Garel, J. P., Mante, A., & Le Neindre, P. (1998). Beef cattle react differently to different handlers according to the test situation and their previous interactions with their caretaker. *Applied Animal Behaviour Science, 55*, 245.

Boivin, X., Le Neindre, P., and Chupin, J. M. (1992a). Establishment of cattle–human relationships. *Applied Animal Behaviour Science, 32*, 325–335.

Boivin, X., Le Neindre, P., Chupin, J. M., Garel, J. P., and Trillat, G. (1992b). Influence of breed and early management on ease of handling and open-field behaviour of cattle. *Applied Animal Behaviour Science, 32*, 313–323.

Boivin, X., Le Neindre, P., Garel, J. P., and Chupin, J. M. (1994). Influence of breed and rearing management on cattle reactions during human handling. *Applied Animal Behaviour Science, 39*, 115–122.

Boivin, X., Nowak, R., Desprès, G., Tournadre, H., and Le Neindre, P. (1997). Discrimination between shepherds by lambs reared under artificial conditions. *Journal of Animal Science, 75*, 2892–2898.

Bokkers, E. A. M. and Koene, P. (2001). Activity, oral behaviour and slaughter data as welfare indicators in veal calves: A comparison of three housing systems. *Applied Animal Behaviour Science, 75*, 1–15.

Borderas, T. F., Pawluczuk, B., de Passillé, A. M., and Rushen, J. (2004). Claw hardness of dairy cows: Relationship to water content and claw lesions. *Journal of Dairy Science, 87*, 2085–2093.

Borg, K. E., Esbenshade, K. L., and Johnson, B. H. (1991). Cortisol, growth hormone, and testosterone concentrations during mating behavior in the bull and boar. *Journal of Animal Science, 69*, 3230–3240.

Boukila, B., Seoane, J. R., and Bernier, J. F. (1995). Effects of dietary hydroxides on intake, digestion, rumen fermentation and acid-base-balance in sheep fed a high-barley diet. *Canadian Journal of Animal Science, 75*, 359–369.

Bradbury, J. W. and Vehrencamp, S. L. (2000). Economic models of animal communication. *Animal Behaviour, 59*, 259–268.

Brambell, F. W. R. (1965). *Report of the technical committee to enquire into the welfare of animals kept under intensive livestock husbandry systems*. London: Her Majesty's Stationary Office.

Braun, U., Rihs, T., and Schefer, U. (1992). Ruminal lactic acidosis in sheep and goats. *Veterinary Record, 130*, 343–349.

Bremel, R. D. and Gangwer, M. I. (1978). Effect of adrenocorticotropin injection and stress on milk cortisol content. *Journal of Dairy Science, 61*, 1103–1108.

Bretschneider, G. (2005). Effects of age and method of castration on performance and stress response of beef male cattle: A review. *Livestock Production Science, 97*, 89–100.

Breuer, K., Hemsworth, P. H., Barnett, J. L., Matthews, L. R., and Coleman, G. J. (2000). Behavioural response to humans and the productivity of commercial dairy cows. *Applied Animal Behaviour Science, 66*, 273–288.

Brightling, P., Grueme, A. M., Malmo, J., and Ryan, D. P. (1998). *Farm guidelines for mastitis control*. Melbourne, Australia: Dairy Research and Development Corporation.

Broom, D. M. (2006). Behaviour and welfare in relation to pathology. *Applied Animal Behaviour Science, 97*, 73–83.

Broom, D. M. and Leaver, J. D. (1978). Effects of group-rearing or partial isolation on later social behaviour of calves. *Animal Behaviour, 26*, 1255–1263.

Broster, W. H. and Broster, V. J. (1998). Body score of dairy cows. *Journal of Dairy Research, 65*, 155–173.

Brown-Brandl, T. M., Eigenberg, R. A., and Nienaber, J. A. (2006). Heat stress risk factors of feedlot heifers. *Livestock Science, 105*, 57–68.

Browning, J. W., Mein, G. A., Barton, M., Nichollas, T. J., and Brightling, P. (1990). Effects of antibiotic therapy at drying-off on mastitis in the dry period and early lactation. *Australian Veterinary Journal, 67*, 440–442.

Bruckmaier, R. M. and Blum, J. W. (1998). Oxytocin release and milk removal in ruminants. *Journal of Dairy Science, 81*, 939–949.

Bruckmaier, R. M., Schams, D., and Blum, J. W. (1993). Milk removal in familiar and unfamiliar surroundings: Concentrations of oxytocin, prolactin, cortisol and beta-endorphin. *Journal of Dairy Research, 60*, 449–456.

Bruckmaier, R. M., Wellnitz, O., and Blum, J. W. (1997). Inhibition of milk ejection in cows by oxytocin receptor blockade, a-adrenergic receptor stimulation and in unfamiliar surroundings. *Journal of Dairy Research, 64*, 315–325.

Bruun, J., Ersbøll, A. K., and Alban, L. (2002). Risk factors for metritis in Danish dairy cows. *Preventive Veterinary Medicine, 54*, 179–190.

Buenger, A., Ducrocq, V., and Swalve, H. H. (2001). Analysis of survival in dairy cows with supplementary data on type scores and housing systems from a region of northwest Germany. *Journal of Dairy Science, 84*, 1531–1541.

Buhman, M. J., Perino, L. J., Galyean, M. L., Wittum, T. E., Montgomery, T. H., and Swingle, R. S. (2000). Association between changes in eating and drinking behaviour and respiratory tract disease in newly arrived calves at a feedlot. *American Journal of Veterinary Research, 61*, 1163–1168.

Burger, J. and Gochfield, M. (1992). Effect of group size on vigilance while drinking in the coati, *Nasua narica* in Costa Rica. *Animal Behaviour, 44*, 1053–1057.

Burrow, H. M. (1997). Measurements of temperament and their relationships with performance traits of beef cattle. *Animal Breeding Abstracts, 65*, 477–495.

Burrow, H. M. (2001). Variances and covariances between productive and adaptive traits and temperament in a composite breed of tropical beef cattle. *Livestock Production Science, 70*, 213–233.

Burrow, H. M. and Dillon, R. D. (1997). Relationship between temperament and growth in a feedlot and commercial carcass traits of bos indicus cross-breeds. *Australian Journal of Experimental Agriculture, 37*, 407–411.

Burton, J. L. and Kehrli, M. E. (1995). Regulation of neutrophil adhesion molecules and shedding of staphylococcus-aureus in milk of cortisol – treated and dexamethasone-treated cows. *American Journal of Veterinary Research, 56*, 997–1006.

Burvenich, C., Paape, M. J., Hoeben, D., Dosogne, H., Massart-Leen, A. M., and Blum, J. (1999). Modulation of the inflammatory reaction and neutrophil defense of the bovine lactating mammary gland by growth hormone. *Domestic Animal Endocrinology, 17*, 149–159.

Busato, A., Steiner, L., Martin, S. W., Shoukri, M. M., and Gaillard, C. (1997). Calf health in cow-calf herds in Switzerland. *Preventive Veterinary Medicine, 30*, 9–22.

Bushe, T. and Oliver, S. P. (1987). Natural protective factors in bovine mammary secretions following different methods of milk cessation. *Journal of Dairy Science, 70*, 696–704.

Campbell, J. B., Boxler, D. J., Adams, D. C., and Applegarth, A. F. (2006). Efficacy of several insecticide ear tags for control of horn flies (diptera: Muscidae) on Nebraska beef cattle. *Journal of the Kansas Entomological Society, 79*, 113–118.

Campbell, J. B., Catangui, M. A., Thomas, G. D., Boxler, D. J., and Davis, R. (2001). Effects of stable flies (diptera: Muscidae) and heat stress on weight gains of grazing yearling cattle. *Journal of Economical Entomology, 94*, 780–783.

Carson, K. and Wood-Gush, D. G. M. (1984). The behaviour of calves at market. *Animal Production, 39*, 389–397.

Carson, T. L. (2000). Current knowledge of water quality and safety for livestock. *Toxicology, 16*, 455–464.

Castillo, C., Benedito, J. L., Méndez, J., Pereira, V., López-Alonso, M., Miranda, M., et al. (2004). Organic acids as a substitute for monensin in diets for beef cattle. *Animal Feed Science and Technology, 115*, 101–116.

Catangui, M. A., Campbell, J. B., Thomas, G. D., and Boxler, D. J. (1997). Calculating economic injury levels for stable flies (diptera: Muscidae) on feeder heifers. *Journal of Economical Entomology, 90*, 6–10.

Ceballos, A. and Weary, D. M. (2002). Feeding small quantities of grain in the parlour facilitates pre-milking handling of dairy cows: A note. *Applied Animal Behaviour Science, 77*, 249–254.

Ceballos, A., Sanderson, D., Rushen, J., and Weary, D. M. (2004). Improving stall design: Use of 3-D kinematics to measure space use by cows when lying down. *Journal of Dairy Science, 87*, 2042–2050.

Chan, S. C., Huber, J. T., Chen, K. H., Simas, J. M., and Wu, Z. (1997). Effects of ruminally inert fat and evaporative cooling on dairy cows in hot environmental temperatures. *Journal of Dairy Science, 80*, 1172–1178.

Chase, C. C. J., Larsen, R. E., Randel, R. D., Hammond, A. C., and Adams, E. L. (1995). Plasma cortisol and white blood cell responses in different breeds of bulls: A comparison of two methods of castration. *Journal of Animal Science, 73*, 975–980.

Chua, B., Coenen, E., van, D. J., and Weary, D. M. (2002). Effects of pair versus individual housing on the behavior and performance of dairy calves. *Journal of Dairy Science, 85*, 360–364.

Clutton-Brock, J. (1999). *A natural history of domesticated mammals* (p. 284). London: Natural History Museum.

Cockcroft, P. D., Henson, F. M. D., and Parker, C. (2000). Thermography of a septic metatarsophalangeal joint in a heifer. *Veterinary Record, 146*, 258–260.

Colam-Ainsworth, P., Lunn, G. A., Thomas, R. C., and Eddy, R. G. (1989). Behaviour of cows in cubicles and its possible relationship with laminitis in replacement dairy heifers. *Veterinary Record, 125*, 573–575.

Colditz, I. (2002). Effects of the immune system on metabolism: Implications for production and disease resistance in livestock. *Livestock Production Science, 75*, 257–268.

Collard, B. L., Boettcher, P. J., Dekkers, J. C. M., Petitclerc, D., and Schaeffer, L. R. (2000). Relationships between energy balance and health traits of dairy cattle in early lactation. *Journal of Dairy Science, 83*, 2683–2690.

Collier, R. J., Dahl, G. E., and VanBaale, M. J. (2006). Major advances associated with environmental effects on dairy cattle. *Journal of Dairy Science, 89*, 1244–1253.

Cook, N. B. (2003). Prevalence of lameness among dairy cattle in Wisconsin as a function of housing type and stall surface. *Journal of the American Veterinary Medical Association, 223*, 1324–1328.

Cook, N. B., Bennett, T. B., and Nordlund, K. V. (2005). Monitoring indices of cow comfort in free-stall-housed dairy herds. *Journal of Dairy Science, 88*, 3876–3885.

Cook, N. B., Nordlund, K. V., and Oetzel, G. R. (2004). Environmental influences on claw horn lesions associated with laminitis and subacute ruminal acidosis in dairy cows. *Journal of Dairy Science, 87, E. supplement*, E36–E46.

Correa, M. T., Erb, H., and Scarlett, J. (1993). Path-analysis for 7 postpartum disorders of Holstein cows. *Journal of Dairy Science, 76*, 1305–1312.

Creel, S. R. and Albright, J. L. (1988). The effects of neonatal social-isolation on the behavior and endocrine function of Holstein calves. *Applied Animal Behaviour Science, 21*, 293–306.

Croney, C. C., Wilson, L. L., Curtis, S. E., and Cash, E. H. (2000). Effects of handling aids on calf behavior. *Applied Animal Behaviour Science, 69*, 1–13.

Curtis, C. R., Erb, H. N., Sniffen, C. J., Smith, R. D., and Kronfield, D. S. (1985). Path analysis of dry period nutrition, postpartum metabolic and reproductive disorders, and mastitis in Holstein cows. *Journal of Dairy Science, 68*, 2347–2360.

Da Silva, R. G., La Scala, N., and Tonhati, H. (2003). Radiative properties of the skin and haircoat of cattle and other animals. *American Society of Agricultural Engineers, 46*, 913–918.

Dallman, M. F. (2001). Stress and sickness decrease food intake and body weight: How does this happen? when does this adaptive response progress to pain and suffering? In: D. M. Broom (ed.), *Coping with challenge: Welfare in animals including humans* (pp. 301–316). Berlin: Dahlem University Press.

Daly, C. C., Gregory, N. G., and Wotton, S. B. (1987). Captive bolt stunning of cattle: Effects on brain function and role of bolt velocity. *British Veterinary Journal, 143*, 574–580.

Daly, C. C., Kallweit, E., and Ellendorf, F. (1988). Cortical function in cattle during slaughter: Conventional captive bolt stunning followed by exsanguination compared with shechita slaughter. *Veterinary Record, 122*, 325–329.

Danbury, T. C., Weeks, C. A., Chambers, J. P., Waterman-Pearson, A. E., and Kestin, S. C. (2000). Self-selection of the analgesic drug carprofen by lame broiler chickens. *Veterinary Record, 146*, 307–311.

Dantzer, R. (2001). Cytokine-induced sickness behavior: Where do we stand? *Brain, Behavior, and Immunity, 15*, 7–24.

Dantzer, R., Mormède, P., Bluthe, R. M., and Soissons, J. (1983). The effect of different housing conditions on behavioural and adrenocortical reactions in veal calves. *Reproduction, Nutrition, Development, 23*, 501–508.

Davis, C. L. and Drackley, J. K. (1998). *The development, nutrition, and management of the young calf.* Ames, IA: State Press.

Davis, M. S., Mader, T. L., Holt, S. M., and Parkhurst, A. M. (2003). Stategies to reduce feedlot cattle heat stress: Effects on tympanic temperature. *Journal of Animal Science, 81*, 649–661.

Dawkins, M. (1983). Battery hens name their price: Consumer demand theory and the measurement of ethological "needs". *Animal Behaviour, 31*, 1195–1205.

Dawkins, M. (1988). Behavioural deprivation: A central problem in animal welfare. *Applied Animal Behaviour Science, 20*, 209–225.

Dawkins, M. (1990). From an animal's point of view: Motivation, fitness, and animal welfare. *Behavioural and Brain Sciences, 13*, 1–61.

Dawkins, M. S. (1998). Evolution and animal welfare. *Quarterly Review of Biology, 73*, 305–328.

Dawkins, M. S. (2004). Using behaviour to assess animal welfare. *Animal Welfare, 13*, S3–S7.

de Mol, R. M. and Ouweltjes, W. (2001). Detection model for mastitis in cows milked in an automatic milking system. *Preventive Veterinary Medicine, 49*, 71–82.

de Mol, R. M. and Woldt, W. E. (2001). Application of fuzzy logic in automated cow status monitoring. *Journal of Dairy Science, 84*, 400–410.

de Passillé, A. M. (2001). Sucking motivation and related problems in calves. *Applied Animal Behaviour Science, 72*, 175–187.

de Passillé, A. M. and Rushen, J. (2005). Can we measure human-animal interactions in on-farm animal welfare assessment? Some unresolved issues. *Applied Animal Behaviour Science, 92*, 193–209.

de Passillé, A. M. and Rushen, J. (2006a). What components of milk stimulate sucking in calves? *Applied Animal Behaviour Science, 101*, 243–252.

de Passillé, A. M. and Rushen, J. (2006b). Calves' behaviour during nursing is affected by feeding motivation and milk availability. *Applied Animal Behaviour Science, 101*, 264–275.

de Passillé, A. M. and Rushen, J. (2006c). Feeding management and stress in calves. In: K. Sejrsen, T. Hvelplund, and M. O. Nielsen (eds), *Ruminant physiology: Digestion, metabolism, and impact of nutrition on gene expression, immunology and stress* (pp. 499–510). Wageningen, The Netherlands: Wageningen Academic Publishers.

de Passillé, A. M., Rushen, J., and Marnet, P. (1997). Effects of nursing a calf on milk ejection and milk yield during milking. *Journal of Dairy Science, 80, supplement 1*, 203.

de Passillé, A. M., Rushen, J., and Martin, F. (1995). Interpreting the behaviour of calves in an open-field test: A factor analysis. *Applied Animal Behaviour Science, 45*, 201–213.

de Passillé, A. M., Rushen, J., and Weary, D. (2004). Designing good environments and management for calves. *Advances in Dairy Technology, 16*, 75–89.

de Passillé, A. M. B., Christopherson, R., and Rushen, J. (1993). Nonnutritive sucking by the calf and postprandial secretion of insulin, CCK and gastrin. *Physiology & Behavior, 54*, 1069–1073.

de Passillé, A. M. B., Metz, J. H. M., Mekking, P., and Wiepkema, P. R. (1992). Does drinking milk stimulate sucking in young calves? *Applied Animal Behaviour Science, 34*, 23–36.

de Passillé, A. M. B., Rushen, J., Ladewig, J., and Petherick, C. (1996). Dairy calves' discrimination of people based on previous handling. *Journal of Animal Science, 74*, 969–974.

De Paula Vieira, A., Guesdon, V., de Passillé, A. M., von Keyserlingk, M. A. G., and Weary, D. M. (2007) Behavioural indicators of hunger in dairy calves. *Applied Animal Behaviour Science*, doi:10.1016/j.applanim.2007.03.006.

de Wilt, J. G. (1985). Behaviour and welfare of veal calves in relation to husbandry systems. Ph.D. thesis, Agricultural University, Wageningen, The Netherlands, p. 138.

Dellmeier, G. R., Friend, T. H., and Gbur, E. E. (1985). Comparison of four methods of calf confinement. II. Behavior. *Journal of Animal Science, 60*, 1102–1109.

Després, G., Veissier, I., and Boissy, A. (2002). Effect of autonomic blockers on heart period variability in calves: Evaluation of the sympatho-vagal balance. *Physiology Research, 51*, 347–353.

deVries, M. J., van der Beek, S., Kaal-Lansbergen, L. M. T. E., Ouweltjes, W., and Wilmink, J. B. M. (1999). Modeling of energy balance in early lactation and the effect of energy deficits in early lactation on first detected estrus postpartum in dairy cows. *Journal of Dairy Science, 82*, 1927–1934.

DeVries, T. J. and von Keyserlingk, M. A. G. (2006). Effects of feeder design on the social and feeding behavior of lactating dairy cows. *Journal of Dairy Science, 89*, 3522–3531.

DeVries, T. J., von Keyserlingk, M. A. G., and Beauchemin, K. A. (2005). Frequency of feed delivery affects the behavior of lactating dairy cows. *Journal of Dairy Science, 88*, 3553–3562.

DeVries, T. J., von Keyserlingk, M. A. G., and Weary, D. M. (2004). Effect of feeding space on the inter-cow distance, aggression, and feeding behavior of free-stall housed lactating dairy cows. *Journal of Dairy Science, 87*, 1432.

Dewell, R. D., Hungerford, L. L., Keen, J. E., Laegreid, W. W., Griffin, D. D., Rupp, G. P., et al. (2006). Association of neonatal serum immunoglobulin G1 concentration with health and performance in beef calves. *Journal of the American Veterinary Medical Association, 228*, 914–921.

Dhabhar, F. S. (2002). Stress-induced augmentation of immune function – the role of stress hormones, leukocyte trafficking, and cytokines. *Brain, Behavior, and Immunity, 16*, 785–798.

Diaz, M. C., Van Amburgh, M., Smith, J. M., Kelsey, J. M., and Hutten, E. L. (2001). Composition of growth of Holstein calves fed milk replacer from birth to 105-kilogram body weight. *Journal of Dairy Science, 84*, 830–842.

Diez-Fraile, A., Meyer, E., Massart-Leën, A. M., and Burvenich, C. (2000). Effect of isoproterenol and dexamethasone on the lipopolysaccharide induced expression of CD11b on bovine neutrophils. *Veterinary Immunology and Immunopathology, 76*, 151–156.

Digesti, R. D. and Weeth, H. J. (1976). A defensible maximum for inorganic sulfate in drinking water of cattle. *Journal of Animal Science, 42*, 1498–1502.

Dingwell, R. T., Duffield, T. F., Leslie, K. E., Keefe, G. P., DesCoteaux, L., Kleton, D. F., et al. (2002). The efficacy of intramammary tilmicosin at drying-off, and other risk factors for the prevention of new intramammary infections during the dry period. *Journal of Dairy Science, 85*, 3250–3259.

Dixit, V. D., Marahrens, M., and Parvizi, N. (2001). Transport stress modulates adrenocorticotropin secretion from peripheral bovine lymphocytes. *Journal of Animal Science, 79*, 729–734.

Dobson, H. and Smith, R. F. (2000). What is stress, and how does it affect reproduction? *Animal Reproduction Science, 61*, 743–753.

Dohoo, I. R. and Martin, S. W. (1984). Subclinical ketosis – prevalence and associations with production and disease. *Canadian Journal of Comparative Medicine, 48*, 1–5.

Drackley, J. K. (1999). Biology of dairy cows during the transition period: The final frontier? *Journal of Dairy Science, 82*, 2259–2273.

Ducrot, C., Calavas, D., Sabatier, P., and Faye, B. (1998). Qualitative interaction between the observer and the observed in veterinary epidemiology. *Preventive Veterinary Medicine, 34*, 107–113.

Duenas, M. I., Paape, M. J., Wettemann, R. P., and Douglass, L. W. (2001). Incidence of mastitis in beef cows after intramuscular administration of oxytetracycline. *Journal of Animal Science, 79*, 1996–2005.

Duffield, T. (2000). Subclinical ketosis in lactating dairy cattle. *Veterinary Clinics of North America – Food Animal Practice, 16*, 231–253.

Duffield, T., Plaizier, J. C., Fairfield, A., Bagg, R., Vessie, G., Dick, P., et al. (2004). Comparison of techniques for measurement of rumen pH in lactating dairy cows. *Journal of Dairy Science, 87*, 59–66.

Duffield, T. F., Kelton, D. F., Leslie, K. E., Lissemore, K. D., and Lumsden, J. H. (1997). Use of test day milk fat and milk protein to detect subclinical ketosis in dairy cattle in Ontario. *Canadian Veterinary Journal, 38*, 713–718.

Duffield, T. F., Sandals, D., Leslie, K. E., Lissemore, K., McBride, B. W., Lumsden, J. H., et al. (1998). Efficacy of monensin for the prevention of subclinical ketosis in lactating dairy cows. *Journal of Dairy Science, 81*, 2866–2873.

Duncan, I. J. H. (2002). Poultry welfare: Science or subjectivity? *British Poultry Science, 43*, 643–652.

Duncan, I. J. H. (2006). The changing concept of animal sentience. *Applied Animal Behaviour Science, 100*, 11–19.

Dunn, C. S. (1990). Stress reactions of cattle undergoing ritual slaughter using two methods of restraint. *Veterinary Record, 126*, 522–525.

EFSA Animal Health and Welfare Panel. (2006). Scientific report on the risk of poor welfare in intensive calf farming systems. An update on the report of the veterinary committee on the welfare of calves. *The EFSA Journal, 366*, 1–36.

Eicher, S. D. (2001). Transportation of cattle in the dairy industry: Current research and future directions. *Journal of Dairy Science, 84*, 19–23.

Eicher, S. D., Cheng, H. W., Sorrells, A. D., and Schutz, M. M. (2006). Short communication: Behavioral and physiological idicators of sensitivity or chronic pain following tail docking. *Journal of Dairy Science, 89*, 3047–3051.

Eicher, S. D., Morrow-Tesch, J. L., Albright, J. L., Dailey, J. W., Young, C. R., and Stanker, L. H. (2000). Tail-docking influences on behavioral, immunological, and endocrine responses in dairy heifers. *Journal of Dairy Science, 83*, 1456–1462.

Eicher, S. D., Morrow-Tesch, J. L., Albright, J. L., and Williams, R. E. (2001). Tail-docking alters fly numbers, fly-avoidance behaviors, and cleanliness, but not physiological measures. *Journal of Dairy Science, 84*, 1822–1828.

Eigenberg, R. A., Hahn, G. L., Nienaber, J. A., Brown-Brandl, T. M., and Spiers, D. E. (2000). Development of a new respiration rate monitor for cattle. *American Society of Agricultural Engineers, 43*, 723–728.

Ekesbo, I., Oltenacu, P. A., Vilson, B., and Nilsson, J. (1994). A disease monitoring-system for dairy herds. *Veterinary Record, 134*, 270–273.

Elbers, A. R. W., Miltenburg, J. D., De Lange, D., Crauwels, A. P. P., Barkema, H. W., and Schukken, Y. H. (1997). Risk factors for clinical mastitis in a random sample of dairy herds from the southern part of the Netherlands. *Journal of Dairy Science, 81*, 420–426.

Elenkov, I. J. and Chrousos, G. P. (1999). Stress hormones, Th1/Th2 patterns, Pro/Anti-inflammatory cytokines and susceptibility to disease. *Trends in Endocrinology and Metabolism, 10*, 359–368.

Elgar, M. A. (1989). Predator vigilance and group size in mammals and birds: A critical review of the empirical evidence. *Biological Reviews, 64*, 13–33.

Elsasser, T. H., Klasing, K. C., Filipov, N., and Thompson, F. (2000). The metabolic consequences of stress: Targets for stress and priorities of nutrient use. In: G. P. Moberg and J. A. Mench (eds), *The biology of animal stress* (pp. 77–110). Wallingford, UK: CAB International.

Embry, L. B., Hoelscher, M. A., Wahlstrom, R. C., Carlson, C. W., Krista, L. M., Brosz, W. R., et al. (1959). *Salinity and livestock water quality*, No. 481. S. Dakota, USA: S. Dakota Agr. Exp. Sta. Bull.

Endres, M. I., DeVries, T. J., von Keyserlingk, M. A. G., and Weary, D. M. (2005). Effect of feed barrier design on the behavior of loose-housed lactating dairy cows. *Journal of Dairy Science, 88*, 2377–2380.

Enevoldsen, C., Grohn, Y. T., and Thysen, I. (1994). Skin injuries on the body and thigh of dairy-cows – associations with season, claw health, disease treatement, and other cow characteristics. *Acta Veterinaria Scandinavica, 35*, 337–347.

Esslemont, R. J. & Kossaibati, M. A. (1996). Incidence of production diseases and other health problems in a group of dairy herds in England. *Veterinary Record, 139*, 486–490.

Espejo, L. A. and Endres, M. I. (2007). Herd-level risk factors for lameness in high-producing Holstein cows housed in freestall barns. *Journal of Dairy Science, 90*, 306–314.

Espejo, L. A., Endres, M. I., and Salfer, J. A. (2006). Prevalence of lameness in high-producing Holstein cows housed in freestall barns in Minnesota. *Journal of Dairy Science, 89*, 3052–3058.

Estep, D. Q. and Hetts, S. (1992). Interactions, relationships, and bonds: The conceptual basis for scientist-animal interaction. In: H. Davis and A. D. Balfour (eds), *The inevitable bond-examining scientist-animal interactions* (pp. 6–26). Cambridge: Cambridge University Press.

Evans, C. S. and Marler, P. (1992). Female appearance as a factor in the responsiveness of male chickens during anti-predator behaviour and courtship. *Animal Behaviour, 43*, 137–145.

Ewbank, R. (1993). Handling cattle in intensive systems. In: T. Grandin (ed.), *Livestock handling and transport* (pp. 59–73). Wallingford, UK: CAB International.

Fairchild, T. P., Mcarthur, B. J., Moore, J. H., and Hylton, W. E. (1982). Coliform counts in various bedding materials. *Journal of Dairy Science, 65*, 1029–1035.

Fallon, R. J., Harte, F. J., and Keane, M. G. (1989). Methods of artificially feeding colostrum to the new-born calf. *Irish Journal of Agricultural Research, 28*, 57–63.

Farm Animal Welfare Council (1997). *Report on the welfare of dairy cattle*. Surbiton, UK: Ministry of Agriculture, Fisheries and Food.

Faulkner, P. M. and Weary, D. M. (2000). Reducing pain after dehorning in dairy calves. *Journal of Dairy Science, 83*, 2037–2041.

Faull, W. B., Hughes, J. W., Clarkson, M. J., Downham, D. Y., Manson, F. J., Merritt, J. B., et al. (1996). Epidemiology of lameness in dairy cattle: The influence of cubicles and indoor and outdoor walking surfaces. *Veterinary Record, 139*, 130–136.

Faye, B., Lescourret, F., Dorr, N., Tillard, E., MacDermott, B., and McDermott, J. (1997). Interrelationships between herd management practices and udder health status using canonical correspondance analysis. *Preventive Veterinary Medicine, 32*, 171–192.

Fell, L. R., Wells, R., and Shutt, D. A. (1986). Stress in calves castrated surgically or by the application of rubber rings. *Australian Veterinary Journal, 63*, 16–18.

Fiems, L., De Campeneere, S., Cottyn, B., and Boucque, C. (1998). Effect of housing on the performance of double-muscled rearing calves. *Annales de Zootechnie, 47*, 207–214.

Filteau, V., Bouchard, E., Fecteau, G., Dutil, L., and DuTremblay, D. (2003). Health status and risk factors associated with failure of passive transfer of immunity in newborn beef calves in Quebec. *Canadian Veterinary Journal, 44*, 907–913.

Fisher, A. D., Crowe, M. A., Alonso de la Varga, M. E., and Enright, W. J. (1996). Effect of castration method and the provision of local anesthesia on plasma cortisol, scrotal circumference, growth, and feed intake of bull calves. *Journal of Animal Science, 74*, 2336–2343.

Fisher, A. D., Crowe, M. A., Ó'Nualláin, E. M., Monaghan, M. L., Larkin, J. A., O'Kiely, P., et al. (1997a). Effects of cortisol on in vitro interferon-m production, acute-phase proteins, growth, and feed intake in a calf castration model. *Journal of Animal Science,* 75, 1041–1047.

Fisher, A. D., Crowe, M. A., O'Nuallain, E. M., Monaghan, M. L., Prendiville, D. J., O'Kiely, P., et al. (1997b). Effects of suppressing cortisol following castration of bull calves on adrenocorticotropic hormone, in vitro interferon-gamma production, leukocytes, acute-phase proteins, growth, and feed intake. *Journal of Animal Science, 75*, 1899–1908.

Fisher, A. D., Verkerk, G. A., Morrow, C. J., and Matthews, L. R. (2002). The effects of feed restriction and lying deprivation on pituitary–adrenal axis regulation in lactating cows. *Livestock Production Science, 73*, 255–263.

Fleischer, P., Metzner, M., Beyerbach, M., Hoedemaker, M., and Klee, W. (2001). The relationship between milk yield and the incidence of some diseases in dairy cows. *Journal of Dairy Science, 84*, 2025–2035.

Flower, F. C. and Weary, D. M. (2001). Effects of early separation on the dairy cow and calf: 2. Separation at 1 day and 2 weeks after birth. *Applied Animal Behaviour Science, 70*, 275–284.

Flower, F. C., and Weary, D. M. (2003). The effects of early separation on the dairy cow and calf. *Animal Welfare, 12*, 339–348.

Flower, F. C. and Weary, D. M. (2006). Effect of hoof pathologies on subjective assessments of dairy cow gait. *Journal of Dairy Science, 89*, 139–146.

Flower, F. C., de Passillé, A. M., Weary, D. M., Sanderson, D. J., and Rushen, J. (2007). Softer, higher-friction flooring improves gait of cows with and without sole ulcers. *Journal of Dairy Science, 90*, 1235–1242.

Flower, F. C., Sanderson, D. J., and Weary, D. M. (2005). Hoof pathologies influence kinematic measures of dairy cow gait. *Journal of Dairy Science, 88*, 3166–3173.

Fordyce, G., Goddard, M. E., Tyler, R., Williams, G., and Toleman, M. A. (1985). Temperament and bruising of *Bos indicus* cross cattle. *Australian Journal of Experimental Agriculture, 25*, 283–288.

Fourichon, C., Beaudeau, F., Bareille, N., and Seegers, H. (2001). Incidence of health disorders in dairy farming systems in western France. *Livestock Production Science, 68*, 157–170.

Franck, A. and De Belie, N. (2006). Concrete floor-bovine claw contact pressures related to floor roughness and deformation of the claw. *Journal of Dairy Science, 89*, 2952–2964.

Frank, N. A. and Kaneene, J. B. (1993). Management risk factors associated with calf diarrhea in Michigan dairy herds. *Journal of Dairy Science, 76*, 1313–1323.

Franklin, S. T. (2004). Enhancing immunity of the calf. *Official Proceedings of the 39th Annual Pacific Northwest Animal Nutrition Conference*, Seattle, WA, 1–7.

Franklin, S. T., Amaral-Phillips, D. M., Jackson, J. A., and Campbell, A. A. (2003). Health and performance of Holstein calves that suckled or were hand-fed colostrum and were fed one of three physical forms of starter. *Journal of Dairy Science, 86*, 2145–2156.

Fraser, D. (2003). Assessing animal welfare at the farm and group level: The interplay of science and values. *Animal Welfare, 12*, 433–443.

Fraser, D. (2006). Animal welfare assurance programs in food production: A framework for assessing the options. *Animal Welfare, 15*, 93–104.

Fraser, D. and Mathews, L. R. (1997). Preference and motivation testing. In: M. C. Appleby and B. O. Hughes (eds), *Animal welfare*. Wallingford, UK: CAB International.

Fraser, D. and Weary, D. M. (2004). Applied animal behavior and animal welfare. In: J. J. Bolhuis and L. A. Giraldeau (eds), *The behavior of animals: Mechanisms, function and evolution* (pp. 345–366). Oxford: Blackwell.

Fraser, D., Weary, D. M., Pajor, E. A., and Milligan, B. N. (1997). A scientific conception of animal welfare that reflects ethical concerns. *Animal Welfare, 6*, 187–205.

Fregonesi, J. A. and Leaver, J. D. (2001). Behaviour, performance and health indicators of welfare for dairy cows housed in strawyard or cubicle systems. *Livestock Production Science, 68*, 205–216.

Fregonesi, J. A. and Leaver, J. D. (2002). Influence or space allowance and milk yield level on behaviour, performance and health of dairy cows housed in strawyard and cubicle systems. *Livestock Production Science, 78*, 245–257.

Fregonesi, J. A., Tucker, C. B., Weary, D. M., Flower, F. C., and Vittie, T. (2004). Effect of rubber flooring in front of the feed bunk on the time budgets of dairy cattle. *Journal of Dairy Science, 87*, 1203–1207.

Fregonesi, J. A., Tucker, C. B., Weary, D. M. (2007). Overstocking reduces lying time in dairy cows. *Journal of Dairy Science. 90*, 3349–3354.

Frei, C., Frei, P. P., Stark, D. C., Pfeiffer, D. U., and Kihm, U. (1997). The production system and disease incidence in a national random longitudinal study of swiss dairy herds. *Preventive Veterinary Medicine, 32*, 1–21.

Frid, A. (1997). Vigilance by female Dall's sheep: Interactions between predation risk factors. *Animal Behaviour, 53*, 799–808.

Friedman, W. A., Garland, T., and Dohm, M. (1992). Individual variation in locomotor behavior and maximal oxygen consumption in mice. *Physiology & Behavior, 52*, 97–104.

Friend, T. H., Dellmeier, G. R., and Gbur, E. E. (1985). Comparison of four methods of calf confinement. I. Physiology. *Journal of Animal Science, 60*, 1095–1101.

Friend, T. H., Gwazdauskas, F. C., and Polan, C. E. (1979). Change in adrenal response from free stall competition. *Journal of Dairy Science, 62*, 768

Friend, T. H., Polan, C. E., Gwazdauskas, F. C., and Heald, C. W. (1977). Adrenal glucocorticoid response to exogenous adrenocorticotropin mediated by density and social disruption in lactating cows. *Journal of Dairy Science, 60*, 1958–1963.

Funk, D. A., Freeman, A. E., and Berger, P. J. (1982). Environmental and physiological factors affecting mastitis at drying-off and post-calving. *Journal of Dairy Science, 65*, 1258–1268.

Galindo, F. and Broom, D. M. (2000). The relationships between social behaviour of dairy cows and the occurrence of lameness in three herds. *Research in Veterinary Science, 69*, 75–79.

Gallo, C., Lizondo, G., and Knowles, T. G. (2003). Effects of journey and lairage time on steers transported to slaughter in Chile. *Veterinary Record, 152*, 361–364.

Galyean, M. L. and Rivera, J. D. (2003). Nutritionally related disorders affecting feedlot cattle. *Canadian Journal of Animal Science, 83*, 13–20.

Ganaba, R., Bengaly, Z., and Ouattara, L. (2002). Calf morbidity, mortality and parasite prevalences in the cotton zone of Burkina Faso. *Preventive Veterinary Medicine, 55*, 209–216.

Ganaba, R., Bigras-Poulin, M. B., Bélanger, D., and Couture, Y. (1995). Description of cow-calf productivity in northwestern Quebec and path models for calf mortality and growth. *Preventive Veterinary Medicine, 24*, 31–42.

Garrett, E. F., Pereira, M. N., Nordlund, K. V., Armentano, L. E., Goodger, W. J., and Oetzel, G. R. (1999). Diagnostic methods for the detection of subacute ruminal acidosis in dairy cows. *Journal of Dairy Science, 82*, 1170–1178.

Gaughan, J. B., Davis, M. S., and Mader, T. L. (2004). Wetting and the physiological responses of grain-fed cattle in a heated environment. *Australian Journal of Agricultural Research, 55*, 253–260.

Gauly, M., Mathiak, H., Hoffmann, K., Kraus, M., and Erhardt, G. (2001). Estimating genetic variability in temperamental traits in german Angus and Simmental cattle. *Applied Animal Behaviour Science, 74*, 109–119.

Gaworski, M. A., Tucker, C. B., and Weary, D. M. (2003). Effects of two free-stall designs on dairy cattle behaviour. *Proceedings of the 5th International Dairy Housing Conference*, 29–31 January, Fort Worth, TX, 139–146; American Society of Agricultural Engineers, St. Joseph, MI.

Genchi, C., Traldi, G., and Locatelli, A. (1986). Influence of transport stress on trichostrongylid infection in feedlot beef cattle. *Veterinary Parasitology, 21*, 211–215.

Giovambattista, G., Ripoli, M. V., Peral-Garcia, P., and Bouzat, J. L. (2001). Indigenous domestic breeds as reservoirs of genetic diversity: The argentinean creole cattle. *Animal Genetics, 32*, 240–247.

Gitau, T., McDermott, J. J., and Mbiuki, S. M. (1996). Prevalence, incidence, and risk factors for lameness in dairy cattle in small-scale farms in kikuyu division, Kenya. *Preventive Veterinary Medicine, 28*, 101–115.

Godden, S., Rapnicki, P., Stewart, S., Fetrow, J., Johnson, A., Bey, R., et al. (2003). Effectiveness of an internal teat seal in the prevention of new intramammary infections during the dry and early-lactation periods in dairy cows when used with a dry cow intramammary antibiotic. *Journal of Dairy Science, 86*, 3899–3911.

Godden, S. M., Fetrow, J. P., Feirtag, J. M., Green, L. R., and Wells, S. J. (2005). Economic analysis of feeding pasteurized nonsaleable milk versus conventional milk replacer to dairy calves. *Journal of the American Veterinary Medical Association, 9*, 1547–1554.

Goff, J. P. (2006). Major advances in our understanding of nutritional influences on bovine health. *Journal of Dairy Science, 89*, 1292–1301.

Goff, J. P. and Horst, R. L. (1997). Physiological changes at parturition and their relationship to metabolic disorders. *Journal of Dairy Science, 80*, 1260–1268.

Goonewardene, L. A., Pang, H., Berg, R. T., and Price, M. A. (1999a). A comparison of reproductive and growth traits of horned and polled cattle in three synthetic beef lines. *Canadian Journal of Animal Science, 79*, 123–127.

Goonewardene, L. A., Price, M. A., Liu, M. F., Berg, R. T., and Erichsen, C. M. (1999b). A study of growth and carcass traits in dehorned and polled composite bulls. *Canadian Journal of Animal Science, 79*, 383–385.

Goonewardene, L. A., & Hand, R. K. (1991). Studies on dehorning steers in Alberta feedlots. *Canadian Journal of Animal Science, 71*, 1249–1252.

Gottardo, F., Mattiello, S., Cozzi, G., Canali, E., Scanziani, E., Ravarotto, L., et al. (2002). The provision of drinking water to veal calves for welfare purposes. *Journal of Animal Science, 80*, 2362–2372.

Graf, B. and Senn, M. (1999). Behavioural and physiological responses of calves to dehorning by heat cauterisation with or without local anaesthesia. *Applied Animal Behaviour Science, 62*, 153–171.

Grafen, A. (1990). Biological signals as handicaps. *Journal of Theoretical Biology, 144*, 517–546.

Grandin, T. (1997). The design and construction of facilities for handling cattle. *Livestock Production Science, 49*, 103–119.

Grandin, T. (1998a). The feasibility of using vocalization scoring as an indicator of poor welfare during cattle slaughter. *Applied Animal Behaviour Science, 56*, 121–128.

Grandin, T. (2000). Special reports: Effect of animal welfare audits of slaughter plants by a major fast food company on cattle handling and stunning practices. *Journal of the American Veterinary Medical Association, 216*, 848–851.

Grandin, T. (2001a). Cattle vocalizations are associated with handling and equipment problems at beef slaughter plants. *Applied Animal Behaviour Science, 71*, 191–201.

Grandin, T. (2001b). Welfare of cattle during slaughter and the prevention of nonambulatory (downer) cattle. *Journal of the American Vetrinary Medical Association, 219*, 1377–1382.

Grandin, T. (2002). Return-to-sensibility problems after penetrating captive bolt stunning of cattle in commercial beef slaughter plants. *Journal of the American Veterinary Medical Association, 221*, 1258–1261.

Grandin, T. (2006). Progress and challenges in animal handling and slaughter in the US. *Applied Animal Behaviour Science, 100*, 129–139.

Grandin, T. and Deesing, M. J. (1998). Genetics and behavior during handling, restraint, and herding. In: T. Grandin (ed), *Genetics and the behavior of domestic animals* (pp. 113–144). Wallingford, UK: CAB International.

Grandin, T., Odde, K. G., Schutz, D. N., and Behrns, L. M. (1994). The reluctance of cattle to change a learned choice may confound preference tests. *Applied Animal Behaviour Science, 39*, 21–28.

Green, L. E., Hedges, V. J., Schukken, Y. H., Blowey, R. W., and Packington, A. J. (2002). The impact of clinical lameness on the milk yield of dairy cows. *Journal of Dairy Science, 85*, 2250–2256.

Gregory, N. (2004). *Physiology and behaviour of animal suffering*. Oxford: Blackwell.

Gregory, N. G. (ed.) (1998). *Animal welfare and meat science*. Wallingford, UK: CAB International.

Griffin, D. R. (1981). *The question of animal awareness*. New York: Rockefeller University Press.

Grigor, P. N., Cockram, S. M., Steele, W. B., Le Sueur, C. J., Forsyth, R. E., Guthrie, J. A., et al. (2001). Effects of space allowance during transport and duration of midjourney lairage period on the physiological, behavioural and immunological responses of young calves during and after transport. *Animal Science, 73*, 341–360.

Gröhn, Y. T., Rajala-Schultz, P. J., Allore, H. G., DeLorenzo, M. A., Hertl, J. A., and Galligan, D. T. (2003). Optimizing replacement of dairy cows: Modelling the effects of diseases. *Preventive Veterinary Medicine, 61*, 27–43.

Grøndahl-Nielsen, C., Simonsen, H. B., Damkjer Lund, J., and Hesselholt, M. (1999). Behavioural, endocrine and cardiac responses in young calves undergoing dehorning without and with use of sedation and analgesia. *The Veterinary Journal, 158*, 14–20.

Grout, A. S., Veira, D. M., Weary, D. M., von Keyserlingk, M. A. G., and Fraser, D. (2006). Differential effects of sodium and magnesium sulfate on water consumption by beef cattle. *Journal of Animal Science, 84*, 1252–1258.

Grunert, K. G., Bredahl, L., and Brunsø, K. (2004). Consumer perception of meat quality and implications for product development in the meat sector: A review. *Meat Science, 66*, 259–272.

Guard, C. (2001). Investigating herds with lameness problems. *Veterinary Clinics of North America – Food Animal Practice, 17*, 175–187.

Gwazdauskas, F. C., Paape, M. J., Peery, D. A., and McGillard, M. L. (1980). Plasma glucocorticoid and circulating blood leukocyte responses in cattle after sequential intramuscular injections of ACTH. *American Journal Veterinary Research, 41*, 1052–1056.

Gwazdauskas, F. C., Thatcher, W. W., Paape, M. J., Kiddy, C. A., and Wilcox, C. J. (1975). Plasma characteristics and adrenal response of heifers to a thermal stress. *Journal of Dairy Science, 58*, 776.

Gygax, L., Neuffer, I., Kaufmann, C., Hauser, R., and Wechsler, B. (2006). Milk cortisol concentration in automatic milking systems compared with auto-tandem milking parlors. *Journal of Dairy Science, 89*, 3447–3454.

Gyger, M. and Marler, P. (1988). Food calling in the domestic-fowl, *Gallus gallus* – the role of external referents and deception. *Animal Behaviour, 36*, 358–365.

Hagen, K. & Broom, D. M. (2003). Cattle discriminate between individual familiar herd members in a learning experiment. *Applied Animal Behaviour Science, 82*, 13–28.

Hagen, K., Langbein, J., Schmied, C., Lexer, D., and Waiblinger, S. (2005). Heart rate variability in dairy cows – influences of breed and milking system. *Physiology & Behavior, 85*, 195–204.

Hahn, G. L. (1999). Dynamic responses of cattle to thermal heat loads. *Journal of Animal Science, 77*, 10–20.

Haley, D. B., Bailey, D. W., and Stookey, J. M. (2005). The effects of weaning beef calves in two stages on their behavior and growth rate. *Journal of Animal Science, 83*, 2205–2214.

Haley, D. B., de Passillé, A. M., and Rushen, J. (2001). Assessing cow comfort: Effects of two floor types and two tie stall designs on the behaviour of lactating dairy cows. *Applied Animal Behaviour Science, 71*, 105–117.

Haley, D. B., Rushen, J., and de Passillé, A. M. (2000). Behavioural indicators of cow comfort: Activity and resting behaviour of dairy cows in two types of housing. *Canadian Journal of Animal Science, 80*, 257–263.

Haley, D. B., Rushen, J., Duncan, I., Widowski, T., and de Passillé, A. M. B. (1998). Effects of resistance to milk flow and provision of hay on non-nutritive sucking by dairy calves. *Journal of Dairy Science, 81*, 2165–2172.

Hall, C. S. (1936). Emotional behavior in the rat. III. The relationship between emotionality and ambulatory activity. *Journal of Comparative Psychology, 22*, 345–352.

Hall, S. J. G. (1986). Chillingham cattle: Dominance and affinities and access to supplementary food. *Ethology, 71*, 201–215.

Hall, S. J. G. (1989). Chillingham cattle: Social and maintenance behaviour in an ungulate that breeds all year round. *Animal Behaviour, 38*, 215–225.

Hall, S. J. G. and Moore, G. F. (1986). Feral cattle of Swona, Orkney islands. *Mammal Review, 16*, 89–96.

Hammon, H. M., Schiessler, G., Nussbaum, A., and Blum, J. W. (2002). Feed intake patterns, growth performance, and metabolic and endocrine traits in calves fed unlimited amounts of colostrum and milk by automate, starting in the neonatal period. *Journal of Dairy Science, 85*, 3352–3362.

Hanna, D., Sneddon, I. A., Beattie, V. E., and Breue, K. (2006). Effects of the stockperson on dairy cow behaviour and milk yield. *Animal Science, 82,* 791–797.

Hänninen, L. (2007). Sleep and rest in calves. Ph.D. thesis, University of Helsinki.

Hänninen, L., de Passillé, A. M., and Rushen, J. (2005). The effect of flooring type and social grouping on the rest and growth of dairy calves. *Applied Animal Behaviour Science, 91*, 193–204.

Hänninen, L., Hepola, H., Rushen, J., de Passillé, A. M., Pursiainen, P., Tuure, V. M., et al. (2003). Resting behaviour, growth and diarrhoea incidence rate of young dairy calves housed individually or in groups in warm or cold buildings. *Acta Agriculturae Scandinavica, Section A, Animal Science, 53*, 21–28.

Hänninen, L., Lovendahl, P., De Passillé, A. M., and Rushen, J. (2006). The effect of floor type or relocation on calves' pulsatile growth hormone and cortisol secretion. *Acta Agriculturae Scandinavica, Section A, Animal Science, 56*, 99–108.

Hänninen, L., Makela, J. P., Rushen, J., de Passille, A. M. & Saloniemi, H. (2007) Assessing sleep state in calves through electrophysiological and behavioural recordings: A preliminary study. *Applied Animal Behaviour Science*, doi:10.1016/j.applanim.2007.06.009.

Hansen, M., Misztal, I., Lund, M. S., Pedersen, J., and Christensen, L. G. (2004). Undesired phenotypic and genetic trend for stillbirth in Danish Holsteins. *Journal of Dairy Science, 87*, 1477–1486.

Hard, D. L., Bhatnagar, R. K., Molina, J. R., and Anderson, L. L. (2001). Secretion of dopamine and norepinephrine in hypophyseal protal blood and prolactin in peripheral blood of Holstein cattle. *Domestic Animal Endocrinology, 20*, 89–100.

Harrison, R. (1964). *Animal machines; the new factory farming industry*. London: Stuart.

Hart, B. L. (1988). Biological basis of the behavior of sick animals. *Neuroscience and Biobehavioral Reviews, 12*, 123–137.

Haskell, M. J., Rennie, L. J., Bowell, V. A., Bell, M. J., and Lawrence, A. B. (2006). Housing system, milk production, and zero-grazing effects on lameness and leg injury in dairy cows. *Journal of Dairy Science, 89*, 4259–4266.

Hemsworth, P. H. (2003). Human-animal interactions in livestock production. *Applied Animal Behaviour Science, 81*, 185–198.

Hemsworth, P. H. and Coleman, G. J. (1998). *Human–livestock interactions*. Wallingford, UK: CAB International.

Hemsworth, P. H., Barnett, J. L., Beveridge, L., and Matthews, L. R. (1995). The welfare of extensively managed dairy cattle: A review. *Applied Animal Behaviour Science, 42*, 161–182.

Hemsworth, P. H., Coleman, G. J., Barnett, J. L., and Borg, S. (2000). Relationships between human-animal interactions and productivity of commercial dairy cows. *Journal of Animal Science, 78*, 2821–2831.

Hemsworth, P. H., Coleman, G. J., Barnett, J. L., Borg, S., and Dowling, S. (2002). The effects of cognitive behavioral intervention on the attitude and behavior of stockpersons and the behavior and productivity of commercial dairy cows. *Journal of Animal Science, 80*, 68–78.

Hemsworth, P. H., Hansen, C., and Barnett, J. L. (1987). The effects of human presence at the time of calving of primiparous cows on their subsequent behavioural response to milking. *Applied Animal Behaviour Science, 18*, 247–255.

Hemsworth, P. H., Verge, J., and Coleman, G. J. (1996). Conditioned approach-avoidance responses to humans: The ability of pigs to associate feeding and aversive social experiences in the presence of humans with humans. *Applied Animal Behaviour Science, 50*, 71–82.

Hepola, H. (2003). Milk feeding systems for dairy calves in groups: Effects on feed intake, growth and health. *Applied Animal Behaviour Science, 80*, 233–243.

Herlin, A. H. (1997). Comparison of lying area surfaces for dairy cows by preference, hygiene and lying down behaviour. *Swedish Journal of Agricultural Research, 27*, 189–196.

Hernandez, J. A., Zanella, E. L., Bogden, R., de Avila, D. M., Gaskins, C. T., and Reeves, J. J. (2005). Reproductive characteristics of grass-fed, luteinizing hormone-releasing hormone-immunocastrated *Bos indicus* bulls. *Journal of Animal Science, 83*, 2901–2907.

Hernandez, L., Barral, H., Halffter, G., and Sanchez Colon, S. (1999). A note on the behavior of feral cattle in the chihuahuan desert of Mexico. *Applied Animal Behaviour Science, 63*, 259–267.

Hernandez-Mendo, O., von Keyserlingk, M. A. G., Veira, D. M., and Weary, D. M. (2007). Effects of pasture on lameness in dairy cows. *Journal of Dairy Science, 90*, 1209–1214.

Higuchi, H. and Nagahata, H. (2001). Relationship between serum biotin concentration and moisture content of the sole horn in cows with clinical laminitis or sound hooves. *Veterinary Record, 148*, 209–210.

Higuchi, H., Maeda, T., Kawai, K., Kuwano, A., Kasamatsu, M., and Nagahata, H. (2003). Physiological changes in the concentrations of biotin in the serum and milk and in the physical properties of the claw horn in Holstein cows. *Veterinary Research Communications, 27*, 407–413.

Hindhede, J., Mogensen, L., and Sorensen, J. T. (1999). Effect of group composition and feeding system on behaviour, production and health of dairy heifers in deep bedding systems. *Acta Agriculture Scandinavica Section A – Animal Science, 49*, 211–220.

Hirvonen, J., Huszenicza, G., Kulcsar, M., and Pyorala, S. (1999). Acute-phase response in dairy cattle with acute post-calving metritis. *Theriogenology, 51*, 1071–1083.

Holloway, K. S. and Suter, R. B. (2004). Play deprivation without social isolation: Housing controls. *Developmental Psychobiology, 44*, 58–67.

Holm, L., Jensen, M. B., and Jeppesen, L. L. (2002). Calves' motivation for access to two different types of social contact measured by operant conditioning. *Applied Animal Behaviour Science, 79*, 175–194.

Holzhauer, M., Hardenberg, C., Bartels, C. J. M., and Frankena, K. (2006). Herd- and cow-level prevalence of digital dermatitis in the Netherlands and associated risk factors. *Journal of Dairy Science, 89*, 580–588.

Hopster, H., Bruckmaier, R. M., van der Werf, J. T. N., and Korte, S. M. (2002). Stress responses during milking; comparing conventional and automatic milking in primiparous dairy cows. *Journal of Dairy Science, 85*, 3206–3216.

Hopster, H., O'Connel M. J., and Blokhuis J. H. (1995). Acute effects of cow-calf separation on heart rate, plasma cortisol and behaviour in multiparous dairy cows. *Applied Animal Behaviour Science, 44*, 1–8.

Hopster, H., ven der werf, J. T. N., and Blokhuis, H. J. (1998). Stress enhanced reduction in peripheral blood lymphocyte numbers in dairy cows during endotoxin-induced mastitis. *Veterinary Immunology and Immunopathology, 66*, 83–97.

Horst, R. L. and Jorgensen, N. A. (1982). Elevated plasma cortisol during induced and spontaneous hypocalcemia in ruminants. *Journal of Dairy Science, 65*, 2332–2337.

Hughes, B. O. and Black, A. J. (1973). Preference of domestic hens for different types of battery cage floor. *British Poultry Science, 14*, 615–619.

Hughes, B. O. and Duncan, I. J. H. (1988). The notion of ethological "need", models of motivation and animal welfare. *Animal Behaviour, 36*, 1696–1707.

Hultgren, J. (2002). Foot/leg and udder health in relation to housing changes in Swedish dairy herds. *Preventive Veterinary Medicine, 53*, 167–189.

Hultgren, J. and Bergsten, C. (2001). Effects of a rubber-slatted flooring system on cleanliness and foot health in tied dairy cows. *Preventive Veterinary Medicine, 52*, 75–89.

Huzzey, J. M., DeVries, T. J., Valois, P., and von Keyserlingk, M. A. G. (2006). Stocking density and feed barrier design affect the feeding and social behavior of dairy cattle. *Journal of Dairy Science, 89*, 126–133.

Huzzey, J. M., Veira, D. M., Weary, D. M., and von Keyserlingk, M. A. G. (2007). Prepartum behavior and dry mater intake identify dairy cows at risk for metritis. *Journal of Dairy Science, 90*, 3220–3233.

Hydbring, E., Madej, A., MacDonald, E., Drugge-Boholm, G., Berglund, B., and Olsson, K. (1999). Hormonal changes during parturition in heifers and goats are related to the phases and severity of labour. *Journal of Endocrinology, 160*, 75–85.

Ingvartsen, K. L. (2006). Feeding- and management-related diseases in the transition cow: Physiological adaptations around calving and strategies to reduce feeding-related diseases. *Animal Feed Science and Technology, 126*, 175–213.

Ingvartsen, K. L. and Andersen, H. R. (1993). Space allowance and type of housing for growing cattle. *Acta Agriculture Scandanvica, Animal Science, 43*, 65–80.

Ingvartsen, K. L. and Andersen, J. B. (2000). Integration of metabolism and intake regulation: A review focussing on periparturient animals. *Journal of Dairy Science, 83*, 1573–1597.

Ingvartsen, K. L., Dewhurst, R. J., and Friggens, N. C. (2003). On the relationship between lactational performance and health: Is it yield or metabolic imbalance that cause production diseases in dairy cattle? A position paper. *Livestock Production Science, 83*, 277–308.

Inui, A. (2001). Cytokines and sickness behavior: Implications from knockout animal models. *Trends in Immunology, 22*, 469–473.

Irwin, M. (2002). Effects of sleep and sleep loss on immunity and cytokines. *Brain, Behaviour and Immunity, 16*, 503–512.

Jacobson, L. H. and Cook, C. J. (1996). Heart rate response as a measure of stress and welfare in cattle. *Proceedings of the New Zealand Society of Animal Production, 56*, 103–106.

Jacobson, L. H. and Cook, C. J. (1998). Partitioning psychological and physical sources of transport-related stress in young cattle. *Veterinary Journal, 155*, 205–208.

Jago, J., Krohn, C. C., and Matthews, L. R. (1999). The influence of feeding and handling on the development of the human-animal interactions in young cattle. *Applied Animal Behaviour Science, 62*, 137–151.

James, S. M. and Kyriazakis, I. (2002). The effect of consumption of foods that differ in energy density and/or sodium bicarbonate supplementation on subsequent diet selection in sheep. *British Journal of Nutrition, 88*, 81–90.

Jarvis, A. M., Harrington, D. W. J., and Cockram, M. S. (1996). Effect of source and lairage on some behavioural and biochemical measurements of feed restriction and dehydration in cattle at a slaughterhouse. *Applied Animal Behaviour Science, 50*, 83–94.

Jarvis, A. M., Selkirk, L., and Cockram, M. S. (1995). The influence of source, sex class and pre-slaughter handling on the bruising of cattle at two slaughterhouses. *Livestock Production Science, 43*, 215–224.

Jasper, J., Budzynska, M., and Weary, D. M. (2007). Weaning distress in dairy calves: Acute behavioural responses by limit-fed calves. *Applied Animal Behaviour Science*, doi:10.1016/j.applanim.2007.03.017.

Jasper, J. and Weary, D. M. (2002). Effects of ad libitum milk intake on dairy calves. *Journal of Dairy Science, 85*, 3054–3058.

Jelinski, M. D., Ribble, C. S., Campbell, J. R., and Janzen, E. D. (1996). Descriptive epidemiology of fatal abomasal ulcers in Canadian beef calves. *Preventitive Veterinary Medicine, 26*, 9–15.

Jensen, M. B. (1999). Effects of confinement on rebounds of locomotor behaviour of calves and heifers, and the spatial preferences of calves. *Applied Animal Behaviour Science, 62*, 43–56.

Jensen, M. B. (2001). A note on the effect of isolation during testing and length of previous confinement on locomotor behaviour during open-field test in dairy calves. *Applied Animal Behaviour Science, 70*, 309–315.

Jensen, M. B. (2004). Computer-controlled milk feeding of dairy calves: The effects of number of calves per feeder and number of milk protions on use of feeder and social behavior. *Journal of Dairy Science, 87*, 3428–3438.

Jensen, M. B. and Holm, L. (2003). The effect of milk flow rate and milk allowance on feeding related behaviour in dairy calves fed by computer controlled milk feeders. *Applied Animal Behaviour Science, 82*, 87–100.

Jensen, M. B. and Kyhn, R. (2000). Play behaviour in group-housed dairy calves, the effect of space allowance. *Applied Animal Behavioral Science, 67*, 35–46.

Jensen, M. B., Munksgaard, L., Mogensen, L., and Krohn, C. C. (1999). Effects of housing in different social environments on open-field and social responses of female dairy calves. *Acta Agriculturae Scandinavica, Section A, Animal Science, 49*, 113–120.

Jensen, M. B., Pedersen, L. J. and Munksgaard, L. (2005). The effect of reward duration on demand functions for rest in dairy heifers and lying requirements as measured by demand functions. *Applied Animal Behaviour Science. 90*, 207–217.

Jensen, M. B., Vestergaard, K. S., and Krohn, C. C. (1998). Play behaviour in dairy calves kept in pens: The effect of social contact and space allowance. *Applied Animal Behaviour Science, 56*, 97–108.

Jensen, M. B., Verstergaard, K. S., Krohn, C. C., and Munksgaard, L. (1997). Effect of single versus group housing and space allowance on responses of calves during open-field tests. *Applied Animal Behaviour Science, 54*, 109–121.

Jensen, P. (2006). Domestication – from behaviour to genes and back again. *Applied Animal Behaviour Science, 97*, 3–15.

Jensen, P. and Toates, F. (1993). Who needs "behavioural needs"? Motivational aspects of the needs of animals. *Applied Animal Behaviour Science, 37*, 161–181.

Jensen, P., Recen, B., and Ekesbo, I. (1988). Preference of loose housed dairy-cows for 2 different cubicle floor coverings. *Swedish Journal of Agricultural Research, 18*, 141–146.

Johanson, J. M. and Berger, P. J. (2003). Birth weight as a predictor of calving ease and perinatal mortality in Holstein cattle. *Journal of Dairy Science, 86*, 3745–3755.

Johnson, R. W. (1998). Immune and endocrine regulation of food intake in sick animals. *Domestic Animal Endocrinology, 15*, 309–318.

Johnson, R. W. (2002). The concept of sickness behavior: A brief chronological account of four key discoveries. *Veterinary Immunology and Iimmunopathology, 87*, 443–450.

Johnstone, R. A. (1999). Signaling of need, sibling competition, and the cost of honesty. *Proceedings of the National Academy of Sciences of the United States of America, 96*, 12644–12649.

Jung, J. and Lidfors, L. (2001). Effects of amount of milk, milk flow and access to a rubber teat on cross-sucking and non-nutritive sucking in dairy calves. *Applied Animal Behaviour Science, 72*, 201–213.

Kadel, M. J., Johnston, D. J., Burrow, H. M., Graser, H. U., and Ferguson, D. M. (2006). Genetics of flight time and other measures of temperament and their value as selection criteria for improving meat quality traits in tropically adapted breeds of beef cattle. *Australian Journal of Agricultural Research, 57*, 1029–1035.

Kadzere, C. T., Murphy, M. R., Silanikove, N., and Maltz, E. (2002). Heat stress in lactating dairy cows: A review. *Livestock Production Science, 77*, 59–91.

Kaneene, J. B. and Miller, R. (1995). Risk factors for metritis in Michigan dairy cattle using herd- and cow-based modeling approaches. *Preventive Veterinary Medicine, 23*, 183–200.

Kehrli, M. E. and Goff, J. P. (1989). Periparturient hypocalcemia in cows: Effects on peripheral blood neutrophil and lymphocyte function. *Journal of Dairy Science, 72*, 1188.

Keil, N. M. and Langhans, W. (2001). The development of intersucking in dairy calves around weaning. *Applied Animal Behaviour Science, 72*, 295–308.

Keister, Z. O., Moss, K. D., Zhang, H. M., Teegerstrom, T., Edling, R. A., Collier, R. J., et al. (2002). Physiological responses in thermal stressed Jersey cows sujected to different management strategies. *Journal of Dairy Science, 85*, 3217–3224.

Kelm, S. C., Freeman, A. E., and NC-2 Technical Committee (2000). Direct and correlated responses to selection for milk yield: Results and conclusions of regional project NC-2, "improvement of dairy cattle through breeding, with emphasis on selection". *Journal of Dairy Science, 83*, 2721–2732.

Kent, J. E. and Ewbank, R. (1983). The effect of road transportation on the blood constituents and behaviour of calves. I. Six months old. *British Veterinary Journal, 139*, 228–235.

Kent, J. E. and Ewbank, R. (1986). The effect of road transportation on the blood constituents and behaviour of calves. III. Three months old. *British Veterinary Journal, 142*, 326–335.

Ketelaar-de Lauwere, C. C. and Smits, A. C. (1991). *Spatial requirements of individually housed veal calves of 175 to 300 kg* (pp. 49–53). Wageningen, The Netherlands: Pudoc.

Keunen, J. E., Plaizier, J. C., Kyriazakis, L., Duffield, T. F., Widowski, T. M., Lindinger, M. I., et al. (2002). Effects of subacute ruminal acidosis model on the diet selection of dairy cows. *Journal of Dairy Science, 85*, 3304–3313.

Khan, M. A., Lee, H. J., Lee, W. S., Kim, H. S., Kim, S. B., Ki, K. S., et al. (2007). Pre- and post-weaning performance of Holstein female calves fed milk through step-down and conventional methods. *Journal of Dairy Science, 90*, 876–885.

Kiley, M. (1972). The vocalizations of ungulates, their causation and function. *Zeitschrift für Tierpsychologie, 31*, 171–222.

Kilgour, R. (1975). The open-field test as an assessment of the temperament of dairy cows. *Animal Behaviour, 23*, 615–624.

Kilgour, R. J., Melville, G. J., and Greenwood, P. L. (2006). Individual differences in the reaction of beef cattle to situations involving social isolation, close proximity of humans, restraint and novelty. *Applied Animal Behaviour Science, 99*, 21–40.

King, B. D., Cohen, R. D. H., Guenther, C. L., and Janzen, E. D. (1991). The effect of age and method of castration on plasma-cortisol in beef-calves. *Canadian Journal of Animal Science, 71*, 257–263.

King, D. A., Pfeiffer, C. E. S., Randel, R. D., Welsh, T. H., Oliphint, R. A., Baird, B. E., et al. (2006). Influence of animal temperament and stress responsiveness on the carcass quality and beef tenderness of feedlot cattle. *Meat Science, 74*, 546–556.

Kirkden, R. D. and Pajor, E. A. (2006). Using preference, motivation and aversion tests to ask scientific questions about animals' feelings. *Applied Animal Behaviour Science, 100*, 29–47.

Kjoestad, H. P. and Simensen, E. (2001). Cubicle refusal and rearing accommodation as possible mastitis risk factors in cubicle-housed dairy heifers. *Acta Veterinaria Scandinavica, 42*, 123–130.

Knight, T. W., Cosgrove, G. P., Death, A. F., Anderson, C. B., and Fisher, A. D. (2000). Effect of method of castrating bulls on their growth rate and liveweight. *New Zealand Journal of Agricultural Research, 43*, 187–192.

Knowles, T. G. (1995). A review of post transport mortality among younger calves. *Veterinary Record, 137*, 406–407.

Knowles, T. G. (1999). A review of the road transport of cattle. *Veterinary Record, 144*, 197–201.

Knowles, T. G., Brown, S. N., Edwards, J. E., Phillips, A. J., and Warriss, P. D. (1999). Effect on young calves of a one-hour feeding stop during a 19-hour road journey. *The Veterinary Record, 144*, 687–692.

Knowles, T. G., Warriss, P. D., Brown, S. N., Edwards, J. E., Watkins, P. E., and Phillips, A. J. (1997). Effects on calves less than one month old of feeding or not feeding them during road transport of up to 24 hours. *Veterinary Record, 140*, 116–124.

Kondo, S., Sekine, J., Okubo, M., and Asahida, Y. (1989). The effect of group-size and space allowance on the agonistic and spacing behavior of cattle. *Applied Animal Behaviour Science, 24*, 127–135.

Kristal, M. B., Thompson, A. C., Abbott, P., Di Pirro, J. M., Ferguson, E. J., and Doerr, J. C. (1990). Amniotic fluid ingestion by parturient rates enhances pregnancy-mediated analgesia. *Life Science, 46*, 693–698.

Krohn, C. C. (2001). Effects of different suckling systems on milk production, udder health, reproduction, calf growth and some behavioural aspects in high producing dairy cows: A review. *Applied Animal Behaviour Science, 72*, 271–280.

Krohn, C. C. and Munksgaard, L. (1993). Behaviour of dairy cows kept in extensive (loose housing/pasture) or intensive (tie stall) environments. II. lying and lying-down behaviour. *Applied Animal Behaviour Science, 37*, 1–16.

Krohn, C. C., Boivin, X., and Jago, J. G. (2003). The presence of the dam during handling prevents the socialization of young calves to humans. *Applied Animal Behaviour Science, 80*, 263–275.

Krohn, C. C., Flodager, J., and Morgensen, L. (1999). Long term effect of colostrum feeding methods on behaviour of female dairy calves. *Acta Agriculturae Scandinavica, 49*, 57–64.

Krohn, C. C., Jago, J. G., and Boivin, X. (2001). The effect of early handling on the socialisation of young calves to humans. *Applied Animal Behaviour Science, 74*, 121–133.

Kung, L., Demarco, S., Siebenson, L. N., Joyner, E., Haenlein, G. F. W., and Morris, R. M. (1997). An evaluation of two management systems for rearing calves fed milk replacer. *Journal of Dairy Science, 80*, 2529–2533.

Ladewig, J. and Smidt, D. (1989). Behavior, episodic secretion of cortisol and adrenocortical reactivity in bulls subjected to tethering. *Hormones and Behavior, 23*, 344–360.

Lago, A., McGuirk, S. M., Bennett, T. B., Cook, N. B., and Nordlund, K. V. (2006). Calf respiratory disease and pen microenvironments in naturally ventilated calf barns in winter. *Journal of Dairy Science, 89*, 4014–4025.

Lambooy, E. and Spanjaard, W. (1981). Effect of the shooting position on the stunning of calves by captive bolt. *Veterinary Record, 109*, 359–361.

Lane, J. (2006). Can non-invasive glucocorticoid measures be used as reliable indicators of stress in animals? *Animal Welfare, 15*, 331–342.

Larson, S. J. and Dunn, A. J. (2001). Behavioral effects of cytokines. *Brain, Behavior, and Immunity, 15*, 371–387.

Lawrence, A. B. and Illius, A. W. (1997). Measuring preferences and the problems of identifying proximate needs. *British Society of Animal Science, 20*, 19–26.

Lay Jr., D. C., Friend, T. H., Randel, R. D., Jenkins, O. C., Neuendorff, D. A., Kapp, G. M., et al. (1996). Adrenocorticotropic hormone dose response and some physiological effects of transportation on pregnant Brahman cattle. *Journal of Animal Science, 74*, 1806–1811.

Lay, D. C., Friend, T. H., Bowers, C. L., Grissom, K. K., and Jenkins, O. C. (1992a). A comparative physiological and behavioral study of freeze and hot-iron branding using dairy cows. *Journal of Animal Science, 70*, 1121–1125.

Lay, D. C., Friend, T. H., Randel, R. D., Bowers, C. L., Grissom, K. K., and Jenkins, O. C. (1992b). Behavioral and physiological-effects of freeze or hot-iron branding on crossbred cattle. *Journal of Animal Science, 70*, 330–336.

Lazo, A. (1994). Social segregation and the maintenance of social stability in a feral cattle population. *Animal Behaviour, 48*, 1133–1141.

Le Neindre, P. (1993). Evaluating housing systems for veal calves. *Journal of Animal Science, 71*, 1345–1354.

LeBlanc, S. J., Duffield, T. F., Leslie, K. E., Bateman, K. G., Keefe, G. P., Walton, J. S., et al. (2002). Defining and diagnosing postpartum clinical endometritis and its impact on reproductive performance in dairy cows. *Journal of Dairy Science, 85*, 2223–2236.

Lefcourt, A. M., Bitman, J., and Wood, D. L. K. S. (1993). Circadian and ultradian rhythms of peripheral cortisol concentrations in lactating dairy cows. *Journal of Dairy Science, 76*, 2607–2612.

Lefcourt, A. M., Kahl, S., and Akers, R. M. (1986). Correlation of indices of stress with intensity of electric shock for cows. *Journal of Dairy Science, 69*, 833–842.

LeGrand, E. K. and Brown, C. C. (2002). Darwinian medicine: Applications of evolutionary biology for veterinarians. *Canadian Veterinary Journal, 43*, 556–559.

Lensink, B. J., Boivin, X., Pradel, P., Le Neindre, P., and Veissier, I. (2000). Reducing veal calves' reactivity to people by providing additional human contact. *Journal of Animal Science, 78*, 1213–1218.

Lensink, B. J., Fernandez, X., Cozzy, G., Florand, L., and Veissier, I. (2001b). The influence of farmers' behavior on calves' reactions to transport and quality of veal meat. *Journal of Animal Science, 79*, 642–652.

Lensink, B. J., Raussi, S., Boivin, X., Pyykkonen, M., and Veissier, I. (2001c). Reactions of calves to handling depend on housing condition and previous experience with humans. *Applied Animal Behaviour Science, 70*, 187–199.

Lensink, B. J., Veissier, I., and Florand, L. (2001a). The farmers' influence on calves' behaviour, health and production of a veal unit. *Animal Science, 72*, 105–116.

Leonard, F. C., O'Connell, J., and O'Farrell, K. (1994). Effect of different housing conditions on behaviour and foot lesions in Friesian heifers. *Veterinary Record, 134*, 490–494.

Leonard, F. C., O'Connell, J. M., and O'Farrell, K. J. (1996). Effect of overcrowding on claw health in first-calved Friesian heifers. *British Veterinary Journal, 152*, 459–472.

Lewis, G. S. (1997). Uterine health and disorders. *Journal of Dairy Science, 80*, 984–994.

Lewis, N. J. and Hurnik, J. F. (1998). The effect of some common management practices on the ease of handling of dairy cows. *Applied Animal Behaviour Science, 58*, 213–220.

Lidfors, L. (1989). The use of getting up and lying down movements in the evaluation of cattle environments. *Veterinary Research Communications, 13*, 307–324.

Lidfors, L. (1993). Cross-sucking in group-housed dairy calves before and after weaning off milk. *Applied Animal Behaviour Science, 38*, 15–24.

Lidfors, L. (1996). Behavioural effects of separating the dairy calf immediately or 4 days post-partum. *Applied Animal Behaviour Science, 49*, 269–283.

Lidfors, L. and Isberg, L. (2003). Intersucking in dairy cattle – review and questionnaire. *Applied Animal Behaviour Science, 80*, 207–231.

Lidfors, L., Jensen, P., and Algers, B. (1994). Suckling in free-ranging beef cattle: Temporal patterns of suckling bouts and effects of age and sex. *Ethology, 98*, 321–332.

Lindström, T. and Redbo, I. (2000). Effect of feeding duration and rumen fill on behaviour in dairy cows. *Applied Animal Behaviour Science, 70*, 84–97.

Lindström, T., Redbo, I., and Uvnäs-Moberg, K. (2001). Plasma oxytocin and cortisol concentrations in dairy cows in relation to feeding duration and rumen fill. *Physiology & Behavior, 72*, 73–81.

Linnane, M. I., Brereton, A. J., and Giller, P. S. (2001). Seasonal changes in circadian grazing patterns of kerry cows (*Bos taurus*) in semi-feral conditions in Killarney national park, Co. Kerry, Ireland. *Applied Animal Behaviour Science, 71*, 277–292.

Linthorst, A. C. E., Flachskamm, C., Hopkins, S. J., Hoadley, M. E., Labeur, M. S., Holsboer, F., et al. (1997). Long-term intracerebroventricular infusion of corticotropin-releasing hormone alters neuroendocrine, neurochemical, autonomic, behavioral, and cytokine responses to a systemic inflammatory challenge. *The Journal of Neuroscience, 17*, 4448–4460.

Livesey, C. T., Harrington, T., Johnston, A. M., May, S. A., and Metcalf, J. A. (1998). The effect of diet and housing on the development of sole haemorrhages, white line haemorrhages and heel erosions in Holstein hiefers. *Animal Science, 67*, 9–16.

Livesey, C. T., Marsh, C., Metcalf, J. A., and Laven, R. A. (2002). Hock injuries in cattle kept in straw yards or cubicles with rubber mats or mattresses. *Veterinary Record, 150*, 677–679.

Loberg, J. and Lidfors, L. (2001). Effect of milkflow rate and presence of a floating nipple on abnormal sucking between dairy calves. *Applied Animal Behaviour Science, 72*, 189–199.

Locatelli, A., Sartorelli, P., Agnes, F., Bondiolotti, G. P., and Picotti, G. B. (1989). Adrenal response in the calf to repeated simulated transport. *The British Veterinary Journal, 145*, 517–522.

Loerch, S. C. and Fluharty, F. L. (1999). Physiological changes and digestive capabilities of newly received feedlot cattle. *Journal of Animal Science, 77*, 1113–1119.

Loerch, S. C. and Fluharty, F. L. (2000). Use of trainer animals to improve performance and health of newly arrived feedlot calves. *Journal of Animal Science, 78*, 539–545.

Loneragan, G. H., Dargatz, D. A., Morley, P. S., and Smith, M. A. (2001). Trends in mortality ratios among cattle in US feedlots. *Journal of the American Veterinary Medical Association, 219*, 1122–1127.

Long, C. R. and Gregory, K. E. (1978). Inheritance of the horned, scurred and polled condition in cattle. *Journal of Heridity, 69*, 395–400.

Losinger, W. C. and Heinrichs, A. J. (1997). Management practices associated with high mortality among preweaned dairy heifers. *Journal of Dairy Research, 64*, 1–11.

Losinger, W. C., Wells, S. J., Garber, L. P., Hurd, H. S., and Thomas, L. A. (1995). Management factors related to salmonella shedding by dairy heifers. *Journal of Dairy Science, 78*, 2464–2472.

Lott, D. F. and Hart, B. J. (1979). Applied ethology in cattle cultures. *Applied Animal Ethology, 5*, 309–319.

Lowe, D. E., Steen, R. W. J., and Beattie, V. E. (2001a). Preferences of housed finishing beef cattle for different floor types. *Animal Welfare, 10*, 395–404.

Lowe, D. E., Steen, R. W. J., Beattie, V. E., and Moss, B. W. (2001b). The effects of floor type systems on the performance, cleanliness, carcass composition and meat quality of housed finishing beef cattle. *Livestock Production Science, 69*, 33–42.

Lucy, M. C., Billings, H. J., Butler, W. R., Ehnis, L. R., Fields, M. J., Kesler, D. J., et al. (2001). Efficacy of an intravaginal progesterone insert and an injection of PGF (2 alpha) for synchronizing estrus and shortening the interval to pregnancy in postpartum beef cows, peripubertal beef heifers, and dairy heifers. *Journal of Animal Science, 79*, 982–995.

Lunam, C. A., de Passillé, A. M., and Rushen, J. (2002). Neuroma formation following tail docking of dairy calves. *Journal of Dairy Science, 85*, (Suppl 1), 371.

Lund, V. and Weary, D. M. (2003). Understanding animal behaviour and its use in organic animal husbandry. In: M. Vaarst, S. Roderick, V. Lund, and W. Lockeretz (eds), *Animal health and welfare in organic agriculture* (pp. 95–110). Wallingford, UK: CAB International.

Lundborg, G. K., Oltenacu, P. A., Maizon, D. O., Svensson, E. C., and Liberg, P. G. A. (2003). Dam-related effects on heart girth at birth, morbidity and growth rate from birth to 90 days of age in Swedish dairy calves. *Preventive Veterinary Medicine, 60*, 175–190.

Lupoli, B., Johansson, B., Uvnäs-Moberg, K., and Svennersten-Sjaunja, K. (2001). Effect of suckling on the release of oxytocin, prolactin, cortisol, gastrin, cholecystokinin, somatostatin and insulin in dairy cows and their calves. *Journal of Dairy Research, 68*, 175–187.

MacKenzie, A. M., Drennan, M., Rowan, T. G., Dixon, J. B., and Carter, S. D. (1997). Effect of transportation and weaning on humoral immune responses of calves. *Research in Veterinary Science, 63*, 227–230.

Macuhova, J., Tancin, V., Kraetzl, W. D., Meyer, H. H. D., and Bruckmaier, R. M. (2002). Inhibition of oxytocin release during repeated milking in unfamiliar surroundings: The importance of opioids and adrenal cortex sensitivity. *Journal of Dairy Research, 69*, 63–73.

Madden, K. S. (2003). Catecholamines, sympathetic innervation, and immunity. *Brain, Behavior, and Immunity, 17*, 5–10.

Mader, T. L. (2003). Environmental stress in confined beef cattle. *Journal of Animal Science, 81*, 110–119.

Mader, T. L., Dahlquist, J. M., and Gaughan, J. B. (1997). Wind protection effects and airflow patterns in outside feedlots. *Journal of Animal Science, 75*, 26–36.

Mader, T. L., Dahlquist, J. M., Hahn, G. L., and Gaughan, J. B. (1999). Shade and wind barrier effects on summertime feedlot cattle performance. *Journal of Animal Science, 77*, 2065–2072.

Mader, T. L. and Davis, M. S. (2004). Effect of management strategies on reducing heat stress of feedlot cattle: Feed and water intake. *Journal of Animal Science, 82*, 3077–3087.

Mader, T. L., Holt, S. M., Hahn, G. L., Davis, M. S., and Spiers, D. E. (2002). Feeding strategies for managing heat load in feedlot cattle. *Journal of Animal Science, 80*, 2373–2382.

Main, D. C. J., Whay, H. R., Green, L. E., and Webster, A. J. F. (2003a). Preliminary investigation into the use of expert opinion to compare the overall welfare of dairy cattle farms in different farms assurance schemes. *Animal Welfare, 12*, 565–569.

Main, D. C. J., Whay, H. R., Green, L. E., and Webster, A. J. F. (2003b). Effect of the RSPCA freedom food scheme on the welfare of dairy cattle. *Veterinary Record, 153*, 227–231.

Mallard, B. A., Dekkers, J. C., Ireland, M. J., Leslie, K. E., Sharif, S., Vankampen, C. L., et al. (1998). Alteration in immune responsiveness during the peripartum period and its ramification on dairy cow and calf health. *Journal of Dairy Science, 81*, 585–595.

Mallard, B. A., Wagter, L. C., Ireland, M. J., and Dekkers, J. C. M. (1997). Effects of growth hormone, insuline-like growth factor-I, and cortisol on periparturient antibody response profiles of dairy cattle. *Veterinary Immunology and Immunopathology, 60*, 61–76.

Manninen, E., de Passillé, A. M., Rushen, J., Norring, M., and Saloniemi, H. (2002). Preferences of dairy cows kept in unheated buildings for different kind of cubicle flooring. *Applied Animal Behaviour Science, 75*, 281–292.

Mannion, M. A., Cowan, C., and Gannon, M. (2000). Factors associated with perceived quality influencing beef consumption behaviour in Ireland. *British Food Journal, 102*, 195–210.

Manske, T., Hultgren, J., and Bergsten, C. (2002). Prevalence and interships of hoof lesions and lameness in Swedish dairy cows. *Preventive Veterinary Medicine, 25*, 247–263.

Manson, F. J. and Leaver, J. D. (1988). The influence of concentrate amount on locomotion and clinical lameness in dairy cattle. *Animal Production, 47*, 185–190.

Manteuffel, G., Puppe, B., and Schön, P. C. (2004). Vocalization of farm animals as a measure of welfare. *Applied Animal Behaviour Science, 88*, 163–182.

Margerison, J. K., Preston, T. R., Berry, N., and Phillips, C. J. C. (2003). Cross-sucking and other oral behaviours in calves, and their relation to cow suckling and food provision. *Applied Animal Behaviour Science, 80*, 277–286.

Markowitz, T. M., Dally, M. R., Gursky, K., and Price, E. O. (1998). Early handling increases lamb affinity for humans. *Animal Behaviour, 55*, 573–587.

Mason, G. and Rushen, J. (2006). *Stereotypic animal behaviour: Fundamentals and applications to welfare*, 2nd edn. Wallingford, UK: CAB International.

Mason, G., Cooper, J., and Garner, J. (1997). Models of motivational decision-making and how they affect the experimental assessment of motivational priorities. *British Society of Animal Science, 20*, 9–17.

Mason, G. J. and Latham, N. R. (2004). Can't stop, won't stop: Is stereotypy a reliable animal welfare indicator? *Animal Welfare, 13*, S57–S69.

Mason, G. J., Cooper, J., and Clarebrough, C. (2001). Frustrations of fur-farmed mink. *Nature, 410*, 35–36.

Matteri, R. L., Carroll, J. A., and Dyer, C. J. (2000). Neuroendocrine responses to stress. In: G. P. Moberg and J. A. Mench (eds), *The biology of animal stress: Basic principles and implications foranimal welfare* (pp. 43–76). Wallingford, UK: CAB International.

Mattiello, S., Canali, E., Ferrante, V., Caniatti, M., Gottardo, F., Cozzi, G., et al. (2002). The provision of solid feeds to veal calves: II. Behavior, physiology, and abomasal damage. *Journal of Animal Science, 80*, 367–375.

Maule, A. G. and Vanderkooi, S. P. (1999). Stress-induced immune-endocrine interaction. In: P. H. M. Balm (ed.), *Stress physiology in animals* (pp. 205–245). Sheffield, UK: Academic Press.

McFarland, D. J. (1989). *Problems of animal behaviour.* Harlow, UK: Longman Scientific and Technical.

McGuirk, S. M. and Collins, M. (2004). Managing the production, storage, and delivery of colostrum. *Veterinary Clinics of North America – Food Animal Practice, 20*, 593–603.

McKenna, D. R., Roeber, D. L., Bates, P. K., Schmidt, T. B., Hale, D. S., Griffin, D. B., et al. (2002). National beef quality audit – 2000: Survey of targeted cattle and carcass characteristics related to quality, quantity, and value of fed steers and heifers. *Journal of Animal Science, 80*, 1212–1222.

McKnight, D. R. (1978). Performance of newborn dairy calves in hutch housing. *Canadian Journal of Animal Science, 58*, 517–520.

McMartin, S., Godden, S., Metzger, L., Feirtag, J., Bey, R., Stabel, J., et al. (2006). Heat treatment of bovine colostrum. I: Effects of temperature on viscosity and immunoglobulin G level. *Journal of Dairy Science, 89*, 2110–2118.

McMeekan, C. M., Stafford, K. J., Mellor, D. J., Bruce, R. A., Ward, R. N., and Gregory, N. G. (1998). Effects of regional analgesia and/or a non-steriodal anti-inflammatory analgesic on the acute cortisol response to dehorning in calves. *Research in Veterinary Science, 64*, 147–150.

McMeekan, C. M., Stafford, K. J., Mellor, D. J., Bruce, R. A., Ward, R. N., and Gregory, N. G. (1999). Effects of a local anaesthetic and a non-steroidal anti-inflammatory analgesic on the behavioural responses of calves to dehorning. *New Zealand Veterinary Journal, 47*, 92–96.

Meischke, H. R. C., Ramsay, W. R., and Shaw, F. D. (1974). The effect of horns on bruising cattle. *Australian Veterinary Journal, 50*, 432–434.

Mellor, D. J. and Gregory, N. G. (2003). Responsiveness, behavioural arousal and awareness in fetal and newborn lambs: Experimental, practical and therapeutic implications. *New Zealand Veterinary Journal, 51*, 2–13.

Mellor, D. J. and Stafford, K. J. (2001). Integrating practical, regulatory and ethical strategies for enhancing farm animal welfare. *Australian Veterinary Journal, 79*, 762–768.

Mellor, D. J. and Stafford, K. J. (2004). Animal welfare implications of neonatal mortality and morbidity in farm animals. *Veterinary Journal, 168*, 118–133.

Mellor, D. J., Stafford, K. J., Todd, S. E., Lowe, T. E., Gregory, N. G., Bruce, R. A., et al. (2002). A comparison of catecholamine and cortisol responses of young lambs and calves to painful husbandry procedures. *Australian Veterinary Journal, 80*, 228–233.

Mench, J. A. (2003). Assessing animal welfare at the farm and group level: A United States perspective. *Animal Welfare, 12*, 493–503.

Menke, C., Waiblinger, S., Fölsch, D. W., and Wiepkema, P. R. (1999). Social behaviour and injuries of horned cows in loose housing systems. *Animal Welfare, 8*, 243–258.

Metz, J. H. M. (1985). The reaction of cows to a short-term deprivation of lying. *Applied Animal Behaviour Science, 13*, 301–307.

Metz, J. and Metz, J. H. M. (1987). Behavioural phenomena related to normal and difficult deliveries in dairy cows. *Netherlands Journal of Agricultural Science, 35*, 87–101.

Meyer, C. L., Berger, P. J., Koehler, K. J., Thompson, J. R., and Sattler, C. G. (2001). Phenotypic trends in incidence of stillbirth for Holsteins in the United States. *Jouranl of Dairy Science, 84*, 515–523.

Miller, K. and Wood-Gush, D. G. M. (1991). Some effects of housing on the social-behavior of dairy-cows. *Animal Prouction, 53*, 271–278.

Minton, J. (1994). Function of the hypothalamic–pituitary–adrenal axis and the sympathetic nervous system in models of acute stress in domestic farm animals. *Journal of Animal Science, 72*, 1891–1898.

Mitlohner, F. M., Galyean, M. L., and McGlone, J. J. (2002). Shade effects on performance, carcass traits, physiology, and behavior of heat-stressed feedlot heifers. *Journal of Animal Science, 80*, 2043–2050.

Mitlohner, F. M., Morrow, J. L., Dailey, J. W., Wilson, S. C., Galyean, M. L., Miller, M. F., et al. (2001). Shade and water misting effects on behavior, physiology, performance, and carcass traits of heat-stressed feedlot cattle. *Journal of Animal Science, 79*, 2327–2335.

Moberg, G. (1985). *Animal stress*. Bethesda, MD: American Physiological Society.

Moberg, G. P. (1996). Suffering from stress: An approach for evaluating the welfare of an animal. *Acta Agriculturae Scandinavica, Section A, Animal Science*, 46–49.

Moberg, G. P. (2000) Biological response to stress: Implications for animal welfare. In: G. P. Moberg and J. A. Mench. *The biology of animal stress: Basic principles and implications for animal welfare*. Wallingford, UK: CAB International.

Moberg, G. P. and Mench, J. A. (2000). *The biology of animal stress: Basic principles and implications for animal welfare*. Wallingford, UK: CAB International.

Mohammed, H. O., Wade, S. E., and Schaaf, S. (1999). Risk factors associated with *cryptosporidium parvum* infection in dairy cattle in southeastern New York State. *Veterinary Parasitology, 83*, 1–13.

Mohr, E., Langbein, J., and Nürnberg, G. (2002). Heart rate variability. A noninvasive approach to measure stress in calves and cows. *Physiology & Behavior, 75*, 251–259.

Molina, A. and Tuero, O. (2000). Effect of the management method and access to the drinking trough on the water intake of grazing cattle. *Cuban Journal of Agricultural Science, 34*, 195–199.

Molony, V., Kent, J. E., and McKendrick, I. J. (2002). Validation of a method for assessment of an acute pain in lambs. *Applied Animal Behaviour Science, 76*, 215–238.

Moore, D. A., Sischo, W. M., Festa, D. M., Reynolds, J. P., Atwill, E. R., and Holmberg, C. A. (2002). Influence of arrival weight, season and calf supplier on survival in Holstein beef calves on a calf ranch in California, USA. *Preventive Veterinary Medicine, 53*, 103–115.

Morisse, J. P., Cotte, J. P., and Huonnic, D. (1995). Effect of dehorning on behavior and plasma-cortisol responses in young calves. *Applied Animal Behaviour Science, 43*, 239–247.

Mormède, P., Andanson, S., Aupérin, B., Beerda, B., Guémené, D., Malmkvist, J., et al. (2007). Exploration of the hypothalamic–pituitary–adrenal function as a tool to evaluate animal welfare. *Physiology & Behavior*, doi:10.1016/j.physbch.2006.12.003.

Morrow, C. J., Kolver, E. S., Verkerk, G. A., and Matthews, L. R. (2002). Fecal glucocorticoid metabolites as a measure of adrenal activity in dairy cattle. *General and Comparitive Endocrinology, 126*, 229–241.

Morton, D. B. and Griffiths, P. H. M. (1985). The recognition and measurement of pain and distress in laboratory-animals. *Laboratory Animal Science, 35*, 540–540.

Möstl, E. and Palme, R. (2002). Hormones as indicators of stress. *Domestic Animal Endocrinology, 23*, 67–74.

Mounier, L., Veissier, I., and Boissy, A. (2005). Behavior, physiology, and performance of bulls mixed at the onset of finishing to form uniform body weight groups. *Journal of Animal Science, 83*, 1696–1704.

Mourits, M. C. M., Van der Fels-Klerx, H. J., Huirne, R. B. M., and Huyben, M. W. C. (2000). Dairy-heifer management in the Netherlands. *Preventive Veterinary Medicine, 46*, 197–208.

Mowbray, L., Vittie, T., and Weary, D. M. (2003). Hock lesions and free-stall design: Effects of stall surface. *Proceedings of the 5th International Dairy Housing Conference. American Society of Agricultural Engineers*, St. Joseph, MI, 288–295.

Mulei, C. M., Gitau, G. K., and Mbuthia, P. G. (1995). Causes of calf mortality in Kabete area of Kenya. *Onderstepoort Journal of Veterinary Research, 62*, 181–185.

Müller, C., Ladewig, J., Thielscher, H. H., and Smidt, D. (1989). Behavior and heart rate of heifers housed in tether stanchions without straw. *Physiology & Behavior, 46*, 751–754.

Muller, C. J. C. and Botha, J. A. (1997). Cow behaviour in relation to different freestall surfaces during winter temperate climate. *Livestock environment 5*, Vol. 2 (pp. 1069–1076). St. Joseph, MI: American Society of Agricultural Engineers.

Müller R. and Schrader, L. (2005). Behavioural consistency during social separation and personality in dairy cows. *Behaviour, 142*, 1295–1312.

Müller, R. and von Keyserlingk, M. A. G. (2006). Consistency of flight speed and its correlation to productivity and to personality in *Bos taurus* beef cattle. *Applied Animal Behaviour Science, 99*, 193–204.

Munck, A., Guyre, P. M., and Holbrook, N. J. (1984). Physiological functions of glucocorticoids in stress and their relation to pharmacological actions. *Endocrine Reviews, 5*, 25–44.

Munksgaard, L., de Passillé, A. M., Rushen, J., Herskin, M. S., and Kristensen, A. M. (2001). Dairy cows' fear of people: Social learning, milk yield and behaviour at milking. *Applied Animal Behaviour Science, 73*, 15–26.

Munksgaard, L., de Passillé, A. M. B., Rushen, J., Thodberg, K., and Jensen, M. B. (1997). Discrimination of people by dairy cows based on handling. *Journal of Dairy Science, 80*, 1106–1112.

Munksgaard, L., Ingvartsen, K. L., Pedersen, L. J., and Nielsen, V. K. (1999). Deprivation of lying down affects behaviour and pituitary–adrenal axis responses in young bulls. *Acta Agriculturae Scandinavica, Section A, Animal Science, 49*, 172–178.

Munksgaard, L., Jensen, M. B., Pedersen, L. J., Hansen, S. W. and Matthews, L. (2005). Quantifying behavioural priorities-effects of time constraints on behaviour of dairy cows, *Bos taurus*. *Applied Animal Behaviour Science. 92*, 3–14.

Munksgaard, L. and Løvendahl, P. (1993). Effects of social and physical stressors on growth hormone levels in dairy cows. *Canadian Journal of Animal Science, 73*, 847–853.

Munksgaard, L. and Simonsen, H. B. (1995). Behavioural and pituitary–adrenal axis responses of tethered cows or cows kept in pens with slatted floors. *Acta Agriculturae Scandinavica, Section A, Animal Science, 45*, 132–138.

Munksgaard, L. and Simonsen, H. B. (1996). Behavioral and pituitary adrenal-axis responses of dairy cows to social isolation and deprivation of lying down. *Journal of Animal Science, 74*, 769–778.

Munoz-Zanzi, C. A., Thurmond, M. C., Johnson, W. O., and Hietala, S. K. (2002). Predicted ages of dairy calves when colostrum-derived bovine viral diarrhea virus antibodies would no longer offer protection against disease or interfere with vaccination. *Journal of the American Veterinary Medical Association, 221*, 678–685.

Myers, S. E., Faulkner, D. B., Ireland, F. A., and Parrett, D. F. (1999). Comparison of three weaning ages on cow-calf performance and steer carcass traits. *Journal of Animal Science, 77*, 323–329.

Nagaraja, T. G. and Chengappa, M. M. (1998). Liver abscesses in feedlot cattle: A review. *Journal of Animal Science, 76*, 287–298.

Nanda, A. S., Dobson, H., and Ward, W. R. (1990). Relationship between an increase in plasma cortisol during transport-induced stress and failure of oestradiol to induce a luteinising hormone surge in dairy cows. *Research in Veterinary Science, 49*, 25–28.

Nanda, A. S., Dobson, H., and Ward, W. R. (1992). Opioid modulation of the hypothalamo–pituitary–adrenal axis in dairy cows. *Domestic Animal Endocrinology, 9*, 181–186.

Nanda, A. S., Ward, W. R., and Dobson, H. (1989). Effects of naloxone on the oestradiol-induced LH surge and cortisol release in transported cows. *Journal of Reproduction and Fertility, 87*, 803–807.

National Research Council. (2000). *Nutrient requirement for beef cattle*. Washington, DC: National Academy of Sciences.

Natzke, R. P., Bray, D. R., and Everett, R. W. (1982). Cow preference for free stall surface material, rubber mats, carpeting, and a layered mat. *Journal of Dairy Science, 25*, 146–153.

Natzke, R. P., Everett, R. W., and Bray, D. R. (1975). Effect of drying-off practices on mastitis infection. *Journal of Dairy Science, 58*, 1828–1835.

Negrao, J. A., Porcionato, M. A., de Passillé, A. M., and Rushen, J. (2004). Cortisol in saliva and plasma of cattle after ACTH administration and milking. *Journal of Dairy Science, 87*, 1713–1718.

Neveux, S., Weary, D. M., Rushen, J., von Keyserlingk, M. A. G., and de Passillé, A. M. (2006). Hoof discomfort changes how dairy cattle distribute their body weight. *Journal of Dairy Science, 89*, 2503–2509.

Nicol, C. J., Davidson, H. P. D., Harris, P. A., Waters, A. J., and Wilson, A. D. (2002). Study of crib-biting and gastric inflammation and ulceration in young horses. *Veterinary Record, 151*(22), 658–662.

Nilsson, C. (1992). Walking and lying surfaces in livestock houses. In: C. Phillips and D. Piggins (eds), *Farm animals and the environment* (pp. 93–110). Wallingford, UK: CAB International.

Nix, J. M., Spitzer, J. C., Grimes, L. W., Burns, G. L., and Plyler, B. B. (1998). A retrospective analysis of factors contributing to calf mortality and dystocia in beef cattle. *Theriogenology, 49*, 1515–1523.

O'Connell, J. M., Giller, P. S., and Meaney, W. J. (1993). Weanling training and cubicle usage as heifers. *Applied Animal Behaviour Science, 37*, 185–195.

O'Driscoll, K., von Keyserlingk, M. A. G., and Weary, D. M. (2006). Effects of mixing on drinking and competitive behavior of dairy calves. *Journal of Dairy Science, 89*, 229–233.

Oetzel, R. R., Nordlund, K. V., and Garret, E. F. (1999). Effect of ruminal pH and stage of lactation on ruminal lactate concentration in dairy cows (abst.). *Journal of Dairy Science, 82*(38).

Offer, J. E., McNulty, D., and Logye, D. N. (2000). Observations of lameness, hoof conformation and development of lesions in dairy cattle over four lactations. *Veterinary Record, 147*, 105–109.

Ok, M., Sen, I., Turgut, K., and Irmak, K. (2001). Plasma gastrin activity and the diagnosis of bleeding abomasal ulcers in cattle. *Journal of Veterinary Medicine Series A – Physiology Pathology Clinical Medicine, 48*, 563–568.

Oliver, S. P. and Sordillo, L. M. (1988). Udder health in the periparturient period. *Journal of Dairy Science, 71*, 2584–2606.

Olsson, S. O., Viring, S., Emanuelsson U., & Jacobsson, S. O. (1993). Calf diseases and mortality in Swedish dairy herds. *Acta Veterinaria Scandinavica, 34*, 263–269.

Oltenacu, P. A., Hultgren, J., and Algers, B. (1998). Associations between use of electric cow-trainers and clinical diseases, reproductive performance and culling in Swedish dairy cattle. *Preventive Veterinary Medicine, 37*, 77–90.

Ortman, K. and Svensson, C. (2004). Use antimicrobial drugs in Swedish dairy calves and replacement heifers. *Veterinary Record, 154*, 136–140.

Østergaard, S. and Larsen, T. (2000). Short communication: Associations between blood calcium status at calving and milk yield in dairy cows. *Journal of Dairy Science, 83*, 2438–2440.

Østergaard, S. and Sorensen, J. T. (1998). A review of the feeding-health-production complex in a dairy herd. *Preventive Veterinary Medicine, 36*, 109–129.

Pacák, K. and Palkovits, M. (2001). Stressor specificity of central neuroendocrine responses: Implications for stress-related disorders. *Endocrine Reviews, 22*, 502–548.

Pajor, E. A., Rushen, J., and de Passillé, A. M. B. (2000). Aversion learning techniques to evaluate dairy cattle handling practices. *Applied Animal Behaviour Science, 69*, 89–102.

Pajor, E. A., Rushen, J., and de Passillé, A. M. B. (2003). Dairy cattle's choice of handling treatments in a Y-maze. *Applied Animal Behaviour Science, 80*, 93–107.

Palechek, N. P., Schoonderwoerd, M., and Perry, A. W. (1987). A case study of respiratory disease in a veal calf operation. *Canadian Veterinary Journal, 28*, 363–365.

Palme, R., Robia, C., Baumgartner, W., and Möstl, E. (2000). Transport stress in cattle as reflected by an increase in faecal cortisol metabolite concentrations. *Veterinary Record, 146*, 108–109.

Pastell, M., Aisla, A. M., Hautala, M., Poikalainen, V., Praks, J., Veermae, I., et al. (2006). Contactless measurement of cow behavior in a milking robot. *Behavior Research Methods, 38*, 479–486.

Pastell, M. E. and Kujala, M. (2007). A probabilistic neural network model for lameness detection. *Journal of Dariy Science, 90*, 2283–2292.

Paton, L. J., Beauchemin, K. A., Veira, D. M., and von Keyserlingk, M. A. G. (2006). Effects of sodium bicarbonate provided free choice or incorporated into a high grain diet on reducing the risk of acidosis in cattle. *Canadian Journal of Animal Science, 86*, 429–437.

Patterson, H. H., Johnson, P. S., Patterson, T. R., and Young, D. B. (2002). Effects of water quality on performance and health of growing steers. *Journal of Animal Science, 80, supplement 2*, 113.

Pedersen, L. J., Jensen, M. B., Hansen, S. W., Munksgaard, L., Ladewig, J., and Matthews, L. (2002). Social isolation affects the motivation to work for food and staw in pigs as measured by operant conditioning techniques. *Applied Animal Behaviour Science, 77*, 295–309.

Peeler, E. J., Green, M. J., Fitzpatrick, J. L., Morgan, K. L., and Green, L. E. (2000). Risk factors associated with clinical mastitis in low somatic cell count British dairy herds. *Journal of Dairy Science, 83*, 2464–2472.

Petherick, J. C. (2005). Animal welfare issues associated with extensive livestock production: The northern Australian beef cattle industry. *Applied Animal Behaviour Science, 92*, 211–234.

Petrie, N. J., Mellor, D. J., Stafford, K. J., Bruce, R. A., and Ward, R. N. (1996a). Cortisol responses of calves to two methods of disbudding used with or without local anaesthetic. *New Zealand Veterinary Journal, 44*, 9–14.

Petrie, N. J., Mellor, D. J., Stafford, K. J., Bruce, R. A., and Ward, R. N. (1996b). Cortisol responses of calves to two methods of tail docking used with or without local anaesthetic. *New Zealand Veterinary Journal, 44*, 4–8.

Petrie, N. J., Stafford, K. J., Mellor, D. J., Bruce, R. A., and Ward, R. N. (1995). The behaviour of calves tail docked with a rubber ring used with or without local anaesthesia. *Proceedings of the New Zealand Society of Animal Production, 55*, 58–60.

Petty, D. B., Hattingh, J., Ganhao, M. F., and Bezuidenhout, L. (1994). Factors which affect blood variables of slaughtered cattle. *Journal of the South African Veterinary Association, 65*, 41–45.

Philipot, J. M., Pluvinage, P., Cimarosti, I., Sulpice, P., and Bugnard, F. (1994). Risk factors of dairy cow lameness associated with housing conditions. *Veterinary Research, 25*, 244–248.

Phillips, C. J. C. (1990). Adverse effects on reproductive performance and lameness of feeding grazing dairy cows partially on silage indoors. *Journal of Agricultural Science, Cambridge, 115*, 253–258.

Phillips, C. J. C. (1993). *Cattle behaviour*. Ipswich, UK: Farming Press.

Phillips, C. J. C. and Morris, I. D. (2000). The locomotion of dairy cows on concrete floors that are dry, wet, or covered with a slurry of excreta. *Journal of Dairy Science, 83*, 1767–1772.

Phillips, C. J. C. and Morris, I. D. (2001). The locomotion of dairy cows on floor surfaces with different frictional properties. *Journal of Dairy Science, 84*, 623–628.

Phillips, C. J. C. and Morris, I. D. (2002). The ability of cattle to distinguish between, and their preference for, floors with diffent levels of friction, and their avoidance of floors contaminated with excreta. *Animal Welfare, 11*, 21–29.

Phillips, C. J. C. and Rind, M. I. (2001). The effects on production and behavior of mixing uniparous and multiparous cows. *Journal of Dairy Science, 84*, 2424–2429.

Phillips, C. J. C. and Rind, M. I. (2002). The effects of social dominance on the production and behavior of grazing dairy cows offered forage supplements. *Journal of Dairy Science, 85*, 51–59.

Phillips, C. J. C., Youssef, M. Y., Chiy, P. C., and Arney, D. R. (1999). Sodium chloride supplements increase the salt appetite and reduce stereotypies in confined cattle. *Animal Behaviour, 98*, 741–747.

Phy, T. S. and Provenza, F. D. (1998). Sheep fed grain prefer foods and solutions that attenuate acidosis. *Journal of Animal Science, 76*, 954–960.

Pinheiro Machado, L. C., Hurnik, J. F., and Burton, J. H. (1997). The effect of amniotic fluid ingestion on the nociception of cows. *Physiology & Behavior, 62*, 1339–1344.

Pinheiro Machado, L. C., Teixeira, D. L., Weary, D. M., von Keyserlingk, M. A. G., and Hötzel, M. J. (2004). Designing better water troughs: Dairy cows prefer and drink more from larger troughs. *Applied Animal Behaviour Science, 89*, 185–193.

Prayaga, K. C. (2007). Genetic options to replace dehorning in beef cattle: A review. *Australian Journal of Agricultural Research, 58*, 1–8.

Preisler, M. T., Weber, P. S. D., Tempelman, R. J., Erskine, R. J., Hunt, H., and Burton, J. L. (2000a). Glucocorticoid receptor down-regulation in neutrophils of periparturient cows. *Americal Journal of Veterinary Research, 61*, 14–19.

Preisler, M. T., Weber, P. S. D., Tempelman, R. J., Erskine, R. J., Hunt, H., and Burton, J. L. (2000b). Glucocorticoid receptor expression profiles in mononuclear leukocytes of periparturient Holstein cows. *Journal of Dairy Science, 83*, 38–47.

Presley, S. M., Knapp, F. W., Boling, J. A., and Burg, J. G. (1996). Effects of the horn fly (diptera: Muscidae) on physiological and nutritional responses of beef steers: Continuous fly population levels. *Journal of Economical Entomology, 89*, 138–143.

Price, E. O. (1998). In: T. Grandin (ed.), *Behavioral genetics and the process of animal domestication; genetics and the behavior of domestic animals* (pp. 31–65). San Diego, CA: Academic Press.

Price, E. O. (1999). Behavioral development in animals undergoing domestication. *Applied Animal Behaviour Science, 65*, 245–271.

Price, E. O. (2003). *Animal domestication and behavior*. Wallingford, UK: CAB International.

Price, E. O. and Wallach, S. J. R. (1990). Physical isolation of hand-reared hereford bulls increases their aggressiveness toward humans. *Applied Animal Behaviour Science, 27*, 263–267.

Price, E. O., Harris, J. E., Borgwardt, R. E., Sween, M. L., and Connor, J. M. (2003). Fenceline contact of beef calves with their dams at weaning reduces the negative effects of separation on behavior and growth rate. *Journal of Animal Science, 81*, 116–121.

Provenza, F. D. (1995). Postingestive feedback as an elementary determinant of food preference and intake in ruminants. *Journal of Range Management, 43*, 2–17.

Pursley, J. R., Silcox, R. W., and Wiltbank, M. C. (1998). Effect of time of artificial insemination on pregnancy rates, calving rates, pregnancy loss, and gender ratio after synchronization of ovulation in lactating dairy cows. *Journal of Dairy Science, 81*, 2139–2144.

Pusillo, G. M., Hoffman, M. P., and Self, H. L. (1991). Effects of placing cattle on feed at two-month intervals and housing on feedlot performance and carcass grades. *Journal of Animal Science, 69*, 443–450.

Quigley, J. D. and Drewry, J. J. (1998). Nutrient and immunity transfer from cow to calf pre- and postcalving. *Journal of Dairy Science, 81*, 2779–2790.

Quigley, J. D., Wolfe, T. A., and Elsasser, T. H. (2006). Effects of additional milk replacer feeding on calf health, growth, and selected blood metabolites in calves. *Journal of Dairy Science, 89*, 207–216.

Quimby, W. F., Sowell, B. F., Bowman, J. G. P., Branine, M. E., Hubbart, M. E., and Sherwood, H. W. (2001). Application of feeding behaviour to predict morbidity in a commercial feedlot. *Canadian Journal of Animal Science, 81*, 315–320.

Radostits, O. M., Blood, D. C., and Gay, C. C. (1994). Acute carbohydrate engorgement of ruminants (rumen overload). In: O. M. Radostits, D. C. Blood, and C. C. Gay (eds), *Veterinary medicine*, 8th edn. (pp. 262–269). Philadelphia, PA: W. B. Saunders.

Rajala, P. and Castrén, H. (1995). Serum immunoglobulin concentrations and health of dairy calves in two management systems from birth to 12 weeks of age. *Journal of Dairy Science, 78*, 2737–2744.

Rajala, P. J. and Gröhn, Y. T. (1998). Disease occurrence and risk factor analysis in Finnish Ayrshire cows. *Acta Veterinaria Scandinavica, 39*, 1–13.

Rajkondawar, P. G., Tasch, U., Lefcourt, A. M., Erez, B., Dyer, R. M., and Varner, M. A. (2002). A system for identifying lameness in dairy cattle. *Applied Engineering in Agriculture, 18*, 87–96.

Ralphs, M. H. and Provenza, F. D. (1999). Conditioned food aversions: Principles and practices, with special reference to social facilitation. *Proceedings of the Nutrition Society, 58*, 813–820.

Ralston, B. J., Cockwill, C. L., Guselle, N. J., Van Herk, F. H., McAllister, T. A., and Olson, M. E. (2003). Prevalence of *Giardia* and *Cryptosporidium andersoni and their effects on* performance in feedlot beef cattle. *Canadian Journal of Animal Science, 83*, 153–159.

Rasmussen, M. D., Wiking, L., Bjerring, M., and Larsen, H. C. (2006). Influence of air intake on the concentration of free fatty acids and vacuum fluctuations during automatic milking. *Journal of Dairy Science, 89*, 4596–4605.

Rauw, W. M., Kanis, E., Noordhuizen-Stassen, E. N., and Grommers, F. J. (1998). Undesirable side effects of selection for high production efficiency in farm animals: A review. *Livestock Production Science, 56*, 15–33.

Rea, D. E., Tyler, J. W., Hancock, D. D., Besser, T. E., Wilson, L., Krytenberg, D. S., et al. (1996). Prediction of calf mortality by use of tests for passive transfer of colostral immunoglobulin. *Journal of the American Veterinary Medical Association, 208*, 2047–2049.

Redbo, I. (1990). Changes in duration and frequency of stereotypies and their adjoining behaviours in heifers, before, during and after the grazing period. *Applied Animal Behaviour Science, 26*, 57–67.

Redbo, I. (1992). The influence of restraint on the occurrence of oral stereotypies in dairy cows. *Applied Animal Behaviour Science, 35*, 115–123.

Redbo, I. (1993). Stereotypies and cortisol secretion in heifers subjected to tethering. *Applied Animal Behaviour Science, 38*, 213–225.

Redbo, I. (1998). Relations between oral stereotypies, open-field behavior, and pituitary–adrenal system in growing dairy cattle. *Physiology & Behavior, 64*, 273–278.

Redbo, I. and Nordblad, A. (1997). Stereotypies in heifers are affected by feeding regime. *Applied Animal Behaviour Science, 53*, 193–202.

Redbo, I., Emanuelson, M., Lundberg, K., and Oredsson, N. (1996). Feeding level and oral stereotypies in dairy cows. *Animal science, 62*, 199–206.

Reinhardt, C., Reinhardt, A., and Reinhardt, V. (1986). Social behaviour and reproductive performance in semi-wild Scottish highland cattle. *Applied Animal Behaviour Science, 15*, 125–136.

Reinhardt, V. and Reinhardt, A. (1981). Natural sucking performance and age of weaning in zebu cattle (*Bos indicus*). *Journal of Agricultural Science, Camridge, 96*, 309–312.

Reinhardt, V., Mutiso, F. M., and Reinhardt, A. (1978). Social behaviour and social relationships between female and male prepubertal bovine calves (*Bos indicus*). *Applied Animal Ethology, 4*, 43–54.

Reinhardt, V., Reinhardt, A., and Mutiso, F. M. (1977). Cow–calf relationship in Massai cattle, 1–7.

Reist, M., Erdin, D., von Euw, D., Tschuemperlin, K., Leuenberger, H., Chilliard, Y., et al. (2003). Estimation of energy balance at the individual and herd level using blood and milk traits in high-yielding dairy cows. *Journal of Dairy Science, 85*, 3314–3327.

Rhind, S. M., Reid, H. W., and McMillen, S. R. (1999). Effects of pulsed or continuous infusion of cortisol on immune function in sheep. *Domestic Animal Endocrinology, 16*, 1–9.

Ribble, C. S., Meek, A. H., Shewen, P. E., Jim, G. K., and Guichon, P. T. (1995). Effect of transportation on fatal fibrinous pneumonia and shrinkage in calves arriving at a large feedlot. *Journal of the American Veterinary Medical Association, 207*, 612–615.

Riley, D. G., Chase, C. C., Olson, T. A., Coleman, S. W., and Hammond, A. C. (2004). Genetic and nongenetic influences on vigor at birth and preweaning mortality of purebred and high percentage Brahman calves. *Journal of Animal Science, 82*, 1581–1588.

Risbrough, V. B. and Stein, M. B. (2006). Role of corticotropin releasing factor in anxiety disorders: A translational research perspective. *Hormones and Behavior, 50*, 550–561.

Ritter, W. and Sorrenson, W. J. (1985). *Cattle production on the highlands of Santa Catarina, Brazil, current situation and prospects*. Eschborn, Germany: Deutsche Gesellschaft für Technishe Zusammenarbeit.

Ritz, L. R., Glowatzki-Mullis, M. L., MacHugh, D. E., and Gaillard, C. (2000). Phylogenetic analysis of the tribe bovini using microsatellites. *Animal Genetics, 31*, 178–185.

Roberts, G. (1996). Why individual vigilance declines as group size increases. *Animal Behaviour, 51*, 1077–1086.

Roche, J. R. and Berry, D. P. (2006). Periparturient climatic, animal, and management factors influencing the incidence of milk fever in grazing systems. *Journal of Dairy Science, 89*, 2775–2783.

Rodriguez-Lainz, A., Melendez-Retamal, P., Hird, D., Read, D., and Walker, R. (1999). Farm- and host-level risk factors for papillomatous digital dermatitis in Chilean dairy cattle. *Preventive Veterinary Medicine, 42*, 87–97.

Roeber, D. L., Mies, P. D., Smith, C. D., Belk, K. E., Field, T. G., Tatum, J. D., et al. (2001). National market cow and bull beef quality audit – 1999: A survey of producer-related defects in market cows and bulls. *Journal of Animal Science, 79*, 658–665.

Roets, E., Burvenich, C., Diez-Fraile, A., and Noordhuizen-Stassen, E. N. (1999). Evaluation of the role of endotoxin and cortisol on modulation of CD18 adhesion receptors in cows with mastitis caused by *Escherichia coli*. *American Journal of Veterinary Research, 60*, 534–540.

Rogers, P. L., Gaskins, C. T., Johnson, K. A., and MacNeil, M. D. (2004). Evaluating longevity of composite beef females using survival analysis techniques. *Journal of Animal Science, 82*, 860–866.

Ronchi, B., Stadaioli, G., Verini Supplizi, A., Bernabucci, U., Lacetera, N., Accorsi, P. A., et al. (2001). Influence of heat stess or feed restriction on plasma progesterone, oestradiol-17$, LH, FSH, prolactin and cortisol in Holstein heifers. *Livestock Production Science, 68*, 231–241.

Rose, L. M. and Fedigan, L. M. (1995). Vigilance in white-faced capuchins, *Cebus capucinus*, in Costa Rica. *Animal Behaviour, 49*, 63–70.

Rouda, R. R., Anderson, D. M., Wallace, J. D., and Murray, L. W. (1994). Free-ranging cattle water consumption in South Central New Mexico. *Applied Animal Behaviour Science, 39*, 29–38.

Ruckebusch, Y. (1972). The relevance of drowsiness in the circadian cycle of farm animals. *Animal Behaviour, 20*, 637–643.

Ruckebusch, Y. (1974). Sleep deprivation in cattle. *Brain Research, 78*, 495–499.

Ruckebusch, Y. (1975). Feeding and sleep patterns of cows prior to and post parturition. *Applied Animal Ethology, 1*, 283–292.

Rugbjerg, H., Nielsen, E. M., and Andersen, J. S. (2003). Risk factors associated with faecal shedding of verocytotoxin-producing *Escherichia* coli O157 in eight known-infected Danish dairy herds. *Preventive Veterinary Medicine, 58*, 101–113.

Rukkwamsuk, T., Kruip, T. A. M., and Wensing, T. (1999). Relationship between overfeeding and overconditioning in the dry period and the problems of high producing dairy cows during the postparturient period. *Veterinary Quarterly, 21*, 71–77.

Rushen, J. (1996). Using aversion learning techniques to assess the mental state, suffering, and welfare of farm animals. *Journal of Animal Science, 74*, 1990–1995.

Rushen, J. (2000). Some issues in the interpretation of behavioural responses to stress. In: G. P. Moberg and J. A. Mench (eds), *The biology of animal stress: Basic principles and implications for animal welfare*. Wallingford, UK: CAB International.

Rushen, J. (2003). Changing concepts of farm animal welfare: Bridging the gap between applied and basic research. *Applied Animal Behaviour Science, 81*, 199–214.

Rushen, J. and de Passillé, A. M. (1995). The motivation of non-nutritive sucking in calves, *Bos taurus. Animal Behaviour, 49*, 1503–1510.

Rushen, J. and de Passillé, A. M. (2006). Effects of roughness and compressibility of flooring on cow locomotion. *Journal of Dairy Science, 89*, 2965–2972.

Rushen, J. and de Passillé, A. M. B. (1992). The scientific assessment of the impact of housing on animal welfare: A critical review. *Canadian Journal of Animal Science, 72*, 721–743.

Rushen, J., Boissy, A., Terlouw, E. M. C., and de Passillé, A. M. B. (1999a). Opioid peptides and behavioral and physiological responses of dairy cows to social isolation in unfamiliar surroundings. *Journal of Animal Science, 77*, 2918–2924.

Rushen, J., de Passillé, A. M. B., and Munksgaard, L. (1999b). Fear of people by cows and effects on milk yield, behavior and heart rate at milking. *Journal of Dairy Science, 82*, 720–727.

Rushen, J., de Passillé, A. M. B., Munksgaard, L., and Tanida, H. (2001a). People as social actors in the world of farm animals. In: H. Gonyou and L. Keeling (eds), *Social behaviour of farm animals*. Wallingford, UK: CAB International.

Rushen, J., Haley, D., and de Passillé, A. M. (2007a). Effect of softer flooring in tiestalls on resting behavior and leg injuries of lactating cows. *Journal of Dairy Science, 90*, 3647–3651.

Rushen, J., Munksgaard, L., de Passillé, A. M. B., Jensen, M. B., and Thodberg, K. (1998). Location of handling and dairy cows' responses to people. *Applied Animal Behaviour Science, 55*, 259–267.

Rushen, J., Munksgaard, L., Marnet, P. G., and de Passillé, A. M. (2001b). Human contact and the effects of acute stress on cows at milking. *Applied Animal Behaviour Science, 73*, 1–14.

Rushen, J., Pombourcq, E., and de Passillé, A. M. (2007b) Validation of two measures of lameness in dairy cows. *Applied Animal Behaviour Science, 106*, 173–177.

Rushen, J., Taylor, A. A., and de Passillé, A. M. B. (1999c). Domestic animals' fear of humans and its effect on their welfare. *Applied Animal Behaviour Science, 65*, 285–303.

Rutherford, K. M. D. (2002). Assessing pain in animals. *Animal Welfare, 11*, 31–53.

Rutter, S. M. (2006). Diet preference for grass and legumes in free-ranging domestic sheep and cattle: Current theory and future application. *Applied Animal Behaviour Science, 97*, 17–35.

Rybarczyk, P., Koba, Y., Rushen, J., Tanida, H., and dePassillé, A. M. (2001). Can cows discriminate people by their faces? *Applied Animal Behaviour Science, 74*, 175–189.

Rybarczyk, P., Rushen, J., and de Passillé, A. M. (2003). Recognition of people by dairy calves using colour of clothing. *Applied Animal Behaviour Science, 81*, 307–319.

Sandem, A. I., Janczak, A. M., Salte, R. & Braastad, B. O. (2006). The use of diazepam as a pharmacological validation of eye white as an indicator of emotional state in dairy cows. *Applied Animal Behaviour Science, 96*, 177–183.

Sanderson, M. W. and Dargatz, D. A. (2000). Risk factors for high herd level calf morbidity risk from birth to weaning in beef herds in the USA. *Preventive Veterinary Medicine, 44*, 97–106.

Sapolsky, R. (1998). *Why zebras don't get ulcers*. New York: W. H. Freeman.

Sapolsky, R. M., Romero, L. M., and Munck, A. U. (2000). How do glucocorticoids influence stress responses? Integrating permissive, suppressive, stimulatory, and preparative actions. *Endocrine Reviews, 21*, 55–89.

Sargeant, J. M., Blackwell, T. E., Martin, S. W., and Tremblay, R. R. (1994). Production practices, calf health and mortality on six white veal farms in Ontario. *Canadian Journal of Veterinary Research, 58*, 189–195.

Sato, S., Nagamine, R., and Kubo, T. (1994). Tongue-playing in tethered Japanese black cattle: Diurnal patterns, analysis of variance and behaviour sequences. *Applied Animal Behaviour Science, 39*, 39–47.

Schaefer, A. L., Cook, N., Tessaro, S. V., Deregt, D., Desroches, G., Dubeski, P. L., et al. (2004). Early detection and prediction of infection using infrared thermography. *Canadian Journal of Animal Science, 84*, 73–80.

Schaefer, A. L., Jones, S. D. M., and Stanley, R. W. (1997). The use of electrolyte solutions for reducing transport stress. *Journal of Animal Science, 75*, 258–265.

Schmidt, G. R., Hossner, K. L., Yemm, R. S., and Gould, D. H. (1999). Potential for disruption of central nervous system tissue in beef cattle by different types of captive bolt stunners. *Journal of Food Protection, 62*, 390–393.

Schnepper, R. L. (2001). Veal calf TLC. *Journal of American Veterinary Medical Association, 219*, 1389–1391.

Schnier, C., Hielm, S., and Saloniemi, H. S. (2002). Comparison of the disease incidences of dairy cows kept in cold and warm loose-housing systems. *Preventive Veterinary Medicine, 53*, 247–261.

Schrader, L. and Todt, D. (1998). Vocal quality is correlated with levels of stress hormones in domestic pigs. *Ethology, 104*, 859–876.

Schrama, J. W., Arieli, A., Brandsma, H. A., Luiting, P., and Verstegen, M. W. (1993). Thermal requirements of young calves during standing and lying. *Journal of Animal Science, 71*, 3285–3292.

Schreiner, D. A. and Ruegg, P. L. (2002). Effects of tail docking on milk quality and cow cleanliness. *Journal of Dairy Science, 85*, 2503–2511.

Schutz, K., Davison, D., and Matthews, L. (2006). Do different levels of moderate feed deprivation in dairy cows affect feeding motivation? *Applied Animal Behaviour Science, 101*, 253–263.

Schwartzkopf-Genswein, K. S., Silasi, R., and McAllister, T. A. (2003). Use of remote bunk monitoring to record effects of breed, feeding regime and weather on feeding behaviour and growth performance of cattle. *Canadian Journal of Animal Science, 83*, 29–38.

Schwartzkopf-Genswein, K. S., Stookey, J. M., Crowe, T. G., and Genswein, B. M. (1998). Comparison of image analysis, exertion force, and behavior measurements for use in the assessment of beef cattle responses to hot-iron and freeze branding. *Journal of Animal Science, 76*, 972–979.

Schwartzkopf-Genswein, K. S., Stookey, J. M., de Passillé, A. M., and Rushen, J. (1997a). Comparison of hot-iron and freeze branding on cortisol levels and pain sensitivity in beef cattle. *Canadian Journal of Animal Science, 77*, 369–374.

Schwartzkopf-Genswein, K. S., Stookey, J. M., and Welford, R. (1997b). Behavior of cattle during hot-iron and freeze branding and the effects on subsequent handling ease. *Journal of Animal Science, 75*, 2064–2072.

Scientific Committee on Animal Health and Animal Welfare (2001). *The welfare of cattle kept for beef production*. SANCO.C.2/AH/R22/2000. European Commission, Health and Consumer Protection Directorate-General.

Scientific Committee on Animal Health and Animal Welfare (2002). The welfare of animals during transport (details for horses, pigs, sheep and cattle). European Commission, Health and Consumer Protection Directorate-General.

Seabrook, M. F. (1984). The psychological interaction between the stockman and his animals and its influence on performance of pigs and dairy cows. *The Veterinary Record, 115*, 84–87.

Seabrook, M. F. and Bartle, N. C. (1992). Environmental factors influencing the production and welfare of farm animals. In: C. J. C. Phillips and D. Piggins (eds.), *Farm animals and the enviornment* (pp. 111–130). Wallingford, UK: CAB International.

Seabrook, M. F. and Wilkinson, J. M. (2000). Stockpersons' attitudes to the husbandry of dairy cows. *Veterinary Record, 147*, 157–160.

Seegers, H., Beaudeau, F., Fourichon, C., and Bareille, N. (1998). Reasons for culling in French Holstein cows. *Preventive Veterinary Medicine, 36*, 257–271.

Selman, I. E., McEwan, A. D., and Fisher, E. W. (1970). Studies on natural suckling in cattle during the first eight hours postpartum. I behavioural studies (calves). *Animal Behaviour, 18*, 284–289.

Senn, M., Gross-Luem, S., Leuenberger, H., and Langhans, W. (2000). Meal patterns and meal-induced metabolic changes in calves fed milk ad lib. *Physiology & Behavior, 70*, 189–195.

Shaver, R. (2002). Rumen acidosis in dairy cattle; bunk management considerations. *Advances in Dairy Technology, 14*, 241–249.

Silanikove, N. (2000). Effects of heat stress on the welfare of extensively managed domestic ruminants. *Livestock Production Science, 67*, 1–18.

Simpson, R. B., Wesen, D. P., Anderson, K. L., Armstrong, J. D., and Harvey, R. W. (1995). Subclinical mastitis and milk production in primiparous Simmental cows. *Journal of Animal Science, 73*, 1552–1558.

Sivula, N. J., Ames, T. R., and Marsh, W. E. (1996). Management practices and risk factors for morbidity and mortality in Minnesota dairy heifer calves. *Preventive Veterinary Medicine, 27*, 173–182.

Skoda, S. R., Thomas, G. D., and Campbell, J. B. (1991). Development sites and relative abundance of immature stages of stable fly (diptera: Muscidae) in beef cattle feedlot pens in eastern Nebraska. *Journal of Economical Entomology, 84*, 191–197.

Slyter, L. L. (1976). Influence of acidosis on rumen function. *Journal of Animal Science, 43*, 910–929.

Smagin, G. N., Howell, L. A., Redmann, Jr., S., Ryan, D. H., and Harris, R. B. S. (1999). Prevention of stress-induced weight loss by third ventricle CRF receptor antagonist. *American Journal of Physiology, 268*, R1461–R1469.

Smith, D., Blackford, M., Younts, S., Moxley, R., Gray, J., Hungerford, L., et al. (2001). Ecological relationships between the prevalence of cattle shedding *Escherichia coli* O157:H7 and characteristics of the cattle or conditions of the feedlot pen. *Journal of Food Protection, 64*, 1899–1903.

Smits, A. C. and de Wilt, J. G. (1991). Group housing of veal calves. In New Trends in Veal Calf Production. Metz, J. H. M. and Groenestein, C. M. (eds.), 49–53. EAAP Publications No 52, Pudoc, Wageningen.

Sneva, F. A. (1970). Behavior of yearling cattle on eastern Oregon range. *Journal of Range Management, 23*, 155–157.

Snowder, G. D., Van Vleck, L. D., Cundiff, L. V., and Bennett, G. L. (2006). Bovine respiratory disease in feedlot cattle: Environmental, genetic, and economic factors. *Journal of Animal Science, 84*, 1999–2008.

Somers, J. G. C. J., Frankena, K., Noordhuizen-Stassen, E. N., and Metz, J. H. M. (2003). Prevalence of claw disorders in Dutch dairy cows exposed to several floor systems. *Journal of Dairy Science, 86*, 2082–2093.

Sowell, B. F., Branine, M. E., Bowman, J. G. P., Hubbert, M. E., Sherwood, H. E., and Quimby, W. (1999). Feeding and watering behavior of healthy and morbid steers in a commercial feedlot. *Journal of Animal Science, 77*, 1105–1112.

Spinka, M. (2006). How important is natural behaviour in animal farming systems? *Applied Animal Behaviour Science, 100*, 117–128.

Spinka, M., Newberry, R. C., and Bekoff, M. (2001). Mammalian play: Training for the unexpected. *Quarterly Review of Biology, 76*, 141–168.

Sprecher, D. J., Hostetler, D. E., and Kaneene, J. B. (1997). A lameness scoring system that uses posture and gait to predict dairy cattle reproductive performance. *Theriogenology, 47*, 1179–1187.

Stafford, K. J. and Mellor, D. J. (2005a). Dehorning and disbudding distress and its alleviation in calves. *Veterinary Journal, 169*, 337–349.

Stafford, K. J. and Mellor, D. J. (2005b). The welfare significance of the castration of cattle: A review. *New Zealand Veterinary Journal, 53*, 271–278.

Stafford, K. J., Mellor, D. J., Todd, S. E., Bruce, R. A., and Ward, R. N. (2002). Effects of local anaesthesia or local anaesthesia plus a non-steroidal anti-inflammatory drug on the acute cortisol response of calves to five different methods of castration. *Research in Veterinary Science, 73*, 61–70.

Stearns, S. C. and Ebert, D. (2001). Evolution in health and disease: Work in progress. *Quarterley Review of Biology, 76*, 417–432.

Stefanowska, J., Swierstra, D., Braam, C. R., and Hendriks, M. M. W. B. (2001). Cow behaviour on a new grooved floor in comparison with a slatted floor, taking claw health and floor properties into account. *Applied Animal Behaviour Science, 71*, 87–103.

Stefanowska, J., Swierstra, D., Smits, A. C., Van den Berg, J. V., and Metz, J. H. M. (2002). Reaction of calves to two flooring materials offered simultaneously in one pen. *Acta Agriculturae Scandinavica, Section A, 52*, 57–64.

Steffens, A. B. and de Boer, S. F. (1999). Impact of stress on animal intermediate metabolism. In: P. H. M. Balm (ed.), *Stress physiology in animals* (pp. 108–129). Sheffield, UK: Academic Press.

Stock, R. A. and Britton, R. (2002). *Cow-calf management guide and cattle producer's library*, 2nd edn. Acidosis No. CL624 Western Beef Resource Committee, University of Idaho Extension.

Stockdale, C. R. (2001). Body condition at calving and the performance of dairy cows in early lactation under Australian conditions: A review. *Australian Journal of Experimental Agriculture, 41*, 823–839.

Stockdale, C. R. and King, K. R. (1983). A note on some of the factors that affect the water consumption of lactating dairy cows at pasture. *Animal Production, 36*, 303–306.

Stott, G. H., Marx, D. B., Menefee, B. E., and Nightengale, G. T. (1979). Colostral immunoglobulin transfer in calves. I. period of absorption. *Journal of Dairy Science, 62*, 1632–1638.

Stull, C. L. and McDonough, S. P. (1994). Multidisciplinary approach to evaluating welfare of veal calves in commercial facilities. *Journal of Animal Science, 72*, 2518–2524.

Stull, C. L., Payne, M. A., Berry, S. L., and Hullinger, P. J. (2002). Evaluation of the scientific justification for tail docking dairy cattle. *Journal of the American Veterinary Medical Association, 220*, 1298–1303.

Sutherland, M. A., Mellow, D. J., Stafford, K. J., Gregory, N. G., Bruce, R. A., and Ward, R. N. (2002). Cortisol responses to dehorning of calves given a 5-h local anaesthetic regimen plus phenylbutazone, ketoprofen, or adrenocorticotropic hormone prior to dehorning. *Research in Veterinary Science, 73*, 115–123.

Svensson, C. and Jensen, M. B. (2007). Short communication: Identification of diseased calves by use of data from automatic milk feeders. *Journal of Dairy Science, 90*, 994–997.

Svensson, C. and Liberg, P. (2006). Effect of group size on health and growth rate of Swedish dairy calves housed in pens with automatic milk-feeders. *Preventive Veterinary Medicine, 73*, 43–53.

Svensson, C., Lundborg, K., Emanuelson, U., and Olsson, S. O. (2003). Morbidity in Swedish dairy calves from birth to 90 days of age and individual calf-level risk factors for infectious diseases. *Preventive Veterinary Medicine, 58*, 179–197.

Swanson, J. C. and Morrow-Tesch, J. (2001). Cattle transport: Historical, research, and future perspectives. *Journal of Animal Science, 79*, 102–109.

Sylvester, S. P., Stafford, K. J., Mellor, D. J., and Ward, R. N. (1998). Acute cortisol responses of calves to four methods of dehorning by amputation. *Australian Veterinary Journal, 76*, 123–126.

Tadich, N., Gallo, C., Bustamante, H., Schwerter, M., and van Schaik, G. (2005). Effects of transport and lairage time on some blood constituents of friesian-cross steers in chile. *Livestock Production Science, 93*, 223–233.

Tancin, V., Schams, D., and Kraetzl, W. D. (2000). Cortisol and ACTH release in dairy cows in response to machine milking after pretreatment with morphine and naloxone. *Journal of Dairy Research, 67*, 467–474.

Tarrant, V. and Grandin, T. (2000). Cattle transport. In: T. Grandin (ed.), *Livestock handling and transport*, 2nd edn. Wallingford, UK: CAB International.

Tarrant, P. V., Kenny, F. J., and Harrington, D. (1988). The effect of stocking density during 4 hour transport to slaughter on behaviour, blood constituents and carcass bruising in Friesian steers. *Meat Science, 24*, 209–222.

Tarrant, P. V., Kenny, F. J., Harrington, D., and Murphy, M. (1992). Long distance transportation of steers to slaughter: Effect of stocking density on physiology, behaviour and carcass quality. *Livestock Production Science, 30*, 223–238.

Taylor, A. and Davis, H. (1998). Individual humans as discriminative stimuli for cattle (*Bos taurus*). *Applied Animal Behaviour Science, 58*, 13–21.

Taylor, A. A., Weary, D. M., Lessard, M., and Braithwaite, L. (2001). Behavioural responses of piglets to castration: The effect of piglet age. *Applied Animal Behaviour Science, 73*, 35–43.

Taylor, L. F., Booker, C. W., Jim, G. K., and Guichon, P. T. (1997a). Epidemiological investigation of the buller steer syndrome (riding behaviour) in a western Canadian feedlot. *Australian Veterinary Journal, 75*, 45–51.

Taylor, L. F., Booker, C. W., Jim, G. K., and Guichon, P. T. (1997b). Sickness, mortality and the buller steer syndrome in a western Canadian feedlot. *Australian Veterinary Journal, 75*, 732–736.

Telezhenko, E., Lidfors, L., and Bergsten, C. (2007). Dairy Cow Preferences for Soft or Hard Flooring when Standing or Walking. *Journal of Dairy Science, 90*, 3716–3724.

Tennessen, T. and Hudson, R. J. (1981). Traits relevant to the domestication of herbivores. *Applied Animal Ethology, 7*, 87–102.

Tennessen, T. and Whitney, D. (1990). Estimating animal space needs with video image analysis. *Canadian Journal of Animal Science, 70*, 1183.

Terosky, T. L., Wilson, L. L., Stull, C. L., and Stricklin, W. R. (1997). Effects of individual housing design and size on special-fed Holstein veal calf growth performance, hematology, and carcass characteristics. *Journal of Animal Science, 75*, 1697–1703.

Thiry, E., Saliki, J., Bublot, M., and Pastoret, P. (1987). Reactivation of infectious bovine rhinotracheitis virus by transport. *Comp Immunology Microbiology Infectious Diseases, 10*, 59–63.

Thomas, G. D., Skoda, S. R., Berkebile, D. R., and Campbell, J. B. (1996). Scheduled sanitation to reduce stable fly (diptera: Muscidae) populations in beef cattle feedlots. *Journal of Economical Entomology, 89*, 411–414.

Thomas, T. P. (1971). Drinking by dairy cows at grass (Abstract). *Animal Prouction, 13*, 399–400.

Thomas, T. J., Weary, D. M., and Appleby, M. C. (2001). Newborn and 5-week-old calves vocalize in response to milk deprivation. *Applied Animal Behaviour Science, 74*, 165–173.

Thompson, J. A., Brimacombe, M., Calvin, J. A. T. M. A., Davidson, T. J., and Magee, D. D. (1999). Effects of environmental management on seasonal decrease in milk production in dairy cattle. *Journal of the American Veterinary Medical Association, 214*, 85–88.

Thornton, P. D. and Waterman-Pearson, A. E. (1999). Quantification of the pain and distress responses to castration in young lambs. *Research in Veterinary Science, 66*, 107–118.

Thorpe, W. H. (1965). The assessment of pain and distress in intensive livestock husbandry systems. In: F. W. R. Brambell (ed.), *Report of the technical committee to enquire into the welfare of animals kept under intensive livestock husbandry systems* (pp. 71–79). London: Her Majesty's Stationary Office.

Thun, R., Eggenberger, E., Zerobin, K., Luscher, T., and Vetter, W. (1981). 24-hour secretoy pattern of cortisol in the bull – evidence of episodic secretion and circadian-rhythm. *Endocrinology, 109*, 2208–2212.

Tinbergen, N. (1969). *The study of instinct*. Oxford: Clarendon Press.

Ting, S. T. L., Earley, B., and Crowe, M. A. (2003). Effect of repeated ketoprofen administration during surgical castration of bulls on cortisol, immunological function, feed intake, growth, and behavior. *Journal of Animal Science, 81*, 1253–1264.

Ting, S. T. L., Earley, B., Veissier, I., Gupta, S., and Crowe, M. A. (2005). Effects of age of Holstein-friesian calves on plasma cortisol, acute-phase proteins, immunological function, scrotal measurements and growth in response to burdizzo castration. *Animal Science, 80*, 377–386.

Toates, F. (1995). *Stress: Conceptual and biological aspects*. New York: Wiley.

Tom, E. M., Duncan, I. J. H., Widowski, T. M., Bateman, K. G., and Leslie, K. E. (2002a). Effects of tail docking using a rubber ring with or without anesthetic on behavior and production of lactating cows. *Journal of Dairy Science, 85*, 2257–2265.

Tom, E. M., Rushen, J., Duncan, I. J. H., and De Passillé, A. M. (2002b). Behavioural, health and cortisol responses of young calves to tail docking using a rubber ring or docking iron. *Canadian Journal of Animal Science, 82*, 1–9.

Troy, C. S., MacHugh, D. E., Balley, J. F., Macgee, D. A., Loftus, R. T., Cunningham, P., et al. (2001). Genetic evidence for near-eastern origins of european cattle. *Nature, 410*, 1088–1091.

Tsigos, C. and Chrousos, G. P. (2002). Hypothalamic–pituitary–adrenal axis, neuroendocrine factors and stress. *Journal of Psychosomatic Research, 53*, 865–871.

Tucker, C. B. and Weary, D. M. (2004). Bedding on geotextile mattresses: How much is needed to improve cow comfort? *Journal of Dairy Science, 87*, 2889–2895.

Tucker, C. B., Fraser, D., and Weary, D. M. (2001). Tail docking dairy cattle: Effects on cow cleanliness and udder health. *Journal of Dairy Science, 84*, 84–87.

Tucker, C. B., Weary, D. M., de Passillé, A. M., Campbell, B., and Rushen, J. (2006). Flooring in front of the feed bunk affects feeding behavior and use of freestalls by dairy cows. *Journal of Dairy Science, 89*, 2065–2071.

Tucker, C. B., Weary, D. M., and Fraser, D. (2003). Effects of three types of free-stall surfaces on preferences and stall usage by dairy cows. *Journal of Dairy Science, 86*, 521–529.

Tucker, C. B., Weary, D. M., and Fraser, D. (2004). Freestall dimensions: Effects of preferences and stall usage. *Journal of Dairy Science, 87*, 1208–1216.

Tucker, C. B., Weary, D. M., and Fraser, D. (2005). Influence of neck-rail placement on free-stall preference, use, and cleanliness. *Journal of Dairy Science, 88*, 2730–2737.

Tucker, C. B., Rogers, A. R., Verkerk, G. A., Kendall, P. E., Webster, J. R., and Matthews, L. R. (2007). Effects of shelter and body condition on the behaviour and physiology of dairy cattle in winter. *Applied Animal Behaviour Science, 105*, 7–13.

Turner, S. P. and Lawrence, A. B. (2007). Relationship between maternal defensive aggression, fear of handling and other maternal care traits in beef cows. *Livestock Science, 106*, 182–188.

Turnpenny, J. R., Wathes, C. M., Clark, J. A., and McArthur, A. J. (2000). Thermal balance of livestock 2. applications of a parsimonious model. *Agricultural and Forest Meteorology, 101*, 29–52.

Tuyttens, F. A. M. (2005). The importance of straw for pig and cattle welfare: A review. *Applied Animal Behaviour Science, 92*, 261–282.

Urton, G., von Keyserlingk, M. A. G., and Weary, D. M. (2005). Feeding behavior identifies dairy cows at risk for metritis. *Journal of Dairy Science, 88*, 2843–2849.

USDA (2000). *Feedlot 99*. Fort Collins, CO; USDA:APHIS:VS, CEAH, National Animal Health Monitoring System.

USDA (2002). *Part I: Reference of dairy health and management in the united states*, No. N377.1202. Fort Collins, CO: National Animal Health Monitoring System, United States Departement of Agriculture, Animal and Plant Health Inspection Service, Veterinary Services.

Uvnäs-Moberg, K., Johansson, B., Lupoli, B., and Svennersten-Sjaunja, K. (2001). Oxytocin facilitates behavioural, metabolic and physiological adaptations during lactation. *Applied Animal Behaviour Science, 72*, 225–234.

Vaarst, M., Hindhede, J., and Enevoldsen, C. (1998). Sole disorders in conventionally managed and organic dairy herds using different housing systems. *Journal of Dairy Research, 65*, 175–186.

Vaarst, M., Thamsborg, S. M., Bennedsgaard, T. W., Houe, H., Enevoldsen, C., Aarestrup, F. M., et al. (2003). Organic dairy farmers' decision making in the first 2 years after conversion in relation to mastitis treatments. *Livestock Production Science, 80*, 109–120.

Valde, J. P., Hird, D. W., Thurmond, M. C., and Osteras, O. (1997). Comparison of ketosis, clinical mastitis, somatic cell count, and reproductive performance between free stall and tie stall barns in Norwegian dairy herds with automatic feeding. *Acta Veterinary Scandinavia, 38*, 181–192.

Valizaheh, R., Veira, D. M., & von Keyserlingk, M. A. G. (2007). Behavioural responses by dairy cows provided two hays of contrasting quality at dry-off. *Applied Animal Behaivour Science*, doi:10.1016/j.applanim.2007.03.001.

Val-Laillet, D., de Passillé, A. M., Rushen, J., von Keyserlingk, M. A. G. (2007). The concept of social dominance and the social distribution of feeding related displacements between cattle. *Applied Animal Behaviour Science*, doi:10.106/j.applanim.2007.06.001.

Van Amburgh, M., Diaz, C., and Smith, J. (1999). Nutrition and management of the milk fed calf. *Proceedings of the 1999 Winter Dairy Management Schools*, Cornell University, Ithaca, NY, 54–63.

van der Fels-Klerx, H. J., Horst, H. S., and Dijkhuizen, A. A. (2000). Risk factors for bovine respiratory disease in dairy youngstock in the Netherlands: The perception of experts. *Livestock Production Science, 66*, 35–46.

Van Donkersgoed, J., Jewison, G., Bygrove, S., Gillis, K., Malchow, D., and McLeod, G. (2001). Canadian beef quality audit 1998–99. *Canadian Veterinary Journal, 42*, 121–126.

Van Putten, G. (1982). Welfare in veal calf units. *The Veterinary record, 111*, 437–440.

Van Reenen, C. G., Engel, B., Ruis-Heutinck, L. F. M., Van der Werf, J. T. N., Buist, W. G., Jones, R. B., et al. (2004). Behavioural reactivity of heifer calves in potentially alarming test situations: A multivariate and correlational analysis. *Applied Animal Behaviour Science, 85*, 11–30.

van Reenen, C. G., Mars, M. H., Leushuis, I. E., Rijsewijk, F. A. M., van Oirschot, J. T., and Blokhuis, H. J. (2000). Social isolation may influence responsiveness to infection with bovine herpes virus 1 in veal calves. *Veterinary Microbiology, 75*, 135–143.

Van Reenen, C. G., O'Connell, N. E., Van der Werf, J. T. N., Korte, S. M., Hopster, H., Jones, R. B., et al. (2005). Responses of calves to acute stress: Individual consistency and relations between behavioral and physiological measures. *Physiology & Behavior, 85*, 557–570.

Vanegas, J., Overton, M., Berry, S. L., and Sischo, W. M. (2006). Effect of rubber flooring on claw health in lactating dairy cows housed in free-stall barns. *Journal of Dairy Science, 89*, 4251–4258.

van der Tol, P. P. J., Metz, J. H. M., Noordhuizen-Stassen, E. N., Back, W., Braam, C. R. & Weijs, W. A. (2003). The Vertical Ground Reaction Force and the Pressure Distribution on the Claws of Dairy Cows While Walking on a Flat Substrate. *Journal of Dairy Science, 86*, 2875–2883.

van der Tol, P. P. J., Metz, J. H. M., Noordhuizen-Stassen, E. N., Back, W., Braam, C. R. & Weijs, W. A. (2005). Frictional Forces Required for Unrestrained Locomotion in Dairy Cattle. *Journal of Dairy Science, 88*, 615–624.

Veenhuizen, M. F. and Shurson, G. C. (1992). Effects of sulfate in drinking water for livestock. *Journal of the American Veterinary Medical Association, 201*, 487–492.

Veissier, I., Boissy, A., de Passillé, A. M., Rushen, J., Van Reenen, C. G., Roussel, S., et al. (2001). Calves' responses to repeated social regrouping and relocation. *Journal of Animal Science, 79*, 2580–2593.

Veissier, I., Chazal, P., Pradel, P., and Le Neindre, P. (1997). Providing social contacts and objects for nibbling moderates reactivity and oral behaviors in veal calves. *Journal of Animal Science, 75*, 356–365.

Veissier, I., de Passillé, A. M., Despres, G., Rushen, J., Charpentier, I., de la Fe, A. R. R., et al. (2002). Does nutritive and non-nutritive sucking reduce other oral behaviors and stimulate rest in calves? *Journal of Animal Science, 80*, 2574–2587.

Veissier, I., Gesmier, V., Le Neindre, P., Gautier, J. Y., and Bertrand, G. (1994). The effects of rearing in individual crates on subsequent social behaviour of veal calves. *Applied Animal Behaviour Science, 41*(3/4), 199–210.

Veissier, I., Le Neindre, P., and Trillat, G. (1989). The use of circadian behaviour to measure adaptation of calves to changes in their environment. *Applied Animal Behaviour Science, 22*, 1–12.

Veissier, I., Van Reenen, C. G., Andanson, S., and Leushuis, I. E. (1999). Adrenocorticotropic hormone and cortisol in calves after corticotropin-releasing hormone. *Journal of Animal Science, 77*, 2047–2053.

Veling, J., Wilpshaar, H., Frankena, K., Bartels, C., and Barkema, H. W. (2002). Risk factors for clinical *Salmonella enterica subsp. enterica* serovar typhimurium infection on Dutch dairy farms. *Preventive Veterinary Medicine, 54*, 157–168.

Verga, M., Pavesi, M., and Cerutti, F. (1985). Behaviour and performance of veal calves under different stabling conditions. *Annales de Zootechnie, 34*, 247–256.

Verkerk, G. A., Phipps, A. M., Carragher, J. F., Matthews, L. R., and Stelwagen, K. (1998). Characterization of milk cortisol concentrations as a measure of short-term stress responses in lactating dairy cows. *Animal Welfare, 7*, 77–86.

Vickers, K. J., Niel, L., Kiehlbauch, L. M., and Weary, D. M. (2005). Calf response to caustic paste and hot-iron dehorning using sedation with and without local anesthetic. *Journal of Dairy Science, 88*, 1454–1459.

Voisinet, B. D., Grandin, T., O'Connor, S. F., Tatum, J. D., and Deesing, M. J. (1997). Bos indicus-cross feedlot cattle with excitable temperaments have tougher meat and a higher incidence of borderline dark cutters. *Meat Science, 46*, 367–377.

Vokey, F. J., Guard, C. L., Erb, H. N., and Galton, D. M. (2001). Effects of alley and stall surfaces on indices of claw and leg health in dairy cattle housed in a free-stall barn. *Journal of Dairy Science, 84*, 2686–2699.

Von Borell, E., Bockish, F., Büscher, W., Hoy, S., Krieter, J., Müller, C., et al. (2001). Critical control points for on-farm assessment of pig housing. *Livestock Production Science, 72*, 177–184.

von Keyserlingk, M. A. G. and Weary, D. M. (2007). Maternal behavior in cattle. *Hormones and Behavior, 52*, 106–113.

von Keyserlingk, M. A. G., Brusius, L., and Weary, D. M. (2004). Competition for teats and feeding behavior by group-housed dairy calves. *Journal of Dairy Science, 87*, 4190–4194.

von Keyserlingk, M. A. G., Wolf, F., Hotzel, M., and Weary, D. M. (2006). Effects of continuous versus periodic milk availability on behavior and performance of dairy calves. *Journal of Dairy Science, 89*, 2126–2131.

Waage, S., Sviland, S., and Ødegard, S. A. (1998). Identification of risk factors for clinical mastitis in dairy heifers. *Journal of Dairy Science, 81*, 1275–1284.

Wagner-Storch, A. M., Palmer, R. W., and Kammel, D. W. (2003). Factors affecting stall use for different freestall bases. *Journal of Dairy Science, 86*, 2253–2266.

Waiblinger, S., Boivin, X., Pedersen, V., Tosi, M. V., Janczak, A. M., Visser, E. K., et al. (2006). Assessing the human-animal relationship in farmed species: A critical review. *Applied Animal Behaviour Science, 101*, 185–242.

Waiblinger, S., Menke, C., and Coleman, G. (2002). The relationship between attitudes, personal characteristics and behaviour of stockpeople and subsequent behaviour and production of dairy cows. *Applied Animal Behaviour Science, 79*, 195–219.

Waiblinger, S., Menke, C., Korff, J., and Bucher, A. (2004). Previous handling and gentle interactions affect behaviour and heart rate of dairy cows during a veterinary procedure. *Applied Animal Behaviour Science, 85*, 31–42.

Walker, K. H., Fell, L. R., Reddacliff, L. A., Kilgour, R. J., House, J. R., Wilson, S. C., et al. (2007). Effects of yard weaning and training on the behavioural adaptation of cattle to a feedlot. *Livestock Science, 106*, 210–217.

Warburton, H. and Mason, G. (2003). Is out of sight out of mind? The effects of resources cues on motivation in mink, *Mustela vison*. *Animal Behaviour, 65*, 755–762.

Warnick, V. D., Arave, C. W., and Mickelsen, C. H. (1977). Effects of group, individual and isolated rearing of calves on weight gain and behavior. *Journal of Dairy Science, 60*, 947–953.

Warriss, P. D., Brown, S. N., Knowles, T. G., Kestin, S. C., Edwards, J. E., Dolan, S. K., et al. (1995). Effects on cattle of transport by road for up to 15 hours. *Veterinary Record, 136*, 319–323.

Washburn, S. P., White, S. L., Green, J. T. J., and Benson, G. A. (2002). Reproduction, mastitis, and body condition of seasonally calved Holstein and Jersey cows in confinement of pasture systems. *Journal of Dairy Science, 85*, 105–111.

Wathes, C. M., Zaidan, W. A. R., Pearson, G. R., Hinton, M., and Todd, N. (1988). Aerosol infection of calves and mice with *Salmonella typhimurium*. *Veterinary Record, 123*, 590–594.

Watson, S., Gallagher, P., Del-estal, D., Hearn, A., Ferrier, I. N., and Young, A. H. (2002). Hypothalamic–pituitary–adrenal axis function in patients with chronic depression. *Psychological medicine, 32*, 1021–1028.

Watts, J. M. and Stookey, J. M. (1999). Effects of restraint and branding on rates and acoustic parameters of vocalization in beef cattle. *Applied Animal Behaviour Science, 62*, 125–135.

Watts, J. M. and Stookey, J. M. (2000). Vocal behaviour in cattle: The animal's commentary on its biological processes and welfare. *Applied Animal Behaviour Science, 67*, 15–33.

Watts, J. M. and Stookey, J. M. (2001). The propensity of cattle to vocalise during handling and isolation is affected by phenotype. *Applied Animal Behaviour Science, 74*, 81–95.

Watts, J. M., Stookey, J. M., Schmutz, S. M., and Waltz, C. S. (2001). Variability in vocal and behavioural responses to visual isolation between full-sibling families of beef calves. *Applied Animal Behavior Science, 70*, 255–273.

Waynert, D. F., Stookey, J. M., Schwartzkopf-Genswein, K. S., Watts, J. M., and Waltz, C. S. (1999). The response of beef cattle to noise during handling. *Applied Animal Behaviour Science, 62*, 27–42.

Weary, D. M. and Chua, B. (2000). Effects of early separation on the dairy cow and calf 1. separation at 6h, 1 day and 4 days after birth. *Applied Animal Behaviour Science, 69*, 177–188.

Weary, D. M. and Fraser, D. (1995). Signalling need: Costly signals and animal welfare assessment. *Applied Animal Behaviour Science, 44*, 159–169.

Weary, D. M. and Taszkun, I. (2000). Hock lesions and free-stall design. *Journal of Dairy Science, 83*, 697–702.

Weary, D. M., Jasper, J., & Hötzel, M. J. (2007). Understanding weaning distress. *Applied Animal Behaviour Science*, doi:10.1016/j.applanim.2007.03.025.

Weary, D. M., Niel, L., Flower, F. C., and Fraser, D. (2006). Identifying and preventing pain in animals. *Applied Animal Behaviour Science, 100*, 64–76.

Weatherell, C., Tregear, A., and Allinson, J. (2003). In search of the concerned consumer: UK public perceptions of food, farming and buying local. *Journal of Rural Studies, 19*, 233–244.

Weaver, D. M., Tyler, J. W., VanMetre, D. C., Hostetler, D. E., and Barrington, G. M. (2000). Passive transfer of colostral immunoglobulins in calves. *Journal of Veterinary Internal Medicine, 14*, 569–577.

Webster, A. J. F. (2001). Farm animal welfare: The five freedoms and the free market. *The Veterinary Journal, 161*, 229–237.

Webster, A. J. F. (2002). Effects of housing practices on the development of foot lesions in dairy heifers in early lactation. *The Veterinary Record, 151*, 9–12.

Webster, A. J. F., Saville, C., Church, B. M., Gnanasakthy, A., and Moss, R. (1985a). The effect of different rearing systems on the development of calf behaviour. *The British Veterinary Journal, 141*, 472–483.

Webster, A. J. F., Saville, C., Church, B. M., Gnanasakthy, A., and Moss, R. (1985b). Some effects of different rearing systems on health, cleanliness and injury in calves. *British Veterinary Journal, 141*, 472–483.

Wechsler, B., Schaub, J., Friedli, K., and Hauser, R. (2000). Behaviour and leg injuries in dairy cows kept in cubicle systems with straw bedding or soft lying mats. *Applied Animal Behaviour Science, 69*, 189–197.

Weeth, H. J. and Hunter, J. E. (1971). Drinking of sulfate-water by cattle. *Journal of Animal Science, 32*, 277–281.

Welchman, D. deB, Whelehan, O. P., and Webster, A. J. F. (1988). Haematology of veal calves reared in different husbandry systems and the assessment of iron deficiency. *Veterinary Record: The Journal of British Veterinary Association, 123*(20), 505–510.

Wellnitz, O. and Bruckmaier, R. M. (2001). Central and peripheral inhibition of milk ejection. *Livestock Production Science, 70*, 135–140.

Wells, S. J., Dargatz, D. A., and Ott, S. L. (1996). Factors associated with mortality to 21 days of life in dairy heifers in the United States. *Preventive Veterinary Medicine, 29*, 9–19.

Wells, S. J., Fedorka-Cray, P. J., Dargatz, D. A., Ferris, K., and Green, A. (2001). Fecal shedding of *Salmonella* spp. by dairy cows on farm and at cull cow markets. *Journal of food protection, 64*, 3–11.

Wells, S. J., Garber, L. P., and Wagner, B. A. (1999). Papillomatous digital dermatitis and associated risk factors in US dairy herds. *Preventive Veterinary Medicine, 38*, 11–24.

Wells, S. J., Ott, S. L., and Hillberg Seitzinger, A. (1998). Key health issued for dairy cattle – new and old. *Journal of Dairy Science, 81*, 3029–3035.

Wells, S. J., Trent, A. M., Marsh, W. E., Williamson, N. B., and Robinson, R. A. (1995). Some risk factors associated with clinical lameness in dairy herds in Minnesota and Wisconsin. *Veterinary Record, 136*, 537–540.

Welp, T., Rushen, J., Kramer, D. L., Festa-Bianchet, M., and de Passillé, A. M. (2004). Vigilance as a measure of fear in dairy cattle. *Applied Animal Behaviour Science, 87*, 1–13.

Wentink, G. H., Rutten, V. P. M. G., van den Ingh, T. S. G. A. M., Hoek, A., Müller, K. E., and Wensing, T. (1997). Impaired specific immunoreactivity in cows with hepatic lipidosis. *Veterinary Immunology and Immunopathology, 56*, 77–83.

Wesselink, R., Stafford, K. J., Mellor, D. J., Todd, S., and Gregory, N. G. (1999). Colostrum intake by dairy calves. *New Zealand Veterinary Journal, 47*, 31–34.

Whay, H. R., Main, D. C. J., Green, L. E., and Webster, A. J. F. (2003a). Animal-based measures for the assessment of welfare state of dairy cattle, pigs and laying hens: Consensus of expert opinion. *Animal Welfare, 12*, 205–217.

Whay, H. R., Main, D. C. J., Green, L. E., and Webster, A. J. F. (2003b). Assessment of the welfare of dairy cattle using animal-based measurements: Direct observations and investigation of farm records. *The Veterinary Record, 153*, 197–202.

Whay, H. R., Waterman, A. E., Webster, A. J. F., and O'Brien, J. K. (1998). The influence of lesion type on the duration of hyperalgesia associated with hindlimb lameness in dairy cattle. *The Veterinary Journal, 156*, 23–29.

Whitaker, D. A., Kelly, J. M., and Smith, S. (2000). Disposal and disease rates in 340 British dairy herds. *Veterinary Record, 146*, 363–367.

White, S. L., Benson, G. A., Washburn, S. P., and Green, J. T. J. (2002). Milk production and economic measures in confinement or pasture systems using seasonally calved Holstein and Jersey cows. *Journal of Dairy Science, 85*, 62–104.

Wierenga, H. K. (1990). Social dominance in dairy cattle and the influences of housing and management. *Applied Animal Behaviour Science, 27*, 201–229.

Wilcox, C. J. and Van Horn, H. H. (1999). *Large dairy herd management*, Champaign, IL: American Dairy Science Association.

Willett, L. B. and Erb, R. E. (1972). Short term changes in plasma corticoids in dairy cattle. *Journal of Animal Science, 34*, 103–111.

Wilson, L. L., Terosky, T. L., Stull, C. L., and Stricklin, W. R. (1999). Effects of individual housing design and size on behavior and stress indicators of special-fed Holstein veal calves. *Journal of Animal Science, 77*, 1341–1347.

Winckler, C. and Willen, S. (2001). The reliability and repeatability of a lameness scoring system for use as an indicator of welfare in dairy cattle. *Acta Agriculturae Scandinavica, Section A, Animal Science, supplement 30*, 103–107.

Winckler, C., Capdeville, J., Gebresenbet, G., Horning, B., Roiha, U., Tosi, M., et al. (2003). Selection of parameters for on-farm welfare-assessment protocols in cattle and buffalo. *Animal Welfare, 12*, 619–624.

Wingfield, J. C. and Ramenofsky, M. (1999). Hormones and the behavioral ecology of stress. In: P. H. M. Balm (ed.), *Stress physiology in animals* (pp. 1–51). Sheffield, UK: CRC Press.

Wittum, T. E., Woollen, N. E., Perino, L. J., and Littledike, E. T. (1996). Relationship among treatment for respiratory tract disease, pulmonary lesions evident at slaughter and rate of weight gain in feedlot cattle. *Journal of the American Veterinary Medical Association, 209*, 814–818.

Wolff, P. H. (1968). Sucking patterns of infant mammals. *Brain, Behavior and Evolution, 1*, 354–367.

Wotton, S. B., Gregory, N. G., Whittington, P. E., and Parkman, I. D. (2000). Electrical stunning of cattle. *Veterinary Record, 147*, 681–684.

Xiccato, G., Trocino, A., Queaque, P. I., Sartori, A., and Carazzolo, A. (2002). Rearing veal calves with respect to animal welfare: Effects of group housing and solid feed supplementation on growth performance and meat quality. *Livestock Production Science, 75*, 269–280.

Yang, W. Z. and Beauchemin, K. A. (2006). Physically effective fiber: Method of determination and effects on chewing, ruminal acidosis, and digestion by dairy cows. *Journal of Dairy Science, 89*, 2618–2633.

Zdanowicz, M., Shelford, J. A., Tucker, C. B., Weary, D. M., and von Keyserlingk, M. A. G. (2004). Sand and sawdust bedding affect bacterial populations on teat ends of dairy cows housed in freestalls. *Journal of Dairy Science, 87*, 1694–1701.

Zhou, J., Yan, X., Ryan, D. H., and Harris, R. B. S. (1999). Sustained effects of repeated restraint stress on muscle and adipocyte metabolism in high-fat-fed rats. *American Journal of Physiology – Regulatory Integrative and Comparative Physiology, 277*, R757–R766.

Zurbrigg, K., Kelton, D., Anderson, N., and Millman, S. (2005). Tie-stall design and its relationship to lameness, injury, and cleanliness on 317 Ontario dairy farms. *Journal of Dairy Science, 88*, 3201–3210.

Index